DISCRETE MATHEMATICS AND ALGEBRAIC STRUCTURES

DISCRETE MATHEMATICS AND ALGEBRAIC STRUCTURES

Larry J. Gerstein

University of California, Santa Barbara

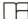

W. H. Freeman and Company

New York

Library of Congress Cataloging-in-Publication Data

Gerstein, Larry J.
  Discrete mathematics and algebraic structures.

  (A series of books in the mathematical sciences)
  Includes index.
  1. Mathematics—1961–   .   2. Algebra, Abstract.
I. Title.   II. Series.
QA39.2.G468   1987   510   86-18417
ISBN 0-7167-1804-9

Printed in the United States of America

1 2 3 4 5 6 7 8 9 0   MP   5 4 3 2 1 0 8 9 8 7

To Susan, David, and Benjamin

Beauty often appears at feasts where only utility or truth has been invited.

FRANÇOIS LE LIONNAIS

Contents

Preface

This book evolved from a set of lecture notes for a course at the University of California, Santa Barbara. That course is taken by the lower-division students intending to major in mathematics or computer science and also by some students in electrical engineering and the physical sciences. The book provides a brief but substantial introduction to ideas, structures, and techniques in discrete mathematics and abstract algebra; it is intended for a one-quarter or one-semester course. This is a mathematics book, not a computer science book, although the selection of topics has been strongly influenced by computer science.

For years many mathematics departments have felt the need for a course to bridge the gap between calculus and upper-division courses on topics such as linear algebra, abstract algebra, combinatorics, and number theory. Instructors of these courses have usually had to start with an extensive review of sets, equivalence relations, and other basic mathematical raw materials before they could get on with the subject at hand. And many students in these courses are alarmed to learn that they are now responsible for the understanding and formulation of mathematical proofs and that mastery of an assortment of technical tricks is no longer a guarantee of academic success.

More recently, computer science departments have been pressing for substantial discrete mathematics courses for their lower-division students as part of the mathematical foundations for upper-division courses in data structures, algorithms, automata and formal languages, program analysis, and so on.

There is therefore a need for a book that explores topics of mutual importance for students in mathematics and computer science at roughly the sophomore level. Students in both disciplines must be comfortable with symbol manipulation and the techniques of mathematical proof, and they need experience in working with abstract ideas at a nontrivial level if they are to achieve the sophisticated blend of creativity, knowledge, and discipline that we call "mathematical maturity." For example, I believe that mathematics students studying Herstein's *Abstract Algebra* and computer science students studying *Elements of the Theory of Computation,* by Lewis and Papadimitriou, have mathematical prerequisites that are virtually identical.

Strictly speaking there are no college-level prerequisites for the material to be found here; indeed, this book could be used as a source of special topics for talented high school students. But in fact I have assumed that this is *not* the student's first encounter with college mathematics, and that some "seasoning" from, say, a year of calculus has already occurred (though calculus is not a prerequisite for anything here). I have also assumed that the student is prepared to pursue ideas with considerable intensity.

The blend of rational thinking and hopeful speculation that leads us to a mathematical truth is usually a far cry from the stately progression of assertions that ultimately constitutes a formal proof. Because this is an introductory text, I have made every effort to give students a broad view of the mathematical experience. Accordingly, the book includes a wide-ranging assortment of examples, exercises, and imagery to motivate the material and to enhance the underlying intuitions. I have tried to strike a balance between rigor and informality, not by operating in some middle region, but by using both styles in what I think is a reasonably balanced way. I have not hesitated to consider a given topic from more than one perspective or at more than one level of rigor.

CONTENTS

Chapter 1 opens with an introduction to the propositional calculus. The notion of mathematical proof is then explored, with guidelines and elementary examples. The chapter concludes with an introduction to logical equivalence and with an application to switching circuits.

Chapter 2 develops the basic operations and constructions of set theory. Quantifiers are introduced. Ordered pairs are defined, leading naturally to Cartesian products and to relations on sets. Equivalence relations and partitions are explored in some detail. The chapter closes with an introduction to mathematical induction and recursion.

Chapter 3 considers functions. Most students will have already had an introduction to this subject; therefore much of the material here consists of examples illustrating aspects of function theory that the students are not likely to have seen before (e.g., the existence of systems of distinct representatives). Sequences are introduced, and Cantor's diagonalization method is considered, anticipating some of the material of the next chapter.

Chapter 4 begins by deriving the fundamental counting principles of finite combinatorics, including the sum and product rules and the pigeonhole principle. The chapter then examines the distinctions between finite and infinite sets as well as the distinctions between countable and uncountable sets. The chapter concludes with a brief introduction to languages and finite automata, a topic that nicely illustrates the use of functions as devices for mathematical modelling and displays the notion of cardinality at work in the context of theoretical computer science.

Chapter 5 centers on finite combinatorics. The chapter opens with the statement of some representative combinatorial problems. The application of the sum and product rules derived in Chapter 4 is then reviewed. Most of the balance of the chapter is concerned with the theory and application of permutations, with special attention given to the concepts of order and parity. The symmetric group is introduced in anticipation of the material on algebraic structures to come in Chapter 6. The chapter concludes with a thorough treatment of binomial and multinomial coefficients, including a proof of the binomial theorem.

Chapter 6 opens by introducing operations and the notion of algebraic structure, including the definitions of semigroup, monoid, and group. The middle of the chapter is devoted to elementary number theory, including the division algorithm, Euclid's theorem on primes, the fundamental theorem of arithmetic, and the irrationality of $\sqrt{2}$. Congruences are introduced and applied to divisibility tests. The chapter concludes with sections on rings and fields, polynomials, complex numbers, and ordered structures (with emphasis on Boolean algebras).

COURSE POSSIBILITIES

There is more than enough material here for a one-semester course. Instructors will need to abbreviate or omit some topics in one-quarter courses, and here are a few guidelines.

CHAPTER 1. The notation introduced in Sections 2, 3, and 5 is used throughout the book, so those sections are essential. Sections 1 and 4 can be treated more briefly, perhaps left as reading assignments. Section 6 can be omitted without loss of continuity.

CHAPTER 2. This chapter is important for everything that follows.

Here and elsewhere in the book the treatment of the material can be shortened by focusing on definitions and the statements of theorems and then considering the examples and the details of proofs as time allows. Section 2 can be safely omitted. Section 6 can be covered quickly, but enough should be said so that the statement and proof of Theorem 2.32 make sense.

CHAPTER 3. A minimal treatment should include the terminology, statement, and proof of Theorems 3.24 and 3.27. The material on sequences in the latter part of Section 2 would be helpful in courses that include Chapter 4, but it can be omitted.

CHAPTER 4. My experience has been that students are generally fascinated (and *surprised* to be fascinated) by the notion that there are different "levels" of infinity. The theorems and examples on countability are good vehicles for the task of sequence construction and hence especially useful for computer science students. Nevertheless, the entire chapter can be omitted without loss of continuity. For an abbreviated treatment, I suggest omitting Sections 4 and 5, skipping most of the proofs in Sections 1, 2, and 3, and proving only Theorems 4.31 and 4.41.

CHAPTER 5. Students generally find this material a welcome relief after the abstraction of Chapter 4. Sections 1 and 4 can be skipped or assigned as outside reading. Section 2 can be skipped if the counting principles in the latter part of Section 1 of Chapter 4 have been covered.

CHAPTER 6. Sections 1, 2, 4, 5, and 6 are basic. Other sections can be covered as time allows and curriculum dictates. If time is really short, I suggest skipping the first three sections, assigning Section 4 as outside reading, and covering only Sections 5 and 6 in detail.

Because this is a book for a one-term course, it makes no claim to be encyclopedic; in particular, it does not include everything the computer scientist should know about discrete mathematics. For example, graph theory is introduced in an ad hoc manner, and finite probability theory makes only a cameo appearance. Instructors designing two-term courses for computer science students would find those two subjects natural extensions of the contents of this book. Similarly, some mathematicians will lament the brevity of the encounter with matrices, for example; and algebraic substructures appear only in the exercises. Therefore, mathematics students would do well to take a course in linear or abstract algebra immediately after completing the material here. Courses in combinatorics or number theory would also be highly appropriate.

ACKNOWLEDGEMENTS

I want to thank my editor, Jerry Lyons, for the wisdom, taste, and sensitivity he has demonstrated in directing this project. And I appreciate the competence of the staff at W. H. Freeman, many of whom are unknown to me by name. I particularly want to mention the excellent work of my project editor, Philip McCaffrey. I thank my colleagues Charles Akemann, Jack Ceder, and Alan Emerson, at UCSB, and Vivian Sewelson and Ernst Snapper, at Dartmouth, for their comments and suggestions. The book was significantly improved as a result of the work of the following reviewers: Ezra Brown (Virginia Polytechnic Institute and State University), David Buchtal (University of Akron), Richard Burden (Youngstown State University), Garret Etgen (University of Houston), Donald Grace (Oklahoma State University), Rick Smith (University of Florida), and Carl Wagner (University of Tennessee). I am especially grateful to consulting editor Victor Klee for his detailed and comprehensive review. I thank the Department of Mathematics and Computer Science at Dartmouth College for its hospitality and support during the latter stages of my work on this book. The manuscript was beautifully typed by Susan Crofoot Nimtz and Christine Willis. Finally, and most of all, I thank the members of my family for the encouragement, affection, and assistance they gave me throughout the book's development. Indeed, to some extent the book was a family project. My son David contributed the splendid drawings on pages 105, 156 (bottom), and 197. My son Benjamin's enthusiasm for big numbers and infinity heightened my belief in the naturalness of mathematical activity. My wife, Susan, has the extraordinary ability to extract order from chaos. Her many suggestions, both mathematical and syntactical, and her imaginative and thorough copyediting have had a dramatic positive impact on the book. My heart throbs when I think of all she has done.

Larry J. Gerstein

CHAPTER 1

Logic

1.1 STATEMENTS, PROPOSITIONS, AND THEOREMS

Mathematics, like ice cream and politics, is discussed in sentences. This is not always immediately apparent, because mathematical sentences may be presented in eccentric formats. For example,

$$\begin{array}{r} 2 \\ +3 \\ \hline 5 \end{array} \quad \text{and} \quad \int_0^{2\pi} \sin x \, dx = 0$$

are sentences. We will not give a precise definition of *sentence* or *statement* (the terms will be used interchangeably), or of the adjectives *true* and *false* used in classifying sentences. But we will adopt the convention that some sentences are true, others are false, no sentence is simultaneously true and false, and some are neither true nor false.

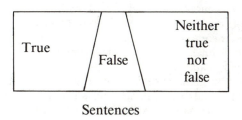

Sentences

Care is essential if we are to make sense out of this scheme of sentence classification. For example, consider the simple statement "Charles snores." Does this mean "Charles snores every night," "Charles has been known to snore," "Charles is presently snoring," or what? To which "Charles" does the sentence refer? We cannot label the statement true or false until we have further clarification.

Distinguishing what is true from what is not true is fundamental in mathematics, and, if we are to proceed without confusion, we must first establish the rules of the game. We begin our activities in a mathematical system by agreeing to recognize certain kinds of expressions as statements in our mathematical language. For example, these might be ordinary English sentences, or statements in a particular computer language, or strings of newly created symbols assembled according to some given guidelines. Next we agree to classify a given collection of one or more mathematical statements as *true;* these statements are called **axioms.** (The word "true" here will not always relate to our ordinary notions of truth, especially since the expressions we accept as statements might not resemble statements in our natural language.) Last, we adopt a system of rules by which we can label certain new statements as "true"; these statements will be related in carefully specified ways to statements that we already know to be true. These rules of truth assignment are called our laws of **logic, deduction, inference,** or **proof.** A proof is a chain of statements leading, implicitly or explicitly, from the axioms to a statement under consideration, compelling us to declare that *that* statement, too, is true. Once we have assumed a system of axioms and logical laws, we become concerned with the consequences of those assumptions, rather than with a more absolute level of truth. A statement that is an axiom or that has been proved is called a **theorem.**

Swallowing a list of axioms as an initial mathematical act may not be easy; the axioms may be in conflict with our intuitions, if "true" has its usual meaning. Consider the following excerpt from Bertrand Russell's autobiography, describing Russell's first encounter with Euclidean geometry. [Russell (1872–1970) was one of this century's great mathematical philosophers.]

> At the age of eleven, I began Euclid, with my brother as my tutor. This was one of the great events of my life, as dazzling as first love. I had not imagined that there was anything so delicious in the world. From that moment until Whitehead and I finished *Principia Mathematica,* when I was thirty-eight, mathematics was my chief interest, and my chief source of happiness. Like all happiness, however, it was not unalloyed. I had been told that Euclid proved things, and was much disappointed that he started with axioms. At first I refused to accept them unless my brother

could offer me some reason for doing so, but he said: "If you don't accept them we cannot go on," and as I wished to go on, I reluctantly admitted them pro tem. The doubt as to the premisses of mathematics which I felt at that moment remained with me, and determined the course of my subsequent work.

The axiom system in the following example is more abstract than that of Euclid's geometry, and this system runs no risk of unsettling our intuitions.

1.1 EXAMPLE.   Consider a mathematical system in which a statement is a string of symbols from the list a, b, S. Some typical statements are

$$abaaS, \quad bSSaba, \quad baaaab, \quad SSS.$$

Take as a system of axioms the statement

$$S$$

The logic consists of two rules: (1) a statement obtained from a true statement by replacing an S with aSb is also true; (2) a statement obtained from a true statement by deleting an S and closing up any resulting space is also true.

In this system the statement $aaabbb$ is a theorem, and here is a proof:

S	(axiom)
aSb	(rule 1)
$aaSbb$	(rule 1)
$aaaSbbb$	(rule 1)
$aaabbb$	(rule 2)

More generally, the theorems in this system are all the statements of the form

$$\underbrace{a \ldots aSb}_{n} \underbrace{\ldots b}_{n} \quad \text{or} \quad \underbrace{a \ldots ab}_{n} \underbrace{\ldots b}_{n},$$

where n is a nonnegative integer. We omit a formal proof.

In Example 1.1 we operated on the initial string S (the axiom system) using successive applications of certain production rules (the logic), and the result was a collection of symbol strings (the theorems) of a special form. This process is an example of what is called *language generation* in computer science. In that context, the rules of logic in Example 1.1 constitute a *grammar,* and the collection of theorems in which no S appears is an example of a *context-free language.* We will have more to say about languages later in the book.

In most mathematical systems, some or all of the statements that are not true are labelled "false," and there may be standard techniques for constructing false statements from true ones, and vice versa. For instance, in arithmetic the false statement $2 + 3 = 7$ has the true companion $2 + 3 \neq 7$. In a mathematical system, the true statements and false statements are the **propositions** of the system, and the label "true" or "false" associated with a given proposition is its **truth value.**

1.2 EXAMPLE.   We take the standard facts and procedures of elementary arithmetic as our axiom system and logic in this example.

(a) The statement "$x + 5 = 7$" is a theorem if it is understood from the context that $x = 2$; it is a false proposition if x represents another numerical value; and it is a statement but not a proposition if no value has been assigned to x.

(b) Recall that a **prime number** is an integer greater than 1 that cannot be expressed as a product of two smaller positive integers. Consider the mammoth integer

$$n = \frac{10^{1031} - 1}{9} = \frac{999 \ldots 9}{9} = \underbrace{111 \ldots 1}_{1031 \text{ digits}}$$

The statement "n is prime" is a proposition, since it is either true or false; but neither "n is prime" nor "n is not prime" is a theorem, because testing n for "primality" is beyond the present capability of our most sophisticated computer technology.

At any level of formality, the search for new truths leads us to link statements together in a variety of ways. To keep mathematical discussions from lapsing into incoherence, we have to standardize the procedures by which truth values are assigned to new propositions. We make these assignments in ways that are **truth functional;** that is, the truth value of a new proposition depends only on the truth values of its component propositions and on the way in which those components have been linked to form the new one. Of course, the truth value of a sentence in our ordinary language will usually depend on far more than this. For example, in conversation we take into account the inflection, facial expressions, and gestures of a speaker in determining his or her meaning. (The sentence "If I have eaten a massive Thanksgiving dinner, then I really feel like eating a bowl of Luigi's spaghetti" probably means the opposite of what the words alone tell us.) For this reason it is unrealistic to expect the formal rules of logic to capture the full richness of our natural language.

1.3 EXAMPLE. Consider the following three statements, and assume them all to be true.

P: This is a good day for a picnic.
W: The weather is beautiful today.
M: Measles is an illness.

We will probably agree that "P because W" is true and "W because M" is false. So when propositions are linked by "because," the truth value of the resulting statement is based on more than the truth values of the component propositions; that is, "because" is *not* truth functional.

Exercises

1. In each case say whether or not the given statement is a proposition. If it is a proposition, indicate its truth value. If it is not a proposition, explain *why* it is not, and then mold it into a proposition by suitable rewriting.

   (a) Lemuel Harrington, of Burbank, California, was President of the United States on July 18, 1897.
   (b) $x/x = 1$.
   (c) $13 + 24 = 35$.
   (d) "The Star Spangled Banner" was played on that occasion.
   (e) If x is a real number, then x is positive, negative, or zero.

2. Consider a mathematical system in which a *statement* is a string of five letters from our usual alphabet. For example, zraav, ththw, and housg

are statements. A *proposition* is a statement found in the latest edition of *Webster's Collegiate Dictionary* (that is, an ordinary five-letter English word); there is only one *axiom:* groan. The *logic* works like this: a proof of a proposition P is a list of propositions $P_1, P_2, P_3, \ldots, P_n$ in which P_1 is a theorem, P_n is just P itself, and for $i > 1$, each P_i is obtained from its immediate predecessor by changing exactly one letter and holding the others fixed. For example, here is a proof of the proposition "frail": groan, groin, grain, grail, frail. (Each word just listed is now a theorem.)

(a) Prove this proposition: cloth.
(b) Discuss the distinction between true propositions and false propositions in this system, and describe some procedure that would convince you that a particular proposition in this system is false. (For instance, suppose you want to demonstrate that the proposition "xylem" is false. How should you proceed?)

3. Consider a mathematical system with the following ingredients: A *statement* is a string of letters, where "letter" means either capital S or a lowercase member of our usual alphabet: a, b, c, . . . , z. The only *axiom:* S; the *logic:* in any theorem, if an occurrence of the letter S is deleted or replaced by one of aSa, bSb, cSc, etc., or by a lowercase letter, the resulting statement will also be a theorem. Describe how to recognize theorems in this system, and write two theorems that are also ordinary English words. Include a proof of each of your theorems.

4. A word that can be inserted between two ordinary English sentences to produce a new sentence is called a **connective.** Consider these connectives: and, hence, or, yet, unless, that. For each connective, give two examples of its use by placing it between two sentences in a way that makes sense. Discuss whether or not the connective is **truth functional;** that is, does the truth or falsity of a new sentence involving a connective depend only on the truth or falsity of the two smaller sentences? If the connective under discussion *is* truth functional, describe its truth-functional action. (That is, which conditions of truth and falsity for the two smaller sentences yield true results, and which yield false results?)

5. (A thought exercise for students with computer programming experience.) Without attempting a written analysis, mentally compare the process of stating and proving a mathematical proposition with the process of writing and debugging a computer program.

1.2 LOGICAL CONNECTIVES AND TRUTH TABLES

We form the **negation** of a statement by using the logical term "not," symbolically denoted by ∼. If P is a statement, then ∼P (verbalized as "not P") denotes its negation. In our ordinary language, ∼P is usually obtained from P by attaching "not" in a grammatically appropriate way to the main verb. If P is "Rosco is smiling," then ∼P is "Rosco is not smiling." ("Rosco is frowning" is *not* the negation of P, since the mere absence of a smile does not force a frown to appear.) There are variations on the ways to phrase a negation, due to the complexities of our language; for example, the negation of "No man is an island" is "At least one man is an island." A safe way to negate P is to precede P by "It is not the case that," though this can lack in linguistic grace what it possesses in logical security. In mathematical contexts, negation is accomplished by a variety of symbolic devices: If P is $2 - 7 = 4$, then ∼P is $2 - 7 \neq 4$. (Here P is a false proposition and ∼P is true.)

The effect of negation on the truth values of propositions is summarized in the following table, called the **truth table** for ∼. Here the letter T in a column indicates that the proposition named at the top of the column is true, and F indicates that it is false. Read each row from left to right. (For example, the bottom row of Table 1.4 tells us that when P is true, ∼P is false.)

(1.4)

P	∼P
F	T
T	F

Two statements P and Q can be joined together with "and" to make a new statement, verbalized as "P and Q," symbolically denoted as $P \wedge Q$, and called the **conjunction** of P and Q. If P and Q are propositions, then asserting that $P \wedge Q$ is true is the same as asserting that P and Q are *both* true. Thus the truth value of $P \wedge Q$ is given by Table 1.5.

(1.5)

P	Q	$P \wedge Q$
F	F	F
F	T	F
T	F	F
T	T	T

The truth value of $P \wedge Q$ depends only on the truth values of P and Q, and there need be no connection between the subject matter of P and Q. Thus the conjuction

$$\underbrace{\text{Most seals swim well}}_{P} \quad \text{and} \quad \underbrace{3 + 14 = 17}_{Q}$$

is a true proposition, although the occasion may seldom arise when we might wish to state it.

The symbolic expression $P \wedge Q$ is not in itself a statement, but it *becomes* a statement when the letters P and Q are replaced by sentences. Expressions can be built up from letters (perhaps with numerical subscripts), logical connective symbols (so far we have seen only $\sim$ and $\wedge$), and parentheses (if necessary for clarity). Those expressions that become statements when the letters are replaced by explicit sentences are called **statement forms** or **sentential forms;** the letters in such forms are called **statement letters** or **sentential variables.** (See Exercise 10 in the next exercise set for further discussion of sentential forms.)

1.6 EXAMPLE. Here are a few statement forms using the connectives $\sim$ and $\wedge$ and the statement letters P, Q, and R:

$$(P \wedge (\sim Q)) \wedge R, \quad \sim(\sim Q), \quad \sim P \wedge ((\sim Q) \wedge R), \quad Q \wedge (\sim(\sim(\sim(\sim P)))).$$

But the following are *not* statement forms:

$$\wedge \wedge P \wedge, \quad P \sim Q, \quad Q \wedge) \sim \sim (R.$$

When the variables in a form are replaced by propositions, the result is a proposition. A truth table for the form displays its truth values in rows, with each row corresponding to a particular combination of truth values for the propositions. To ensure that the truth tables include all possible truth-value combinations, imagine replacing the symbols F and T by 0 and 1, respectively. Then a string of F's and T's takes the form of an integer in binary (base 2) notation: FTTFTFT becomes 0110101. To get all possible strings of length n using the digits 0 and 1, we just count from

$$\underbrace{000. \, . \, .0}_{n \text{ digits}} \quad \text{to} \quad \underbrace{111. \, . \, .1}_{n \text{ digits}}$$

in base 2. That is, we count from 0 to $2^n - 1$. Accordingly, the truth-value combinations in our table will range from

$$\underbrace{\text{FFF. . .F}}_{n \text{ digits}} \quad \text{to} \quad \underbrace{\text{TTT. . .T,}}_{n \text{ digits}}$$

giving 2^n rows in all.

1.7 EXAMPLE. If a statement form S has five variables, P_1, P_2, P_3, P_4, and P_5, then the left part of the truth table will look like this:

P_1	P_2	P_3	P_4	P_5	
F	F	F	F	F	
F	F	F	F	T	
F	F	F	T	F	$2^5 = 32$ rows
F	F	F	T	T	
F	F	T	F	F	
⋮	⋮	⋮	⋮	⋮	
T	T	T	T	T	

In the discussion up to now, the term "statement" has remained undefined; hence the logical framework developed thus far can be applied to any system in which there are two possible truth values. For example, when we say "and" we think of the usual English conjunction if our statements are standard English sentences. But if statements have no intuitive meaning (as in Example 1.1), then what we are presently saying is just this: the expression obtained by inserting ∧ between two statements (or statement forms) in our system will also be viewed as a statement (or statement form). If the two component statements are propositions, then so is the new one, and its truth value is dictated by Table 1.5. A similar observation carries over to the other logical symbols that will appear in this chapter.

The word "or" has at least two meanings in standard English, as the following examples illustrate: (1) At the airport while waiting to board our plane we hear, "Parties with young children or physically handicapped members will be seated first." (2) At the restaurant we learn that "the price of dinner includes soup or salad." In example (1) we understand that if our party satisfies *at least* one of the specified conditions, then we will be seated first; in particular, if our party includes both a physically handicapped person and a child, or if it includes a physically handicapped child, we will still qualify to be seated first. This "or" is called **inclusive,** since our fulfillment of one condition includes the possibility that we may also fulfill the other. In example (2) we understand that we are entitled to *exactly* one of the two possibilities; hence this "or" is called **exclusive.** In mathematics the convention is to use "or" in the inclusive sense unless we explicitly state otherwise; thus "*P* or *Q*" is true if *P* is true, *Q* is true, or both are true. For example,

$$\underbrace{\text{Lions are birds}}_{P} \quad \text{or} \quad \underbrace{6 = 2 + 4}_{Q}$$

and

$$\underbrace{3 + 2 = 5}_{P} \quad \text{or} \quad \underbrace{5 - 7 = -2}_{Q}$$

are both true propositions. And you should henceforth assume that a host who offers the ambiguous invitation, "We hope you or your sister will come to our party" will be pleased to have both of you attend if you can.

The proposition "*P* or *Q*" is called the **disjunction** of *P* and *Q*. It is denoted symbolically by $P \vee Q$; its truth table is

(1.8)

P	*Q*	$P \vee Q$
F	F	F
F	T	T
T	F	T
T	T	T

Here is a mnemonic for distinguishing $\wedge$ from $\vee$: notice that $\wedge$ resembles the A in AND; also, classicists will be happy to learn that the symbol $\vee$ is derived from the Latin *vel* (or).

Now that we have negation, conjunction, and disjunction in hand, we can determine the truth value of more complicated propositions. The strategy is as follows: Decompose the given proposition into **atomic sentences** (that is, sentences that cannot themselves be broken into smaller sentences) linked by logical connectives. (This may require some judicious rephrasing of the original proposition.) Then label each atomic sentence with a sentential variable. The result will be a statement form that represents the given proposition. A truth table will complete the analysis.

1.9 EXAMPLE.   When is the following statement false?

I am either a polo player or the Queen of Zorbia.

The answer, obtained from pure thought (and without the use of truth tables or a logic chapter), is that the statement is false under exactly one pair of circumstances: I am not a polo player *and* I am not the Queen of Zorbia. Although this is clear upon reading the given statement, let's go through the formal analysis anyway to get a feeling for the technique. Introduce sentential variables for the atomic sentences as follows:

P: I am a polo player.
Q: I am the Queen of Zorbia.

The proposition is represented by $P \vee Q$, so Table 1.8 gives the asserted result. (The top row is the only row with "F" under $P \vee Q$, and in that row both P and Q are false.)

1.10 EXAMPLE.   Represent the statement

I will go to the movies on Monday or Tuesday, but not on both days.

by a sentential form.

SOLUTION.   Make the assignments

P: I will go to the movies on Monday.
Q: I will go to the movies on Tuesday.

The first clause of the given statement is represented by $P \vee Q$, and the last by $\sim(P \wedge Q)$, so the complete answer is

$$(P \vee Q) \wedge \sim(P \wedge Q).$$

Here parentheses guide us in our use of the logical connectives. Notice that the effect of the phrase "but not on both days" is to make the "or" exclusive. Thus the essence of "exclusive or" has been captured by a conspiracy of $\vee$, $\wedge$, and $\sim$.

1.11 EXAMPLE.   Let K be the following sentential form:

$$K = \sim(P \wedge Q) \wedge \underbrace{(P \wedge (\sim Q \vee (\sim P \vee Q)))}_{J}.$$

Under what truth values for propositions P and Q is the proposition represented by K true?

SOLUTION.   Construct a truth table for K:

P	Q	$P \wedge Q$	$\sim Q$	$\sim P$	$\sim P \vee Q$	$\sim Q \vee (\sim P \vee Q)$	J	$\sim(P \wedge Q)$	K
F	F	F	T	T	T	T	F	T	F
F	T	F	F	T	T	T	F	T	F
T	F	F	T	F	F	T	T	T	T
T	T	T	F	F	T	T	T	F	F

Here the two columns on the left display all possible truth-value combinations for the sentential variables, and each subsequent column corresponds to a sentential form built from one or two of its predecessors by applying $\sim$, $\wedge$, or $\vee$ and invoking Table 1.4, 1.5, or 1.8, respectively. We conclude that K is true only when P is true and Q is simultaneously false.

1.12 REMARK.   Another sentential form that is true precisely when P is true and Q is simultaneously false is $P \wedge \sim Q$. So it would be reasonable to say that statement form K in Example 1.11 is "logically equivalent" to $P \wedge \sim Q$. We will return to this point later.

Exercises for this section are included in the exercise set at the end of Section 1.3.

1.3   CONDITIONAL STATEMENTS

"All your troubles are due to those 'ifs'," declared the Wizard.

L. FRANK BAUM
The Emerald City of Oz

If P and Q are propositions (whose truth values may or may not be known), we may wish to assert this:

(1.13)   Truth of P and falsity of Q do not coexist.

1.14 EXAMPLE. Your mother tells you

If yesterday's weather was nice in Boise, then Uncle Harry went fishing yesterday.

$\underbrace{\phantom{\text{If yesterday's weather was nice in Boise}}}_{P}$ $\underbrace{\phantom{\text{Uncle Harry went fishing yesterday}}}_{Q}$

In other words, nice weather in Boise didn't coexist with a nonfishing day for Uncle Harry. There is only one combination of circumstances under which Mom would be wrong: Boise's weather was nice, yet Uncle Harry failed to fish. In particular, if the weather was poor, your mother spoke the truth whether or not your uncle went fishing.

Assertion 1.13 is represented by the sentential form $\sim(P \wedge \sim Q)$ (stated formally: it is not the case that P and not Q), which we now abbreviate by

$$P \Rightarrow Q.$$

(Other common notations are $P \rightarrow Q$, $P \supset Q$, $Q \Leftarrow P$). The effect of the new logical connective $\Rightarrow$ can be summarized by Table 1.15:

	P	Q	$P \Rightarrow Q$
	F	F	T
(1.15)	F	T	T
	T	F	F
	T	T	T

A proposition of the form $P \Rightarrow Q$ is called a **conditional proposition** or a **material implication.** The proposition P is the **antecedent** or **hypothesis** of the conditional proposition, and Q is the **consequent** or **conclusion.** We usually verbalize "$P \Rightarrow Q$" as "If P then Q" or "P implies Q," although these usages can lead to confusion by suggesting a relationship (perhaps of obscure origin) between the contents of P and Q that goes beyond their truth values. (See Example 1.16.)

To avoid monotony it is common practice to have several English statements corresponding to one mathematical one. For example, the conditional $P \Rightarrow Q$ can also be read "P is **sufficient** (or a **sufficient condition**) for Q," and "Q is **necessary** (or a **necessary condition**) for P." Other phrasings are "P only if Q," and "Q if P."

1.16 EXAMPLE. Consider the following conditional statements:

(a) If Utah is a state then $2 + 2 = 4$.

(b) If the moon is made of cheese, then an average duck weighs 3 tons.

(c) $23 \cdot 6 = 17$ implies $1 + 23 \cdot 6 = 18$.

(d) $23 \cdot 6 = 17$ implies $1 + 23 \cdot 6 = 2$.

(e) If common digits in the numerator and denominator of a fraction can be cancelled, then $\frac{16}{64} = \frac{1}{4}$.

(f) If $2 + 2 = 4$ then $6 \cdot 3 = 19$.

Statement (a) has the form $T \Rightarrow T$. (We are abbreviating furiously by replacing each component proposition by its truth value, since that is our only concern here.) Hence statement (a) is true, though there is no evident natural pathway from P to Q. Statements (b) through (e) all have false antecedents and are therefore true conditionals; of these only statement (e) has a true consequent. Also note that in statements (c) and (e) the antecedent appears to lead "naturally" to the consequent, whereas in statements (b) and (d) no such connection is evident. Statement (f) has the form $T \Rightarrow F$ and is therefore the only false proposition on the list.

1.17 EXAMPLE. Construct a truth table for the following propositional form J:

$$P \Rightarrow (\underbrace{(\sim Q \Rightarrow P) \wedge (Q \vee \sim P)}_{K})$$

SOLUTION.

P	Q	$\sim P$	$\sim Q$	$\sim Q \Rightarrow P$	$Q \vee \sim P$	K	J
F	F	T	T	F	T	F	T
F	T	T	F	T	T	T	T
T	F	F	T	T	F	F	F
T	T	F	F	T	T	T	T

Thus J is true unless P and Q have truth values T and F, respectively. Suggestion: now reread Remark 1.12.

Conditional statements are an important feature of most computer programming languages. They are used in choosing among two or more procedures at some stage in a program's execution, and they can prevent a program from going into spasm when confronted with an impossible task.

For example, suppose an integer n has been defined and a program that handles only real numbers demands "Let $m = \sqrt{n}$." Should it turn out that $n < 0$, the program will tie up. But a conditional statement of the form "If $n \geq 0$, then let $m = \sqrt{n}$, else . . . [insert some other appropriate procedure]" allows the program to proceed regardless of the truth value of the hypothesis $n \geq 0$. Thus a computer program accepts a conditional statement whose hypothesis is false, just as we label as "true" a conditional statement whose hypothesis is false.

If P and Q are propositions, we abbreviate the proposition

$$(P \Rightarrow Q) \wedge (P \Leftarrow Q)$$
$$(P \text{ only if } Q) \quad \text{and} \quad (P \text{ if } Q)$$

by

$$P \Leftrightarrow Q$$
$$(P \text{ if and only if } Q).$$

(The implication $P \Leftarrow Q$ is called the **converse** of $P \Rightarrow Q$.) Such a statement is called a **biconditional proposition** or a **material equivalence;** we say that P and Q are **equivalent propositions** if $P \Leftrightarrow Q$ is true. From the truth table

	P	Q	$P \Rightarrow Q$	$P \Leftarrow Q$	$P \Leftrightarrow Q$
	F	F	T	T	T
(1.18)	F	T	T	F	F
	T	F	F	T	F
	T	T	T	T	T

we see that the statement $P \Leftrightarrow Q$ is true only when P and Q have the same truth value. Thus

$$\text{George Washington was a President} \quad \Leftrightarrow \quad 3 + 5 = 8$$

and

$$2 + 2 = 5 \quad \Leftrightarrow \quad 2 + 2 = 6$$

are both true.

Our intuitive notion of equivalent statements in ordinary language is concerned with *meaning,* which is a philosophical concept, and not just with truth value. Two statements that are equivalent in that philosophical sense are also equivalent in the sense that has been defined here. (That is, both are true or both are false.) But our equivalence is a broader concept, since if P and Q are propositions, then a statement of the form $P \Leftrightarrow Q$ is either true or false; whereas outside the strictly logical realm, two statements whose meanings have no evident bearing on one another are regarded as incomparable.

We have defined $P \Rightarrow Q$ as an abbreviation for $\sim(P \wedge \sim Q)$, so everything we can express with $\Rightarrow$ can also be expressed with $\sim$, $\wedge$, and parentheses. But conditional statements are ubiquitous in mathematics, and whatever we write is likely to be read (otherwise why write it?), so we retain the connective $\Rightarrow$ in order to avoid the vertigo that can result from fusillades of symbols.

Exercises

1. Let P be the statement "Howard fell" and let Q be the statement "Howard broke his leg."

   (a) Write English statements corresponding to each of these:

   $$P \wedge Q, \quad \sim P \wedge \sim Q, \quad \sim(P \wedge Q), \quad Q \vee \sim P.$$

   When is one of the last two of these statements true and the other one false?

   (b) With P and Q as given, under what conditions is the statement "P or Q" true when "or" is *exclusive? Inclusive?*

2. One Tuesday morning your friend says, "If today is Wednesday, then today is Thursday." Discuss the truth value of your friend's statement.

3. Let $\underline{\vee}$ denote "exclusive or," and exhibit a truth table for the sentential form $P \underline{\vee} Q$.

4. The connective "unless" can be ambiguous, and this exercise will pinpoint the ambiguity. If P and Q are propositions, let $P \text{ u } Q$ denote "P unless Q." (This is not a standard notation.)

   (a) Complete as much of a truth table as possible for $P \text{ u } Q$. Of its four lines of truth values, three will be straightforward. Fill in those three lines, and discuss the ambiguity of the fourth line.

(It may be helpful to consider the following statement as an example:

$$\underbrace{\text{We will have a picnic today}}_{P} \quad \text{unless} \quad \underbrace{\text{it is raining.}}_{Q})$$

(b) Show that the ambiguity of "unless" is essentially the same as the ambiguity between the inclusive and exclusive usages of "or" in ordinary English. (Truth table comparisons may help your discussion.)

5. Suppose four cards are given, each of which has a letter on one side and a number on the other. They are displayed as follows:

Which card(s) *must* be turned over in order to determine the truth value of the following statement: If a card has *B* on one side, then it has 2 on the other side. (This is a standard problem on psychological tests.)

6. Your uncle tells you, "Next Sunday if the weather is nice we will either go on a picnic or go fishing, unless my car needs repair." Represent this statement symbolically and determine those conditions under which you will declare your uncle a liar. (You may wish to refer to Exercise 4.)

7. Represent the following two statements as statement forms (they are both implications) and discuss their truth values.
   (a) I go to the movies in the afternoon if it is rainy.
   (b) I go to the movies in the afternoon only if it is rainy.

8. Represent each of the following as a statement form.
   (a) Claudia will run in the marathon if she has trained properly or is injury-free at race time.
   (b) Claudia will run in the marathon only if she is injury-free at race time.
   (c) Claudia will run in the marathon if and only if she has trained properly and is injury-free at race time.

9. (a) Exhibit a truth table that shows the truth values for the sentential form $P \Rightarrow Q$ and its converse.

(b) Replace the variables P and Q by English sentences so that $P \Rightarrow Q$ becomes a true proposition and $P \Leftarrow Q$ a false one.

10. The notion of *statement form* (also called *well-formed formula*) can be described more carefully than in the text. Start with a collection of symbols that we agree to call **statement letters.** [Do not allow any of the symbols), (, $\sim$, $\vee$, $\wedge$, $\Rightarrow$, $\Leftrightarrow$ to be in this collection.] A statement letter is a statement form of the simplest kind; and if S_1 and S_2 are statement forms, then so are $\sim(S_1)$, $(S_1) \vee (S_2)$, $(S_1) \wedge (S_2)$, $(S_1) \Rightarrow (S_2)$, and $(S_1) \Leftrightarrow (S_2)$. Parentheses can be omitted when confusion is unlikely. [But they are sometimes essential: if S_1 is $P \Rightarrow Q$ and S_2 is $Q \Rightarrow R$, then $S_1 \Rightarrow S_2$ is written $(P \Rightarrow Q) \Rightarrow (Q \Rightarrow R)$.] Taking our usual alphabet as the collection of statement letters, determine which of the following are statement forms. In each case when the given expression is *not* a statement form, insert parentheses so that it becomes such a form.

(a) $(P \Rightarrow \sim Q) \Rightarrow ((\sim(\sim P)) \Leftrightarrow Q)$
(b) $(P \Rightarrow Q) \Rightarrow R) \Rightarrow S) \Rightarrow T$
(c) $(P \Rightarrow (\sim Q \Rightarrow R))) \Rightarrow Q$
(d) $((P \Rightarrow Q \Rightarrow R) \wedge Q) \vee (\sim(P \vee Q))$
(e) $(\sim P \vee Q) \Leftrightarrow (((P \wedge Q) \vee (P \wedge R)) \Rightarrow (S \vee \sim P))$

11. Exhibit truth tables for each of the following statement forms:

(a) $P \Rightarrow (Q \vee \sim Q)$
(b) $P \Rightarrow (P \Rightarrow P)$
(c) $(P \Rightarrow P) \Rightarrow P$
(d) $(P \Rightarrow Q) \Leftrightarrow (\sim Q \Rightarrow \sim P)$
(e) $(P \Rightarrow \sim Q) \Rightarrow (R \Rightarrow (\sim Q \vee P))$

12. List the integers from zero through seventeen in binary notation.

13. A student prepares a truth table for a statement form that has seven sentential variables but omits six rows by mistake. How many rows are in the student's table?

14. How many rows are in the truth table for the form

$$(P \wedge Q) \Rightarrow (((\sim P) \vee R) \Leftrightarrow (R \Rightarrow S))?$$

15. Let S be a statement form in which the sentential variables P, Q, R do not occur. Consider the new statement form

$$S \vee (P \Rightarrow ((\sim Q) \wedge R)).$$

How does the number of rows in the truth table for the new form compare with the number of rows in the table for S?

16. Suppose S is a statement form with n variables $P_1, P_2, \ldots, P_n$, where $n \geq 3$. In exactly how many rows of the truth table for S is P_1 false? In how many rows is P_1 false while P_2 is true?

17. We have defined $P \Rightarrow Q$ as an abbreviation for $\sim(P \wedge \sim Q)$. Express the sentential form

$$P \Rightarrow (Q \Rightarrow (R \Rightarrow S))$$

using $\sim$ and $\wedge$ as the only logical connectives. (That is, pretend the symbol "$\Rightarrow$" has not been introduced.)

18. A **tautology** is a sentential form that becomes a true proposition whenever the letters in the expression are replaced by actual propositions. For example, the expressions $P \vee \sim P$ and $(P \vee Q) \leftrightarrow (Q \vee P)$ are both tautologies. Use truth tables to determine which of the following are tautologies:

(a) $P \Rightarrow (\sim P \Rightarrow Q)$
(b) $(P \wedge Q) \Rightarrow (P \vee R)$
(c) $((P \Rightarrow Q) \leftrightarrow Q) \Rightarrow P$
(d) $P \Rightarrow (Q \Rightarrow (Q \Rightarrow P))$

If a given form is *not* a tautology, display a line of the truth table corresponding to one instance in which the form becomes a false proposition.

1.4   PROOFS: STRUCTURES AND STRATEGIES

Truth tables can be applied systematically to determine the truth values of certain new propositions constructed from old ones. For example, if P_1 and P_2 are theorems, then $\sim P_2$ is false and $\sim P_2 \vee P_1$ is true. But mathematics extends far beyond the uninspired linking of one randomly chosen proposition to another. Mathematics also involves enlightened speculation about what *might* be true, given what is already known to be true; the creation of interesting new propositions on the basis of the calculations, hunches, and fantasies that constitute that speculation; and the proving or disproving of those propositions. In this section some techniques of mathematical proof are discussed, but nowhere will it be suggested that creating and proving interesting new theorems is an automatic consequence of certain rules of procedure. Consider the following remarks on proof technique. (The quotation is from *The Mathematical Experience,* by P. J. Davis and R. Hersh;† it refers to a proof of the Pythagorean theorem in which the key to the argument is the supplementation of an initial diagram with certain "construction lines.")

> Now, how does one know where to draw these lines so as to reason with them? It would seem that these lines are accidental or fortuitous. In a sense this is true and constitutes the genius or the trick of the thing. Finding the lines is part of finding a proof, and this may be no easy matter. With experience come insight and skill at finding proper construction lines. One person may be more skillful at it than another. There is no guaranteed way to arrive at a proof. This sad truth is equally rankling to schoolchildren and to skillful professionals. Mathematics as a whole may be regarded as a systematization of just those questions which have been pursued successfully.

Once again: *There is no guaranteed way to arrive at a proof.*

What are the criteria for validity and goodness of a proof? This is not a simple issue. In an interesting mathematical system, the formal requirements of the laws of deduction are likely to be so awesome in their complexity and rigidity that it is a common practice to adopt a less formal style of discussion, consisting of a palatable mixture of mathematical expressions and ordinary sentences in our natural language. We will follow that practice here. So what we customarily call a proof is usually only an outline of the genuine article, and we must be on guard for irrational leaps that cannot be justified by the available body of axioms, theorems, and logical rules. The alternative to the informal approach is strict adherence to the formal laws, and this can involve intuition-free arguments of colossal length and complexity; any resulting gains in accuracy can easily be offset by losses in insight. For instance, consider this story from mathematician John Kemeny (excerpted from his article "Rigor versus intuition in mathematics"):††

> There was an advanced seminar . . . in which the lecturer devoted the entire hour to writing out a proof with complete rigor. After having filled all the blackboards, he had everyone in the room completely lost, including one of his own colleagues, who jumped up and said, "Look, I

† Boston: Houghton Mifflin, 1981.
†† In Douglas Campbell and John C. Higgens (eds.), *Mathematics: People. Problems. Results.* (Belmont, Calif.: Wadsworth, 1984).

just don't understand this proof at all. I tried to follow you, but I got lost somewhere. I just didn't get it at all." The lecturer stopped for a moment, looked at him, and said, "Oh, didn't you see it? You see, it's just that the two spaces connect like this," intertwining his two arms in a picturesque fashion. And then his colleague exclaimed, "Oh, now I get the whole proof."

The notion of proof is discussed with charm and eloquence by Yu. I. Manin in his book, *A Course in Mathematical Logic,* and here is an excerpt:†

> A proof only becomes a proof after the social act of "accepting it as a proof." This is as true for mathematics as it is for physics, linguistics, or biology. The evolution of commonly accepted criteria for an argument's being a proof is an almost untouched theme in the history of science. In any case, the ideal for what constitutes a mathematical demonstration of a "nonobvious truth" has remained unchanged since the time of Euclid: we must arrive at such a truth from "obvious" hypotheses, or assertions which have already been proved, by means of a series of explicitly described, "obviously valid" elementary deductions.

Manin points out that the "ideal" mentioned in the preceding paragraph is rarely achieved:

> The absence of errors in a mathematical paper (assuming that none are discovered), as in other natural sciences, is often established indirectly: how well the results correspond to what was generally expected, the use of similar arguments in other papers, examination of small sections of the proof "under the microscope," even the reputation of the author.

The history of mathematics includes many instances in which propositions that were widely believed (sometimes for years) to have been proved were later found to be false. We now ask: can we rely on *computers* to lead us safely through the Valley of the Shadow of Falsehood and thereby let us know with certainty what is true and what is not? In considering the role of computers in mathematical proofs, Manin supplies this quotation from a research article by H. P. F. Swinnerton-Dyer:

> When a theorem has been proved with the help of a computer, it is impossible to give an exposition of the proof which meets the traditional

† New York: Springer-Verlag, 1977.

test—that a sufficiently patient reader should be able to work through the proof and verify that it is correct. Even if one were to print all the programs and all the sets of data used (which in this case would occupy some forty very dull pages) there can be no assurance that a data tape has not been mispunched or misread. Moreover, every modern computer has obscure faults in its software and hardware—which so seldom cause errors that they go undetected for years—and every computer is liable to transient faults. Such errors are rare, but a few of them have probably occurred in the course of the calculations reported here.

After further discussion, Manin concludes with this moral:

A good proof is one that makes us wiser.

End of philosophy.

Now, suppose we have at hand a proposition Q that we want to prove. We ask ourselves: Does it seem *likely* to be true? Why? If we suspect that Q is true, we are probably aware (perhaps subconsciously) of a theorem P whose truth seems to be incompatible with the falsity of Q. By explicitly recalling P and establishing that incompatibility, we will have proved Q (this will be shown). After some reflection, if no such P presents itself, we can consider as *candidates* for P some known truths intuitively related to the subject matter of Q. Perhaps one of these or a combination of several will, when subjected to suitable logical maneuvers, lead to a proof of Q.

Let's take a closer look at the logic. We claim that to prove a proposition Q it suffices to isolate a true proposition P such that the implication $P \Rightarrow Q$ is also true. To see this, consider the first three columns of Table 1.19.

	P	Q	$P \Rightarrow Q$	$\sim P$	$\sim Q$	$\sim Q \Rightarrow \sim P$	$Q \Rightarrow P$
	F	F	T	T	T	T	T
(1.19)	F	T	T	T	F	T	F
	T	F	F	F	T	F	T
	T	T	T	F	F	T	T

Notice that only on the bottom line are both P and $P \Rightarrow Q$ true, and in this situation Q is also true. Therefore, to prove Q it is enough to prove $P \wedge (P \Rightarrow Q)$, as claimed. In words: if P is true and if it is also true that P implies Q, then Q is true. This fundamental logical principle is called the **rule of *modus ponens*** or the **law of detachment,** and it dates back at least to Aristotle (384–322 B.C.). An argument of the pattern described here is called a **direct proof** of Q.

1.20 EXAMPLE. Let $\triangle$ denote a triangle with one vertex at the center of a circle C and the other two vertices on C itself. Consider the following proposition Q: Triangle $\triangle$ has two equal angles. Let's quickly sketch a proof and then analyze its logical structure. Let O denote the vertex at the center of C, and let A and B be the other two vertices.

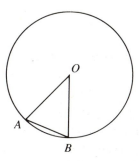

Lines OA and OB have the same length, since they are both radii of C. Therefore $\triangle$ is isosceles. However, base angles of an isosceles triangle are equal, hence $\triangle$ has two equal angles, and we're done.

The heart of the argument is that base angles of an isosceles triangle are equal, which we can rewrite as an implication:

$$\underbrace{\triangle \text{ is an isosceles triangle}}_{P} \Rightarrow \underbrace{\text{base angles of } \triangle \text{ are equal}}_{Q}$$

The proof assumes that the truth of $P \Rightarrow Q$ is known; then we use the definition of *circle* to deduce the truth of P for our triangle $\triangle$, and finally *modus ponens* yields the truth of Q.

1.21 EXAMPLE. Take the facts of elementary arithmetic as our body of "known truths," and consider a proof of this proposition: The square of an odd integer has the form $8k + 1$ for some integer k.

PROOF *(rambling style).* Consider an odd integer n. Our task involves the close scrutiny of n^2 without knowing anything about n except for its oddness. However, "odd" means "not divisible by 2"; so when we attempt to divide n by 2, we get a quotient q and a remainder 1. That is,

$n = 2q + 1$. Notice a consequence of what we have done so far: every integer has the property that either it or its successor (the next largest integer) is even. (*Why* is this a consequence?) We now have

$$n^2 = (2q + 1)^2 = 4q^2 + 4q + 1 = 4q(q + 1) + 1.$$

Either q or $q + 1$ is even, as we noted a moment ago, so $q(q + 1)$ is even and $4q(q + 1)$ is divisible by 8, say $4q(q + 1) = 8k$. Thus $n^2 = 8k + 1$, as desired.

Let's be frank. In practice, printed proofs are almost never as expansive as this one. First, space is generally limited; second, there is much to be said for the reader's pushing through the details independently; and, last, a mountain of details can bury the central ideas beyond excavation. A textbook version of the above proof is more likely to read as follows:

PROOF *(compact style)*.   An odd integer can be written $n = 2q + 1$ for some integer q. Thus $n^2 = 4q(q + 1) + 1$, and putting $q(q + 1) = 2k$ gives $n^2 = 8k + 1$, as desired.

The underlying logic is the same in both proofs. Elementary arithmetic provides the axiomatic foundation. The proof opens when we choose the symbol n for "an odd integer." The key to the proof lies in using arithmetic to translate the statement "n is odd" (taken as true) into its equational formulation, $n = 2q + 1$, then proving the implication

$$\underbrace{(n = 2q + 1)}_{P} \Rightarrow \underbrace{(n^2 = 8k + 1)}_{Q}$$

by showing that truth of P must inevitably be accompanied by truth of Q. Finally we apply *modus ponens* (without saying so explicitly) to conclude that Q is true.

It would be comforting to know that any proposition that can be stated can be proved or disproved (a *disproof* of P is a proof of $\sim P$), but in the 1930s the logician Kurt Gödel showed that certain kinds of mathematical systems in which we can do ordinary arithmetic must inevitably contain propositions that can be neither proved nor disproved. Determining exactly which propositions fall into this murky category is a subject of continuing mystery. Consider, for example, that over 300 years ago Pierre de Fermat claimed to have a proof (which he did not publish) of this statement: If n is an integer, and if $n \geq 3$, there are no positive integers x, y, and z for which $x^n + y^n = z^n$. Extraordinary efforts have been made to prove this proposition, known as Fermat's Last Theorem; much of the recent work has involved the heavy use of computers. Recently the work of the mathematician Gerd Faltings has given new hope that a proof is obtainable. (Faltings's research is in an area known as *arithmetic algebraic geometry*.) The fact remains that if you prove or disprove Fermat's theorem, your career anxieties will be over; on the other hand, it may be that the proposition is unprovable. (For the record, note that Fermat also conjectured that the integers of the form $2^{(2^n)} + 1$, with n any nonnegative integer, are prime. His evidence: the first few such integers are prime, namely, 3, 5, 17, 257, and 65537. But in 1732 Leonhard Euler observed that the next such integer, $2^{32} + 1$, has the factorization 641×6700417, thereby wiping out the conjecture.)

Although not all propositions can be proved or disproved, there are some general psychological guidelines that can be useful in the search for proofs. Imagine that you are standing at the foot of a mountain and you want to reach the top. Perhaps you will be lucky and see a path in front of you that seems to be headed in the right direction; so you take the path and you reach the top. Great. But you may see no immediate path, or you may see several and have to choose one. What then? By looking through binoculars you may see that the top is accessible only from a few directions, and this may help you eliminate certain paths at the start, or it may encourage you to look for a path in a location you haven't yet fully checked. You may succeed by being continually aware of your location, of the directions in which you can travel from your location, and of the places near the mountaintop from which the top is accessible. If you are prepared to move forward along many different paths from the starting point, and if you can imagine backing down from the top in many different ways, you may discover a route from the bottom to the top. Of course, once you know the way you can draw a map, and then the next person can follow your path directly.

Now suppose you want to prove proposition Q, and you think you know all that is needed for the job; say you know that a proposition P is true. Think: If I know P, what *else* do I know as a consequence? (This corresponds to looking for a path to start up the mountain.) Also, what does it mean to say that Q is true? Are there one or more statements, perhaps simpler than Q, and perhaps separately provable, from which Q will follow readily? (This is like looking for locations near the mountaintop from which the top is accessible.)

In mathematical proofs, definitions are often crucial. A **definition** is a statement introducing a new symbol or word that abbreviates a package of statements or expressions (or both) whose meanings or uses are already understood. Asserting that an object "satisfies" a definition asserts that the statements in the package are true for that object. For instance, saying that an object T "satisfies the definition of trapezoid" says that T is a closed polygon, that it has four sides, and that two of its sides are parallel. In the proofs we have already discussed, the definitions of "odd integer" and "isosceles triangle" play a central role. If a proposition P claims that an object has a certain property then, in the absence of inspiration, try playing with one or more of the properties that characterize the object under consideration; also toy with the definitions of the terms in the desired conclusion.

Strive for flexibility: if an argument leads you in the wrong direction, and if you have been careful with the details, abandon your dedication to it and try another approach. (If a path toward the mountain dead-ends at a swamp, will you take it again and again?) An interesting proposition might not have a proof that comes easily, but if you get it you will rejoice. As you struggle with a proof, these two questions should dominate the proceedings at all times: *What do I know? What do I want to show?*

We will now consider a proof technique called **indirect proof** (also called **proof by contradiction,** or *reductio ad absurdum*). The technique involves proving a proposition Q by showing that the negation $\sim Q$ implies a statement that we know to be false (that is, $\sim Q$ leads to a contradiction of a known truth). Thus $\sim Q$ must itself be false, and therefore Q is true. The logician W. Quine described the method this way: "It consists in assuming the contradictory of what is to be proved and then looking for trouble." As we did with direct proofs, we can justify this procedure by means of Table 1.19. Only on the bottom line of the table are $\sim P$ false and $\sim Q \Rightarrow \sim P$ true; and also on this line Q is true, which is what we want to show.

1.22 EXAMPLE. Let's give an indirect proof of proposition Q in Example 1.20 using the notation introduced in our first proof. If $\triangle$ does *not* have two equal angles, then $\triangle$ cannot be isosceles (by what we know about isosceles triangles); thus OA and OB must have different lengths. This contradicts the fact that all radii of the same circle have equal length, and the argument is finished. Symbolically, let P be the true proposition, "All radii of a given circle have equal length." We have just verified the statement $\sim Q \Rightarrow \sim P$, so by our earlier discussion Q is true.

The basic strategy in proofs by contradiction is to ask this question: What would go wrong if the proposition I am trying to prove were actually *false*?

You may ask: For a given proposition Q, what is better, a direct proof or an indirect proof? The answer varies with the proposition, but in specific cases the information provided by the the two kinds of proofs can differ substantially. For example, suppose Q asserts the existence of a number with a certain property. A direct proof of Q may actually produce the number or lead to its discovery, but an indirect proof will merely rule out the possibility that there is *no* such number and leave untouched the problem of actually finding it. Why would anyone *ever* choose to construct an indirect proof? Because we do what we can do. The facts at hand may lack the clout required for a direct proof of a proposition Q, whereas adding $\sim Q$ to our arsenal of assumptions *may* lead to some interesting conclusions, including the negation of a known theorem.

A special form of indirect proof for a proposition Q consists of proving an implication $P \Rightarrow Q$ in which P is known to be true, by proving the implication $\sim Q \Rightarrow \sim P$ instead. As in the general case for indirect proofs, we assume $\sim Q$ and seek a false consequence; however, here we specifically work to deduce $\sim P$ as that false consequence. The sentential form $\sim Q \Rightarrow \sim P$ is called the **contrapositive** of the form $P \Rightarrow Q$.

1.23 EXAMPLE. Let n denote an integer. Prove the implication

$$\text{If } \underbrace{n^2 \text{ is even}}_{P} \text{ then } \underbrace{n \text{ is even.}}_{Q}$$

SOLUTION. We prove the contrapositive, assuming the facts of elementary arithmetic. Suppose n is odd (this is $\sim Q$), say $n = 2k + 1$. (Glance back at the proof of Example 1.21 for justification.) Then

$$n^2 = (2k + 1)^2 = 4k^2 + 4k + 1 = \underbrace{4(k^2 + k)}_{\text{even}} + 1.$$

This number is odd, so we have proved $\sim Q \Rightarrow \sim P$, and we are through. (It may be useful for you to attempt a *direct* proof of $P \Rightarrow Q$. Such a proof follows easily from properties of factorization into primes, which we will look at in Section 6.5.)

Exercises

1. Suppose the following propositions are known to be true:

   If Dracula seizes power then democracy is lost.
   If democracy is lost then the use of food additives increases.

If the use of food additives increases, then mutations can be expected.

Dracula seizes power.

Represent each proposition by a sentential form, then show that repeated application of *modus ponens* proves the truth of "mutations can be expected."

2. For each of the following propositions, answer the question "What must I prove?" To do this, use the definition of each mathematical term to rephrase the given proposition. In each case assume that a planar geometrical configuration G has been presented to you. (That is, G is a collection of points or lines or both in a given plane.)

   (a) G is a rectangle.
   (b) G is a circle.
   (c) G is a parabola.

3. Prove that 5 is a prime number.

4. Prove that the sum of two odd numbers or two even numbers is even, and that the sum of an odd number and an even number is odd. (You are really being asked to prove three propositions here.)

5. A diamond is inscribed in a rectangle that is inscribed in a circle, as pictured here:

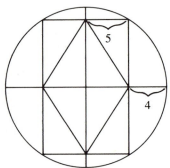

Determine the length of a side of the diamond, and then write a proof of your claim. (In the figure, the point that looks like the center *is* the center; and you may assume that the two lines through the center are mutually perpendicular.)

6. In this section the process of proving a theorem is compared to the process of climbing to the top of a mountain. Compare the notions of direct and indirect proof from this "mountain-theoretic" point of view.

1.5   LOGICAL EQUIVALENCE

Imagine someone with the bizarre habit of stating all his atomic sentences twice, with "and" wedged in between. Thus his days are filled with the likes of "the toast is hot and the toast is hot," "$2 + 3 = 5$ and $2 + 3 = 5$," and so on. In response to this, your instinct tells you that if P is a proposition, the statement symbolized by $P \wedge P$ carries the same information as P alone; in particular, P and $P \wedge P$ have the same truth value. You may even develop the urge to say that P and $P \wedge P$ are "logically equivalent." Giving a precise definition of this concept is our goal here. In the next section we will see a useful application to electronic circuit design.

1.24 DEFINITION.   A statement form is said to be a **tautology** if every substitution of propositions for its sentential variables yields a true proposition.

To verify a statement form's exalted status as a tautology we inspect the form's truth table: the column under the form itself must be an unbroken column of T's.

1.25 EXAMPLE.   Each of the following statement forms is a tautology:

$$P \Leftrightarrow \sim(\sim P), \quad P \wedge Q \Rightarrow P, \quad \sim(P \wedge \sim P), \quad ((P \Rightarrow Q) \wedge P) \Rightarrow Q.$$

When we replace the variables in tautologies by propositions, the new propositions thus obtained are true as a consequence of their form, not because of their content. For example, here are some incarnations of the tautologies in Example 1.25:

Jones likes waffles if and only if it is not the case that Jones dislikes waffles.

If Smith is dead and Doe likes melon, then Smith is dead.

It is not the case that today is Tuesday and not Tuesday.

If Washington is President implies Disney is Vice President, and Washington *is* President, then Disney is Vice President.

None of these statements creates major excitement, because in each case the actual content of the components has no bearing on the truth value of the complete statement.

1.26 DEFINITION. Suppose S_1 and S_2 are sentential forms. We say that S_1 and S_2 are **logically equivalent** if the sentential form $S_1 \leftrightarrow S_2$ is a tautology; the form $S_1 \leftrightarrow S_2$ is then called a **logical equivalence**. Thus a logical equivalence is a sentential form that becomes a true equivalence whenever its variables are replaced by propositions.

1.27 EXAMPLE. Let S_1 be the sentential form $\sim(P \wedge Q)$, and let S_2 denote the form $\sim P \vee \sim Q$. We claim that S_1 and S_2 are logically equivalent. To see this, consider the truth table for $S_1 \leftrightarrow S_2$:

P	Q	$\sim P$	$\sim Q$	$P \wedge Q$	S_1	S_2	$S_1 \leftrightarrow S_2$
F	F	T	T	F	T	T	T
F	T	T	F	F	T	T	T
T	F	F	T	F	T	T	T
T	T	F	F	T	F	F	T

The T's in the right-hand column show that $S_1 \leftrightarrow S_2$ is a tautology, and this proves the claim. A similar argument will show that $\sim(P \vee Q)$ is logically equivalent to $\sim P \wedge \sim Q$. The two logical equivalences in this example are called **De Morgan's laws**.

Example 1.27 illustrates the standard test for logical equivalence of two forms S_1 and S_2: display them both on the same truth table and ask, "Are the truth-value columns under S_1 and S_2 identical?"

For two statement forms S_1 and S_2, the assertion "S_1 is logically equivalent to S_2" is sometimes abbreviated

$$S_1 \equiv S_2.$$

For instance, in this notation De Morgan's laws look like this:

$$(1.28) \qquad \begin{aligned} \sim(P \wedge Q) &\equiv \sim P \vee \sim Q, \\ \sim(P \vee Q) &\equiv \sim P \wedge \sim Q. \end{aligned}$$

In elementary arithmetic we are accustomed to replacing part of an algebraic expression by something equal to it, thereby getting a new expression equal to the original. Thus the equality $3 + 10 = 13$ enables us to write $5 - \sqrt{3 + 10} = 5 - \sqrt{13}$. A similar principle holds in our present context:

REPLACEMENT PRINCIPLE.   If $S_1 \equiv S_2$ and if in some sentential form S an occurrence of S_1 is replaced by S_2, the resulting sentential form is logically equivalent to S.

1.29 EXAMPLE.   A truth table will show that the statement form $A \leftrightarrow A \wedge (A \vee B)$ is a tautology; hence the logical equivalence $A \equiv A \wedge (A \vee B)$ holds. The replacement principle then says that replacement of $A \wedge (A \vee B)$ by A in any sentential form yields a logically equivalent form. For instance,

$$\sim((A \wedge (A \vee B)) \Rightarrow C) \equiv \sim(A \Rightarrow C).$$

A formal proof of the replacement principle will not be given here, but the principle is clearly reasonable. For suppose S' is the sentential form resulting from the substitution of S_2 for S_1 in S, where $S_1 \equiv S_2$. Then, in the truth table for S', the column of truth values under S_2 is identical to the column under S_1 in the table for S. Thus, linking S_2 with the other components to get S' produces the same pattern of truth values that linking S_1 with those components produces in the construction of S.

Here is a list of some of the most basic logical equivalences, with their names. In each case it is left for you to verify the result with a truth table.

(1.30)
$$\left. \begin{array}{l} P \wedge (Q \wedge R) \equiv (P \wedge Q) \wedge R \\ P \vee (Q \vee R) \equiv (P \vee Q) \vee R \end{array} \right\} \quad \textbf{associative laws}$$

(1.31)
$$\left. \begin{array}{l} P \leftrightarrow Q \equiv Q \leftrightarrow P \\ P \wedge Q \equiv Q \wedge P \\ P \vee Q \equiv Q \vee P \end{array} \right\} \quad \textbf{commutative laws}$$

(1.32)
$$\left. \begin{array}{l} P \wedge P \equiv P \\ P \vee P \equiv P \end{array} \right\} \quad \textbf{idempotency laws}$$

(1.33)
$$\left. \begin{array}{l} P \wedge (P \vee Q) \equiv P \\ P \vee (P \wedge Q) \equiv P \end{array} \right\} \quad \textbf{absorption laws}$$

(1.34) $\quad \begin{aligned} P \wedge (Q \vee R) &\equiv (P \wedge Q) \vee (P \wedge R) \\ P \vee (Q \wedge R) &\equiv (P \vee Q) \wedge (P \vee R) \end{aligned} \Bigg\}$ **distributive laws**

(1.35) $\qquad\qquad \sim(\sim P) \equiv P$ **law of double negation**

Our interest in a sentential form centers on its truth-functional behavior. That is, we view the truth values of inserted propositions as "inputs"; and then the outputs are the truth values of the resulting propositions after the insertions. The form's "behavior" is determined by the pattern of inputs and outputs. (So the leftmost columns of the truth table, in which truth values are assigned to the variables, and the rightmost column, in which the truth values of the entire form appear, are the "behavioral" columns.) Logically equivalent forms have the same behavior in this sense. Notice that in each of laws 1.32 through 1.35 and in De Morgan's laws (1.28), one of the two equivalent forms is shorter than the other. Our natural desire for notational economy will usually lead us to choose the shorter of two equivalent forms for our work, just as in numerical calculations we prefer to use the numeral 1 in place of $(\cos \pi)^3/(\sin \frac{3}{2}\pi)^{77}$. There are payoffs. Shortening the sentential forms that represent statements in a mathematical proof (or a computer program) may provide simplifications that lead to greater understanding (or greater efficiency) and therefore perhaps to further developments. Sentential forms can be used as symbolic representations of certain kinds of electronic circuits (we will have a brief look at this in the next section), and simpler forms can lead to simpler and less expensive circuits.

1.6   INTRODUCTION TO SWITCHING CIRCUITS

Sentential forms are commonly used as mathematical representations of electronic switching circuits. The general scheme: replacing a sentential variable by a true proposition corresponds to the idea of a circuit wire's being electrically "alive" or "on" (that is, voltage is positive); replacement by a false proposition corresponds to the electricity being "off" in the wire. From this perspective, a propositional form corresponds to an electronic circuit in which one or more wires connected to independent power sources have been linked together with switching devices called *gates*. If we determine that a propositional replacement transforms a given sentential form into a true proposition, we then interpret that to mean that the corresponding pattern of power inputs and gates results in a flow of current. A false proposition implies the absence of current flow.

Three of the most basic gates are pictured as follows:

$$P \quad P \wedge Q \qquad P \quad P \vee Q \qquad P \quad \sim P$$

AND   OR   NOT

In each case, current (if any) is understood to flow through the gate from left to right. The AND gate is designed so that if the wires labelled P and Q are simultaneously "on," then the wire labelled $P \wedge Q$ is also "on"; otherwise $P \wedge Q$ is "off." This pattern can be read directly from the truth table for $P \wedge Q$ by reading "on" and "off" in place of T and F, respectively. Similarly, the actions for the OR and NOT gates can be read from the truth tables for $P \vee Q$ and $\sim P$, respectively.

1.36 EXAMPLE.   A room light is to be connected to switches in two locations, and a flick of either switch should change the state of the light (turns it off when on, and vice versa). Design a circuit for this purpose, using gates of the kind that have been described here.

SOLUTION.   Let P and Q denote wires leading from the two wall switches. Then, according to the circuit design specifications, any sentential form S that will represent a satisfactory circuit must have the following truth table:

P	Q	S
F	F	F
F	T	T
T	F	T
T	T	F

Thus S is true when either $\sim P \wedge Q$ is true (this is from line 2 of the truth table) or when $P \wedge \sim Q$ is true (from line 3); otherwise S is false. Thus S has the same truth-value assignments as $(\sim P \wedge Q) \vee (P \wedge \sim Q)$. That is,

$$S \equiv (\sim P \wedge Q) \vee (P \wedge \sim Q).$$

The solution to the problem therefore has the following design:

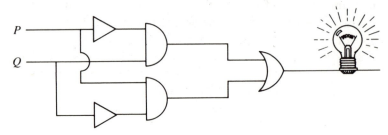

Seeing the light.

We note that if an EXCLUSIVE OR gate

is available, it will achieve the same goal, because the forms $P \veebar Q$ and $(\sim P \wedge Q) \vee (P \wedge \sim Q)$ are logically equivalent.

1.37 EXAMPLE.   A zoo has two special cages for endangered species. An alarm will be installed in the zoo office that will clang if either (or both) of the special cages is opened, assuming that the alarm's power supply is on. Give a diagram of an appropriate circuit.

SOLUTION.   Let A, C_1, and C_2 denote the power supplies for the alarm and the two cages, respectively. We could proceed with a truth table, as in the previous example, but instead let's think it through informally.

For the alarm to sound, either A and C_1 must both be activated or A and C_2 must both be activated. Thus $(A \wedge C_1) \vee (A \wedge C_2)$ represents a solution, and this corresponds to the following circuit diagram.

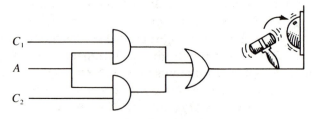

An alarming situation.

Now notice the logical equivalence

$$(A \wedge C_1) \vee (A \wedge C_2) \equiv A \wedge (C_1 \vee C_2).$$

[This is one of the distributive laws; see (1.34).] That is, the alarm will sound if it is on and either C_1 or C_2 is activated. The shorter sentential form $A \wedge (C_1 \vee C_2)$ yields a simpler circuit:

Still alarming, but simpler.

The new circuit is said to be **equivalent** to the original.

If we were to proceed by studying a truth table, as we did in Example 1.36, we would arrive at the grotesque sentential form

$$((A \wedge (C_1 \wedge C_2)) \vee (A \wedge (C_1 \wedge \sim C_2))) \vee (A \wedge (\sim C_1 \wedge C_2)),$$

whose corresponding circuit is a labyrinthine atrocity. (Draw it if you feel the compulsion.)

As Example 1.37 illustrates, the theory of switching circuits leads naturally to the *minimization problem* for sentential forms: given a sentential form S, determine the simplest sentential form among all those logically equivalent to S. The issue is of major importance in the design of complex electronic systems. We will not pursue this topic here. (An important first step would be to give a precise definition of "simple," and this may vary with the situation. For instance, in the design of the extraordinarily complex circuitry for a computer, reducing the amount of interchip communication can outweigh the benefits of reducing the total number of gates in the circuit.) Instead we refer the interested reader to the book *Boolean Algebra and Switching Circuits,* by Elliott Mendelson (New York: McGraw-Hill, 1970) for further discussion; in particular, see the extensive list of references given there. Also see *Introduction to Switching Theory & Logical Design,* by Frederick Hill and Gerald Peterson (New York: Wiley, 1981).

Exercises

1. Do Exercise 18 (if you have not already done so) on page 19.

2. A statement form S is said to be a **contradiction** if every substitution of propositions for its sentential variables yields a false proposition. Rewrite this definition using the term "tautology."

3. (a) Write a sentential form logically equivalent to

$$P \Rightarrow (Q \Rightarrow R)$$

   in which the symbol "$\Rightarrow$" does not occur.
   (b) Write a sentential form logically equivalent to $P \vee Q$ in which the only logical connectives are $\sim$ and $\wedge$.
   (c) Deduce that the work of the connectives $\wedge, \vee, \sim, \Rightarrow$ can be done by $\wedge$ and $\sim$ alone. (We say, therefore, that $\wedge$ and $\sim$ together constitute an **adequate set of connectives.**)

4. The purpose of this exercise is to improve upon Exercise 3(c) by introducing a *single* logical connective that can serve by itself as an adequate set of connectives. We write

$$P \uparrow Q$$

   as an abbreviation for $\sim(P \wedge Q)$; this new connective $\uparrow$ is called the **Sheffer stroke.** (The corresponding circuitry gate is called NAND, for "not and.") Use truth tables to verify the following logical equivalences.
   (a) $\sim P \equiv P \uparrow P$
   (b) $P \vee Q \equiv (P \uparrow P) \uparrow (Q \uparrow Q)$
   (c) $P \wedge Q \equiv (P \uparrow Q) \uparrow (P \uparrow Q)$
   (d) $(P \Rightarrow Q) \equiv$
   (Fill the blank with a form having only $\uparrow$ as a connective, and verify.)

5. (Companion to Exercise 4) We write $P \downarrow Q$ as an abbreviation for $\sim(P \vee Q)$. (The symbol $\downarrow$ is the **dagger;** its gate is called NOR, for "not or.") Represent each of the sentential forms $\sim P$, $P \vee Q$, $P \wedge Q$, $P \Rightarrow Q$ by forms using $\downarrow$ as the only connective, and verify your results.

6. Let S_1 and S_2 be sentential forms. Briefly discuss the distinctions between the expressions $S_1 \leftrightarrow S_2$ and $S_1 \equiv S_2$.

7. Show that the sentential form

$$((P \wedge Q) \vee (P \wedge \sim Q)) \wedge (R \vee \sim R)$$

is logically equivalent to a form consisting of just a single sentential variable.

8. Let S_1 and S_2 be sentential forms. The statement "S_1 **logically implies** S_2," written $S_1 \models S_2$, is defined to mean that the sentential form $S_1 \Rightarrow S_2$ is a tautology. (*Example: $P \wedge Q \models P$. Proof.* The truth table

P	Q	$P \wedge Q$	$P \wedge Q \Rightarrow P$
F	F	F	T
F	T	F	T
T	F	F	T
T	T	T	T

shows that $P \wedge Q \Rightarrow P$ is a tautology.) The terminology "logically implies" is modelled on common usage. A randomly chosen educated individual who has never studied logic will readily agree that "I like jam and Benjamin is charming" is a statement that "logically implies" that Benjamin is charming, even with no personal knowledge of Benjamin.

   (a) How does logical implication relate to material implication? (See page 13.)
   (b) How does logical implication relate to logical equivalence? (See page 30.)
   (c) Suppose $S_1 \models S_2$, and suppose further that S_1 and S_2 are both used as models for circuits with the same power sources. (Say S_1 represents circuit C_1 and S_2 represents C_2.) How are the performances of C_1 and C_2 related?

9. A room light is to be connected to three wall switches, each of which has a "down" position and an "up" position. In each case, design a circuit satisfying the stated specifications.

   (a) The light will be on if *at least* one switch is up; otherwise it will be off.
   (b) The light will be on if *exactly* one switch is up; otherwise it will be off.
   (c) If all switches are down the light is off; and flicking any switch at any time changes the state of the light.

10. Replace the following circuit with an equivalent circuit having only four gates. (Use only the NOT, AND, and OR gates.)

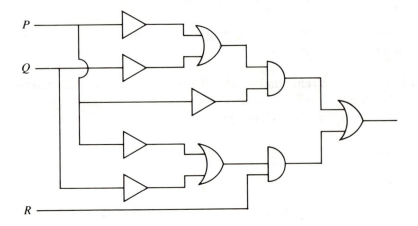

CHAPTER 2

Sets

He had . . . twelve marbles, part of a jew's-harp, a piece of blue bottle glass to look through, a spool cannon, a key that wouldn't unlock anything, a fragment of chalk, a glass stopper of a decanter, a tin soldier, a couple of tadpoles, six firecrackers, a kitten with only one eye, a brass doorknob, a dog collar—but no dog—the handle of a knife, four pieces of orange peel, and a dilapidated old window sash.

MARK TWAIN
The Adventures of Tom Sawyer

2.1 FUNDAMENTALS

We learn the meaning of a word either by seeing it defined in terms of words already known to us or by experiencing examples of its use and sensing the common properties of the examples. For instance, consider these statements: "Snerd is a member of the country club," "This saucer belongs to my dinnerware set," "*Fromage* is one of the French nouns," "Points A and B are on the perpendicular bisector of angle α," "The assault on the western flank was a key element of Varnish's battle plan." In each of these assertions something is said to be a member or element of some collection or set of things. As the examples suggest, sets are of wildly different sorts and, accordingly, "membership" can have a variety of meanings. (Dictionaries provide no substantial help: a standard dictionary

39

defines a set as a "collection of objects," a collection as an "aggregate," and an aggregate as a "collection"; and a member is "one of the elements of which an aggregate is composed." Check *your* dictionary.) But mathematics has to start somewhere, so we agree to accept the words **set** and **member** as undefined (or *primitive*) terms, though we take comfort from the meaningful interpretations we give the terms in our daily lives. For linguistic variety we sometimes use the words **family** and **collection** as synonyms for set, we use the word **element** for a member of a set, and we say that an element **belongs** to any set of which it is a member.

Our goals in this chapter are to learn to describe and manipulate sets and to develop techniques for the construction of new sets from given ones: we'll glue sets together, examine their overlap, take products and differences of them, and chop them up. Justification for our constructions is contained in the axioms of set theory, which we will usually not state formally; we prefer instead a more casual approach to the subject. Set theory is at the foundation of mathematics, and nearly every mathematical object of interest is a set of some kind. [Even the real numbers that we all grew up with (like 1, π, $-\frac{2}{3}$, and $\sqrt{5}$) have definitions as sets, but we will not go into the details here, and we will assume a familiarity with the basic arithmetic of real numbers.]

If x is an element and A is a set, the expression $x \in A$ denotes the statement "x is a member of A," which is either true or false. (In the language of Chapter 1, the statement "$x \in A$" is a *proposition*.) The symbol $\in$ is called the **membership symbol**. The negation of $x \in A$ is usually written $x \notin A$ instead of $\sim(x \in A)$. Another common abbreviation: the statement "$a \in S$ and $b \in S$" is shortened to "$a, b \in S$," and similarly for longer expressions (such as "$x, y, z \in S$"). To minimize confusion in writing about sets and their elements, it is useful to use uppercase letters for sets and lowercase letters for their elements ($x \in X$ instead of $X \in x$). However, we cannot be consistent about this, since, as we will see, sets are themselves members of sets, and larger type is not always a possibility.

$$x \in X \quad \text{and} \quad X \in X \quad \text{and} \quad X \in X \quad \text{and} \ldots$$

An element of an element of an element of an element of . . .

2.1 DEFINITION.   Sets A and B are said to be **equal**, written $A = B$, if they have the same members, that is, if every member of A is a member of B and vice versa. More formally, the statement $A = B$ means that for every element x the following equivalence holds:

$$x \in A \iff x \in B.$$

If A and B are not equal we write $A \neq B$.

2.2 EXAMPLE. Let A be the set of nonnegative real numbers, and let B be the set of all real numbers that are squares of real numbers. Then by Definition 2.1 we have $A = B$, even though the descriptions of A and B are quite different. (The first describes the set's location on the real line, the second has to do with solvability of equations.)

Since a set is completely determined by its members, if we wish to define a set (that is, to specify it completely) it suffices to indicate what its members are, and there are several standard ways to do this. If a set is small we may be able to list its elements between braces: $\{1, \sqrt{2}\}$ has as its elements the numbers 1 and $\sqrt{2}$. When we use this notational device, it is understood that the order in which we list the terms is irrelevant, and also that symbol repetition has no effect on the set being described. Thus

$$\{1, \sqrt{2}\} = \{\sqrt{2}, 1, \sqrt{2}, 1, 1\} = \{\sqrt{2}, 1, \sqrt{2}\} = \{\sqrt{2}, 1\}.$$

Sometimes it is more convenient to *suggest* a listing of the elements: the expression

$$\{0, 1, 2, \ldots, 1000\}$$

denotes the set whose members are the nonnegative integers from 0 through 1000. It is important that we be on guard for ambiguities: does the expression

$$\{3, 5, 7, \ldots, 23\}$$

denote the set of eleven odd numbers from 3 through 23,

$$\{3, 5, 7, 9, 11, 13, 15, 17, 19, 21, 23\},$$

does it represent the set consisting of the first eight odd prime numbers,

$$\{3, 5, 7, 11, 13, 17, 19, 23\},$$

or does it represent some other set?

Very often there is a property P that completely characterizes the elements of a set A, in which case we can write

$$A = \{x \mid P(x)\},$$

which reads, "A is equal to the set of all x such that property P holds for x." (We will not explicitly define what a "property" is.) For example, the set $B = \{0, 1, 2, \ldots, 1000\}$ can be written

$$B = \{x \mid \underbrace{x \text{ is an integer and } 0 \leq x \leq 1000}_{P(x)}\}.$$

In this format the letters chosen to represent elements are unimportant:

$$\{x \mid P(x)\} = \{y \mid P(y)\}.$$

For example,

$$\{x \mid x \text{ is an integer and } 1 < x \leq 3\} = \{y \mid y \text{ is an integer and } 1 < y \leq 3\}$$
$$= \{2, 3\}.$$

Suppose we want to instruct a computer to print out the elements of the set $B = \{0, 1, 2, \ldots, 1000\}$. For this purpose the description $\{0, 1, 2, \ldots, 1000\}$ is inadequate, since the computer will not assign a meaning to the expression ". . ." that we interpret to mean "and so on." (Computers are faster than we are, but not smarter.) We can of course type a one-thousand-line program like this (we ignore format details):

```
PRINT 0
PRINT 1
PRINT 2
    ⋮
PRINT 1000
END
```

(No dots would appear in the actual program.) This is grossly inefficient, because we would actually have typed the desired list while typing the program, and we would also have typed "PRINT" one thousand times! Far better would be something like

```
LET K = 0                          LET K = 0
→ PRINT K                          → DO WHILE K ≤ 1000
⌈ ADD 1 TO K             or        ⌈ PRINT K
⌊ IF K ≤ 1000 THEN LOOP            │ ADD 1 TO K
  END                             ⌊ LOOP
                                    END
```

Such a procedure is called *recursive*. Roughly speaking, a procedure is **re-**

cursive if one or more initial steps are given, and then a general method is given for determining each subsequent step on the basis of steps already taken. The brief program

```
    LET K = 0
┌─► DO
│      PRINT K
│      ADD 1 TO K
└── LOOP
    END
```

recursively instructs a computer to print out *all* the nonnegative integers, a task that no machine can ever finish.

We will return to the topic of recursion later on. The ability to specify enormous and sophisticated entities (for example, computer programs with many steps, sets with many elements) in a few lines of print is clearly desirable.

REMARKS.

(1) Here and elsewhere in the book, our programs will need to be adapted to fit the technical requirements of a specific programming language if they are actually to run on a computer. They are written here in a less formal style in order to avoid entanglement in issues that are primarily programming concerns rather than mathematical concerns.

(2) Computer scientists are usually more restrictive in their use of the label "recursive" than we have been: they require that a recursive procedure make reference to itself. (In the standard terminology, a recursive procedure "calls" itself into action.) For example, a procedure for printing the integers from 0 to N would need to have a format something like the following if it is to be called recursive in this narrower sense:

```
PROCEDURE: NUMBERS (N)
IF N = 0 THEN PRINT 0
ELSE
    CALL NUMBERS (N − 1)
    PRINT N
```

The essential idea is the same as before: we have reduced a highest-level task to a lower level. In this book we will continue to use the term "recursive" without insisting on the self-referencing feature.

Certain sets are used so often that they have been given standard names, and here are a few of them:

$\mathbb{N} = \{1, 2, 3, 4, 5, \ldots\}$,   the **natural numbers** or **positive integers**

$\mathbb{Z} = \{\ldots, -3, -2, -1, 0, 1, 2, 3, \ldots\}$,   the **integers** or **whole numbers**

$\mathbb{Q} = \{x \mid x = a/b$ for some $a, b \in \mathbb{Z}, b \neq 0\}$,   the **rational numbers**

$\mathbb{R}$,   the **real numbers**

These number collections are familiar to everyone who has done arithmetic, and we will use them freely. Carefully justifying the assertion that these are sets is not an elementary matter, and we will leave this task for more advanced courses. (Consider this: if terrorists were to demand that you give a precise definition of "positive integer," would you be able to satisfy their demand?)

We have seen that some sets can be defined with a characterizing property $P(x)$. We do *not* mean to suggest that there is always a sort of camaraderie among the members of a given set that enables us to view them all as variations on a single theme. For instance, if

$$S = \{\pi, \text{Franklin Delano Roosevelt}, -\sqrt{53}, \text{spaghetti}, 12, 0\},$$

then we would be hard put to specify a property characterizing the elements of S other than the property of being on the given list. Similar considerations hold for the set of Tom Sawyer's goods listed at the start of this chapter.

2.3 EXAMPLE.   Consider the following three expressions.

$A = \{x \mid x$ is a left shoe$\}$

$B = \{y \mid y$ is a left sock$\}$

$C = \{z \mid z$ is a Tuesday before January 1983 and after February 1400 on which it rained at noon on what is now the site of the Washington Monument$\}$

Here A is a set, assuming that we have an unambiguous definition of "shoe" to work with. But B is not a set since the term "left sock" is ambiguous; we say B is not **well-defined**. Finally, C *is* a well-defined set, since on any Tuesday in the given time interval it either did or did not rain at the indicated time and place. (So if x is an element, the statement $x \in C$ is a proposition.) But we have no way to determine all the members of C. For example, is the second Tuesday of the year 1417 a member of C?

Once again let us note the distinction between B and C. No set B has been defined because the criterion for membership is unclear. On the other hand, C *is* a set because there is no ambiguity in the membership requirement; yet in practice there is no way of actually determining its members. In the language of Chapter 1, B fails to be a set because the statement that seems to define it is not a proposition; C *achieves* set-hood but is unmanageable because we cannot reliably determine the truth value of its defining statement for all membership candidates.

The axioms of set theory tell us that sets exist. These axioms also provide us with some explicit examples, and they guide us in the formation of new sets from given ones by a variety of abstract cutting-and-pasting maneuvers. The axiom that follows is the set-theoretic version of this common kitchen wisdom: an individual is not obliged to consume an entire roast turkey at one sitting; instead, the diner can extract an appropriate portion with skillful carving.

AXIOM OF SEPARATION. Given a set X and a property P, there is a set whose elements are the elements of X that have property P. That is, $\{x \mid x \in X \text{ and } P(x)\}$ is a set.

A convenient abbreviation: if S is a set and P is a property, the set

$$\{x \mid x \in S \text{ and } P(x)\}$$

is usually written

$$\{x \in S \mid P(x)\}.$$

2.4 EXAMPLE. We have already introduced the set of integers, $\mathbb{Z} = \{0, \pm 1, \pm 2, \pm 3, \ldots\}$. Therefore, by the axiom of separation, the following are also sets:

$\{x \in \mathbb{Z} \mid x = 2y \text{ for some } y \in \mathbb{Z}\}$, the set of even integers;

$\{x \in \mathbb{Z} \mid x = 2y + 1 \text{ for some } y \in \mathbb{Z}\}$, the set of odd integers;

$\{n \in \mathbb{Z} \mid n > 0 \text{ and } x^n + y^n = z^n \text{ is solvable for some } x, y, z \in \mathbb{Z}$
$$\text{with } xyz \neq 0\}.$$

The numbers 1 and 2 are members of the last set; as of this writing no one knows whether there are others.

In general, from now on we will feel free to discuss sets of real numbers that satisfy virtually any condition without officially announcing our dependence on the axiom of separation. We will refer, for instance, to the set of prime numbers, the set of irrational numbers, the set of negative real numbers, and so on.

Just as it is useful to have the number zero available to indicate the absence of items to count, it is useful to have the concept of a set that has no elements. Thus we view a stamp collector as having a collection even if he has just declared himself a collector and as yet has acquired no stamps; we say his collection is *empty*. We can formally justify this idea by appealing to the axiom of separation, which assures us that the following object is a set:

$$\emptyset = \{x \in \mathbb{Z} \mid x \neq x\}.$$

This set clearly has no elements, since no integer has the property $x \neq x$; we call $\emptyset$ an **empty set.**†

We can use the axiom of separation in other ways to define empty sets; for example, these are also empty sets:

$$\{x \in \mathbb{Z} \mid x^2 = 2\} \quad \text{and} \quad \{x \in \mathbb{R} \mid x^2 = -1\}.$$

We claim: *all empty sets are equal.* To prove this, suppose $\emptyset$ and $\emptyset'$ are both empty sets; we must prove $\emptyset = \emptyset'$. (That is, if Joe collects stamps and has no stamps, and Pat collects coins and has no coins, then their collections are equal.) By definition of set equality (2.1), we must verify the equivalence $x \in \emptyset \Leftrightarrow x \in \emptyset'$ for every element x. But for every x the statements $x \in \emptyset$ and $x \in \emptyset'$ are both false, since $\emptyset$ and $\emptyset'$ have no elements; therefore the desired equivalence is true for every x. (Recall Table 1.18). To sum up, we have proved the following:

2.5 THEOREM.   There is exactly one empty set.

---

† In formal studies of the foundations of mathematics, the existence of an empty set is asserted at or near the very beginning of set theory, and this set is a fundamental building block in the subsequent development of number sets. In our less formal approach we have taken the number sets for granted and deduced the existence of an empty set as a consequence.

We will continue to use the symbol $\emptyset$ for the empty set.

The complete set of photographs of ninth-century life.
(Alias $\emptyset$)

2.6 EXAMPLE.   We assert:

$$\emptyset \neq \{\emptyset\}.$$

PROOF.   The set on the left has no members. The set on the right has $\emptyset$ as a member. Therefore the inequality holds.

Notice that in stating Example 2.6 we have used an unannounced axiom of set theory: If S is a set then so is $\{S\}$; that is, there is a set that has S as its only member. We will generally omit the formal statements of the axioms of set formation if confusion seems unlikely.

Exercises

1. At Ace Elementary School the basketball team consists of Martha, Chet, Fred, Mervin, and Peggy. The softball team has exactly the same members. Are they the same set?

2. Describe each of the following sets in the format $\{x \mid P(x)\}$.
   (a) $A = \{0, 2, 4, 6, 8, 10, 12, 14, 16, 18, \ldots\}$
   (b) $B = \{1, 2, 5, 10, 17, 26, 37, 50, \ldots\}$
   (c) $C = \{1, 5, 9, 13, 17, 21, \ldots\}$
   (d) $D = \{1, \frac{1}{2}, \frac{1}{3}, \frac{1}{4}, \frac{1}{5}, \ldots\}$
   (e) $E = \{\text{lemon}, 1, 2, 3, 4, 5, \ldots\}$

3. Describe each of the following sets by listing all its elements between a pair of brackets.
   (a) $\{x \in \mathbb{N} \mid x \text{ is prime and } x^2 < 30\}$
   (b) $\{x \in \mathbb{Z} \mid x \text{ is prime and } (x \text{ is even or } 3 \leqslant x \leqslant 10)\}$

   (c) $\{x \in \mathbb{Q} \mid x \in \mathbb{N}$ and $1 \leqslant x \leqslant 3\}$

   (d) $\{x \in \mathbb{Z} \mid$ There is an element $y \in \mathbb{N}$ such that $x^2 + y^2 \leqslant 25\}$

4. Which of the following are well-defined sets? Explain briefly.

   (a) The set of large real numbers.

   (b) The set of all integers that are not integer multiples of 7.

   (c) The set of all four-digit strings that appear in the decimal expansion of π to the right of the decimal point. (Example: Since $\pi = 3.1415926535. \ldots$, the first four strings of this kind are 1415, 4159, 1592, and 5926.)

5. Label each statement *true* or *false*.

   (a) $\{1, 2, 3\} = \{3, 1, 2\}$

   (b) $\{1, 2, 3\} = \{2, 1, 3, 3, 2\}$

   (c) $\{5, \emptyset\} = \{5\}$

   (d) $\{5\} \in \{2, 5\}$

   (e) $\emptyset \in \{1, 2\}$

   (f) $\{1, 2\} \in \{1, 2\}$

   (g) $\{1, 2\} \in \{3, \{1, 2\}\}$

   (h) $2 \in \{\{1, 2\}\}$

   (i) $\{x \in \mathbb{R} \mid x^2 = -2\} = \emptyset$

   (j) $\emptyset \in \emptyset$

6. In each part give an example of sets A, B, C that simultaneously satisfy the stated conditions.

   (a) $A \in B, \ B \in C, \ A \notin C$

   (b) $A \in B, \ B \in C, \ A \in C$

   (c) $A \notin B, \ B \in C, \ A \in C$

7. Take the following statement as an axiom: No set is a member of itself.

   (a) Show that if A is any set then $A \neq \{A\}$.

   (b) Argue convincingly that if A is a set and

$$\underbrace{\{\{\{ \cdots \{\{A\}\} \cdots \}\}\}}_{n \text{ brackets}} = \underbrace{\{\{\{ \cdots \{\{A\}\} \cdots \}\}\}}_{m \text{ brackets}},$$

then $n = m$. (A precise proof requires *mathematical induction,* a technique to be considered later.)

8. (a) Write a recursive procedure whose object is to print all the integers. Use the format of a computer program, as was done in the text.

   (b) Do the same for the even integers.

   (c) Do the same for the odd integers.

9. (Challenging) Write a recursive procedure in the format of a computer program whose object is to print all the positive rational numbers. (Suggestion: think of a way to begin listing the positive rational numbers by hand in such a way that any preselected positive rational number will eventually be reached.)

2.2   RUSSELL'S PARADOX AND THE HALTING PROBLEM

We may find it reasonable to expect that every property P defines a set, namely, the set of all objects that satisfy that property: $\{x \mid P(x)\}$. Let us assume for the moment that this is the case. Consider the set

$$S = \{A \mid A \text{ is a set and } A \notin A\}.$$

For instance, if A is a set of pencils, then A is not itself a pencil, so $A \notin A$, and therefore $A \in S$. On the other hand, the set B of all abstract ideas is itself an abstract idea, hence $B \in B$, and so $B \notin S$.

Question: Is the set S under discussion a member of itself? If $S \in S$, then the description of the elements of S tells us that $S \notin S$. On the other hand, if $S \notin S$, then S satisfies the property required of elements of S, and so $S \in S$. We have therefore verified the equivalence

$$S \in S \Leftrightarrow S \notin S.$$

This is lunacy, because the statement $S \in S$ is a proposition and therefore has *opposite* truth value from that of its negation, $S \notin S$. What has gone wrong? Answer: The assumption that S is a set has led to a contradiction. Therefore *we must abandon the notion that S is a set, and also the wider assumption that for every property P the expression $\{x \mid P(x)\}$ defines a set.*

Bertrand Russell presented this disheartening news to the mathematical world in 1901, and Russell's paradox (the apparent set that is and is not a member of itself) caused widespread fear that the foundations of mathematics were built in quicksand. The axiom of separation, which we discussed earlier, enables us to avoid Russell's paradox by limiting our expectations: instead of asserting that every property determines a set, the axiom says that *inside a given set whose existence has already been guaranteed,* the elements satisfying a given property constitute a set. For example, once we have accepted the existence of the set $\mathbb{R}$ of real numbers, we are then free to define sets of real numbers that satisfy whatever further property our hearts desire. (Of course, if the property is outrageous the resulting set may be empty.)

2.7 EXAMPLE. If X is a set then

$$S = \{A \in X \mid A \text{ is a set and } A \notin A\}$$

is a set, by the axiom of separation. At first glance this set resembles the subject of Russell's paradox. Again we ask: Is S a member of itself? If $S \in S$ then our description of elements of S tells us these facts: $S \in X$, and S is a set, and $S \notin S$. But the last of these facts contradicts the statement $S \in S$. Therefore the hypothesis $S \in S$ is false, so we have $S \notin S$. Thus S must fail to satisfy some criterion for membership in S. But S is a set and $S \notin S$, so S must fail the only other requirement for membership in S; that is, we conclude that $S \notin X$. *This is no paradox; nowhere did we assume that $S \in X$.*

Example 2.7 shows that given any set X, there is something (in the example it is S) that is not a member of X. So:

There is no set that contains everything.

The spirit of Russell's paradox carries over to theoretical computer science. Just as Russell asked whether every property determines a set, we can reasonably ask whether every programming objective can be achieved by a sufficiently clever program, given enough time and memory. We will now briefly consider this issue. [A more detailed treatment requires a precise definition of the notion of *algorithm* (which, roughly speaking, means "organized procedure") by means of *Turing machines.* We leave this for courses in theoretical computer science.]

A standard problem that every programmer must consider is this: Given a program Q and a set I of data, will the execution of Q come to a halt when I is used as input? Briefly stated: Does Q halt on I? For example, if I is in a format unacceptable for Q, then Q will halt soon after starting. At the other extreme, if I is accepted by Q and the program goes into an endless loop, then Q does not halt on I.

It would by wonderful to have a program P that tests for halting in all cases. (We would call P a *decision procedure for the halting problem.*) Such a program P would accept as input any program Q and data set I and perform the following action:

P prints YES if *Q* halts on *I*, and then *P* halts.

P prints NO if *Q* does not halt on *I*, and then *P* halts.

Let us assume for now that such a program P exists.

A program itself can be used as data on which a program already in the computer can act, possibly by halting immediately. That is, there is no *mechanical* distinction between a program *Q* and a data set *I*: each can be represented by a stack of punched cards, a strip of magnetized tape, a string of 0s and 1s, and so on. It is therefore appropriate to consider what will happen when a program *Q* is loaded into the computer, and then *Q* is again loaded as the data set *I* on which *Q* is to act: either *Q* will halt on *Q* or it won't.

Now let's modify *P* to construct a new program *P** that accepts any program *Q* as input and acts as follows:

P* prints YES if *Q* halts on *Q*, and then *P** goes into the loop:

$$\begin{array}{l} \rightarrow \text{PRINT 0} \\ \llcorner \text{LOOP} \end{array}$$

P* prints NO if *Q* does not halt on *Q*, and then *P** halts.

Question: Does *P** halt on *P**? That is, does *P** halt when it acts on the data set *Q* = *P**? If the answer is "yes" then the answer is "no," and vice versa. Therefore such a program *P** is impossible, and therefore *P* was an impossibility.

Our conclusion: there are fundamental limitations on what a computer program can achieve, just as there are fundamental limitations on the properties that can be used to characterize the members of a set.

The unsolvability of the halting problem is more serious for computer science than Russell's paradox is for mathematics. As we have seen, Russell's paradox can be avoided by introducing the axiom of separation. But the halting problem's unsolvability tells us that there are fundamental limitations inherent in the nature of algorithmic computations.

2.3 QUANTIFIERS

The statement $x^2 + 8 = 17$ is not a proposition, because until we know what x is, we cannot assign a truth value to the statement. But it becomes a proposition when x is replaced by a number: $5^2 + 8 = 17$ is false,

$(-3)^2 + 8 = 17$ is true. Here we say that x is a **variable:** it is a symbol used in a statement P in such a way that P becomes a proposition when the symbol is replaced by a specific element, whereas in its original form P is not a proposition.

Sometimes we may want to say that a given statement P with a variable x is true for *at least one* substitution of an element for x, or for *every* such substitution. For this purpose we introduce the notation

$$(\exists x)P$$

which we verbalize as "There exists an x such that P" or "For some x, P" or "There is at least one x for which P is true." We also introduce the notation

$$(\forall x)P$$

which is verbalized "for every x, P" or "for all x, P." The symbol $\exists$ is called the **existential quantifier,** and $\forall$ is the **universal quantifier.**

2.8 EXAMPLES.

(a) The statement

$$(\exists x)(x \in \mathbb{Z} \text{ and } x^2 - 7x + 10 = 0)$$

is a true proposition since the substitution of 2 for x yields a true equation. In practice, abbreviations are common. For instance we may shorten the preceding statement to

$$(\exists x \in \mathbb{Z})(x^2 - 7x + 10 = 0)$$

or, if we understand that $\mathbb{Z}$ is the underlying set of interest, to

$$(\exists x)(x^2 - 7x + 10 = 0).$$

(b) To assert that the product of any real number with zero is zero, we write

$$(\forall x)(x \in \mathbb{R} \implies x \cdot 0 = 0)$$

or, if $\mathbb{R}$ is understood as our element source,

$$(\forall x)(x \cdot 0 = 0).$$

(c) The statement that every nonnegative real number has a square root is represented by the two-variable proposition

$$(\forall x)(x \in \mathbb{R} \text{ and } x \geqslant 0 \Rightarrow (\exists y)(y \in \mathbb{R} \text{ and } y^2 = x)).$$

More informally (mixing words and symbols, and with the understanding that we are discussing real numbers), we can write

$$\forall x \geqslant 0, \quad \exists y \text{ such that } y^2 = x,$$

which in turn has the compactification

$$(\forall x \geqslant 0)(\exists y)(y^2 = x).$$

Notice that the order in which the quantifying phrases occur is crucial. The present example has the form, "For every nonnegative x something happens"; accordingly, $\forall x$ starts off the symbolic representation. The "something" that happens is that "there is an element with a certain property," and therefore the next part of the symbolic representation starts with $\exists y$. By contrast, the expression

$$(\exists y)(\forall x \geqslant 0)(y^2 = x)$$

translates into "There is a number y that serves as a simultaneous square root for all nonnegative real numbers," clearly a falsehood.

(d) For variety, sometimes instead of $(\forall x)P$ we write P, $\forall x$. For example, in analytic geometry we assert that every point (x, y) in a given set A lies outside the unit circle (the circle with center $(0, 0)$ and radius 1) as follows:

$$x^2 + y^2 > 1, \qquad \forall (x, y) \in A.$$

Now let's consider the negation of quantified statements. If P is a statement in which the variable x occurs, then $(\exists x)P$ asserts that there is an element for which P is true. Therefore the negation $\sim(\exists x)P$ asserts that there is *no* such element. (Read more literally: It is not the case that there is an element for which P holds.) That is, for every element the statement $\sim P$ is true; so we have

(2.9) $$\qquad\qquad\qquad \sim(\exists x)P \iff (\forall x)(\sim P).$$

Similarly, the statement $(\forall x)P$ tells us "P is true for every x," so the negation $\sim(\forall x)P$ says, "There is some x for which P is false." Thus

(2.10) $\sim(\forall x)P \;\Leftrightarrow\; (\exists x)(\sim P)$.

2.11 EXAMPLES.   The purpose of this example is to familiarize you quickly and informally with quantifier manipulation, without getting entangled in substantive mathematical issues. Here are some standard English sentences and their negations in the quantification format. Each variable is named with the initial of the word it represents.

(a) "There is a cat in that house" can be represented

$$(\exists c)(c \text{ is in that house}).$$

By 2.9 the negation is

$$(\forall c)(c \text{ is outside that house}),$$

which means, "Every cat is outside that house."
(b) "Every animal eats some food" is written

$$(\forall a)(\exists f)(a \text{ eats } f).$$

By 2.10 the negation can be written

$$(\exists a)(\sim(\exists f)(a \text{ eats } f)).$$

By 2.9 this in turn is equivalent to

$$(\exists a)(\forall f)(a \text{ does not eat } f),$$

meaning "There is an animal that eats no food."
(c) "There is an animal that eats every food" becomes

$$(\exists a)(\forall f)(a \text{ eats } f).$$

By 2.9 and 2.10, this negates to

$$(\forall a)(\exists f)(a \text{ does not eat } f),$$

which we may interpret as "Associated with every animal is some food it doesn't eat."

(d) The order in which variables are quantified can affect the meaning. For example, if instead of (b) we write

$$(\exists f)(\forall a)(a \text{ eats } f),$$

we get "There is a food that every animal eats." Thus, statement (b) is a statement about each animal; and we would presumably label it "true," assuming that "eats" is interpreted to mean "consumes as a part of its natural diet." But the present statement asserts the existence of a certain food, and we must clearly label it "false" (if we interpret "eats" in the same way). Is there any food eaten by both lobsters and giant pandas?

Similarly, modifying (c) to

$$(\forall f)(\exists a)(a \text{ eats } f)$$

yields "Every food is eaten by some animal." The negations of these two statements are left for you to consider.

(e) "You can fool all of the people all of the time" is represented

$$(\forall p)(\forall t)(\text{You can fool } p \text{ at time } t).$$

By a double application of 2.10, this has the negation

$$(\exists p)(\exists t)(\text{You can't fool } p \text{ at time } t),$$

which translates to "There is a person who sometimes can't be fooled." Alternatively, we could have started with the representation

$$(\forall t)(\forall p)(\text{You can fool } p \text{ at time } t).$$

This negates to

$$(\exists t)(\exists p)(\text{You can't fool } p \text{ at time } t),$$

which means, "There is a time when at least one person can't be fooled."

The analysis of examples of the kind we have been considering here illustrates the vagueness that permeates our usual language. "There is a song that everyone is singing" and "Everyone is singing a song" have different intended meanings that can be captured by different quantifications, whereas in ordinary conversation we may require further discussion for clarification. Ambiguities like this will keep lawyers in business forever.

It is appropriate to pause for a moment to appreciate the benefits of our notation. If we are given a statement of staggering complexity, of the form

$$(\forall x_1)(\exists x_2)(\exists x_3)(\forall x_4) \cdots (\exists x_n)P$$

(where P is a proposition), we can now negate it mechanically to

$$(\exists x_1)(\forall x_2)(\forall x_3)(\exists x_4) \cdots (\forall x_n)(\sim P)$$

by repeated application of rules 2.9 and 2.10, without struggling to grasp the meaning of the given statement. Indeed, a computer can be programmed to carry out the symbol manipulation. Here is a commentary by Alfred North Whitehead, a distinguished mathematical philosopher (and a colleague of Bertrand Russell, our friend from an earlier section), in response to a similar notational triumph:

> This example shows that, by the aid of symbolism, we can make transitions in reasoning almost mechanically by the eye, which otherwise would call into play the higher faculties of the brain.
> It is a profoundly erroneous truism, repeated by all copybooks and by eminent people when they are making speeches, that we should cultivate the habit of thinking of what we are doing. The precise opposite is the case. Civilization advances by extending the number of important operations which we can perform without thinking about them. Operations of thought are like cavalry charges in a battle—they are strictly limited in number, they require fresh horses, and must only be made at decisive moments.†

There is one more useful quantifier, really a souped-up existential quantifier: the prefix

$$(\exists!x)$$

† From A. N. Whitehead, *An Introduction to Mathematics* (Oxford: Oxford University Press, 1959).

means "There is exactly one x" or "There is a unique x." A proof of a statement of the form $(\exists!x)P$ has two parts: the *existence* part, in which we prove $(\exists x)P$; and the *uniqueness* part, in which we prove that if statement P is true for both x and y then necessarily $x = y$.

Exercises

1. Express the following statements symbolically, using quantifiers, logical connectives [$\wedge$ (and), $\vee$ (or), $\sim$ (not), $\Rightarrow$ (implies)], and standard mathematical symbols from arithmetic, but using no English words. [Example: The false proposition "every real number's square is less than 8" can be written $(\forall x)(x \in \mathbb{R} \Rightarrow x^2 < 8)$, or in the more abbreviated form $(\forall x \in \mathbb{R})(x^2 < 8)$.]

    (a) There is a positive integer whose cube, when added to 15, yields a sum of 22.

    (b) Every positive integer has the property that when its cube is added to 15, the result is 22.

    (c) It is not the case that every real number is a square of a real number.

    (d) There is a real number that is not the square of a real number.

    (e) Every real number has a unique cube root.

    (f) For every integer n there is a set A that has a proper subset B with more than n elements. (Here you may use the English phrase "is a set" and the abbreviation $\#B$ for the number of elements in B.)

    (g) If every element of A is an integer then there is a unique negative real number that belongs to B.

2. Let $A = \{1, 2, \pi\}$, and let P be the statement "$x \in A$ and $x \in \mathbb{Z}$." Determine the truth value of each of the following implications, and justify briefly.

    (a) $(\exists x)P \Rightarrow (\forall x)P$

    (b) $(\forall x)P \Rightarrow (\exists x)P$

3. (a) Redo Exercise 2, using $A = \emptyset$ instead.

    (b) Is there a set A for which the truth value of statement (b) of Exercise 2 is *false?* Explain.

4. Write the following statements in more abbreviated form, using quantifiers. Here the short phrases "is prime" and "is a line" are allowed, and the symbol Π may be used for "the plane."

    (a) 17 is not the largest prime number.

    (b) There is no largest prime number.

(c) Every real number has a fifth root.

(d) Every pair of distinct points in the plane lies on a unique line.

5. The following definition can be found in most calculus books. A function f is **continuous** at $x \in \mathbb{R}$ if the following condition holds:

> For every $\epsilon > 0$ there is a number $\delta > 0$ such that
> $|f(x') - f(x)| < \epsilon$ whenever $|x' - x| < \delta$.

(a) Write the condition in abbreviated form, using quantifiers.

(b) Write the negation of this condition in a quantified form, using no negation symbols.

(c) Write out part (b) mostly in words: "A function f is not continuous if. . . ."

6. A **prime number** is an integer greater than 1 that cannot be written as a product of smaller positive integers. Write the definition of prime number using symbols and no words. That is, complete the statement

p is a prime number $\Leftrightarrow$ _____

with a suitable symbolic expression.

7. Suppose you are working with real numbers. Devise equations P and Q, in which x appears as the only variable, such that the propositions

$$(\exists x)(P \wedge Q) \quad \text{and} \quad (\exists x)P \wedge (\exists x)Q$$

have opposite truth values.

8. Show the ambiguity of the statement "All men are not liars" by giving two reasonable interpretations that have different meanings. Use quantifiers, as in Example 2.11; and to convince the reader that your two interpretations are different, give a condition under which one is true while the other is false.

9. Label each of the following statements *true* or *false*. All statements here refer to real numbers.

(a) $(x \geqslant 0) \Rightarrow (\exists y)(y^2 = x)$ (e) $(\forall x)(\exists! y)(xy = 0)$

(b) $(x \geqslant 0) \Rightarrow (\exists! y)(y^2 = x)$ (f) $(\exists x)(\exists! y)(xy = 0)$

(c) $(\forall x)(\exists! y)(y^3 = x)$ (g) $(\forall x)(\exists! y)(x + y = 0)$

(d) $(\exists! x)(\forall y)(xy = y)$

2.4   SET INCLUSION

We often recognize that a given collection of objects is part of a broader class of things: my books are among my prized possessions; owls are nocturnal creatures; the integers are real numbers.

2.12   DEFINITION.   Let A and B be sets. We say A is a **subset** of B (or A is **contained** in B, or B **contains** A, or B **includes** A), written $A \subseteq B$ (or $B \supseteq A$), if every member of A is a member of B. More formally,

$$A \subseteq B \Leftrightarrow (\forall x)(x \in A \Rightarrow x \in B).$$

The symbol "$\subseteq$" is called the **set inclusion** symbol. If A is not a subset of B we write $A \nsubseteq B$.

2.13   EXAMPLES.

(a) $\{1, 3\} \subseteq \{1, \pi, 3\}$
(b) $\{\emptyset\} \subseteq \{\emptyset, 5\}$
(c) $\mathbb{N} \subseteq \mathbb{Z} \subseteq \mathbb{Q} \subseteq \mathbb{R}$

Suppose we wish to prove $A \subseteq B$. We first recall that an implication $P \Rightarrow Q$ is always true if P is false. In particular, the statement $x \in A \Rightarrow x \in B$ is automatically a true proposition whenever x is replaced by an element that is not a member of A. Therefore we need only be concerned with the truth value when x is replaced by a member of A. So we start a *direct* proof of the statement $A \subseteq B$ with "Let $x \in A$. Then . . ." or "If $x \in A$, then . . ." and we show that it follows that x must be a member of B. Or we could exhibit a chain of true implications of the form

$$x \in A \Rightarrow \cdots \Rightarrow x \in B,$$

giving reasons whenever the justification is not obvious. We begin an *indirect* proof of $A \subseteq B$ by supposing that $x \notin B$ and showing that consequently $x \notin A$. [Reason: $(x \notin B \Rightarrow x \notin A)$ is the contrapositive of $(x \in A \Rightarrow x \in B)$.]

Suppose we want to prove $A \not\subseteq B$. We have

$$
\begin{aligned}
A \not\subseteq B &\Leftrightarrow \sim(\forall x)(x \in A \Rightarrow x \in B) & \text{(by 2.12)} \\
&\Leftrightarrow (\exists x)\sim(x \in A \Rightarrow x \in B) & \text{(by 2.10)} \\
&\Leftrightarrow (\exists x)(x \in A \text{ and } x \notin B). & \text{(use a truth table)}
\end{aligned}
$$

Thus, *to prove $A \not\subseteq B$ it suffices to show that A contains an element that is not in B.*

2.14 THEOREM.   Let A be a set. Then $A \subseteq A$ and $\emptyset \subseteq A$.

PROOF.   The inclusion $A \subseteq A$ follows from the truth of the insultingly trivial implication

$$x \in A \Rightarrow x \in A.$$

To see that $\emptyset \subseteq A$, notice that the implication

$$x \in \emptyset \Rightarrow x \in A$$

is always true, since its hypothesis is false no matter how x is chosen.†   □

EXERCISE.   Give an *indirect* proof of the inclusion $\emptyset \subseteq A$.

2.15 THEOREM.   If $A \subseteq B$ and $B \subseteq C$ then $A \subseteq C$.

PROOF.   The hypothesis gives the implication chain

$$x \in A \Rightarrow x \in B \Rightarrow x \in C,$$

from which the result follows.   □

We will follow custom and loosen our grip on the notational reins when an abbreviation is convenient and confusion seems unlikely. For example, the symbolic expression in the proof of Theorem 2.15 is casual shorthand for this:

$$(x \in A \Rightarrow x \in B) \quad \text{and} \quad (x \in B \Rightarrow x \in C).$$

† From now on we will use the symbol □ to mark the end of a proof or partial proof.

2.16 EXAMPLE.   We can use the symbolism of Theorem 2.15 to express many standard logical arguments. For example, consider the sentence

All cats eat fish, and every fish-eater sings; therefore cats sing.

Translation: Let C, E, and S denote (respectively) the sets of cats, fish-eaters, and singers. Then $C \subseteq E \subseteq S$, and so $C \subseteq S$; that is, all cats sing.

2.17 THEOREM. $A = B \Leftrightarrow A \subseteq B$ and $B \subseteq A$.

PROOF OUTLINE.   Compare Definitions 2.1 and 2.12.   □

2.18 EXAMPLE.   In the set T of all triangles, define

$S = \{\triangle \in T |\ \triangle$ has two sides of equal length$\}$,
$A = \{\triangle \in T |\ \triangle$ has two equal angles$\}$.

Show that $S = A$. (Here assume that the standard criteria for triangle congruence are known.)

SEMISOLUTION.   By Theorem 2.17, it suffices to show that $S \subseteq A$ and $A \subseteq S$. Let $\triangle \in S$, and let P, Q, R denote the vertices of $\triangle$, labelled so that $\overline{PQ} = \overline{QR}$. Construct a line from Q to the midpoint M of $\overline{PR}$. Then there is a congruence $\triangle PMQ \cong \triangle RMQ$, since these two triangles have sides of the same lengths. Therefore $\measuredangle MPQ = \measuredangle MRQ$ (why?), and $\triangle \in A$. This shows that $S \subseteq A$. Exercise: Prove that $A \subseteq S$ to finish the solution.   □

2.19 DEFINITION.   For sets A and B we define the **proper** (or **strict**) **inclusion** relation, denoted $\subset$, as follows:

$$A \subset B \Leftrightarrow A \subseteq B \text{ and } A \neq B.$$

We say that A is a **proper subset** of B (or **properly contained** in B) if this condition holds. In view of 2.12 and 2.17, we can reformulate this definition as follows:

$$A \subset B \Leftrightarrow A \subseteq B \text{ and } (\exists x)(x \in B \text{ and } x \notin A).$$

If A is not a proper subset of B we write $A \not\subset B$.

Many authors use "⊂" for ordinary set inclusion instead of our "⊆." Therefore check the notation carefully when reading material on set theory in other sources.

2.20 EXAMPLES.

(a) $\{1, 7\} \not\subset \{1, 7\} \subset \{1, 2, 7\}$
(b) If A is any nonempty set then $\emptyset \subset A$.

   PROOF.   The implication

$$x \in \emptyset \implies x \in A$$

   is true, since its hypothesis is false for every x; therefore $\emptyset \subseteq A$. Also, the statement $\emptyset \neq A$ is true by the assumption that A is not empty. Therefore $\emptyset \subset A$.   □

(c) If $A \subset B$ and $B \subset C$ then $A \subset C$.
(d) Define $A = \{1, 2\}$, $B = \{1, 2, 3\}$, $C = \{3, \{1, 2\}\}$, $D = \{1, 2, \{1, 2\}\}$. Then the following assertions are true, and they should be verified by the reader.

$$A \subset B, \quad A \notin B, \quad A \in C, \quad A \not\subset C, \quad A \in D, \quad A \subset D$$

   Notice that A and C each have two elements, while B and D each have three elements.
(e) The set $S = \{$Boston Red Sox, New York Yankees$\}$ is a set with two members, each of which is itself a set. Thus if Jones is a member of the Boston Red Sox then

$$\text{Jones} \in \text{Boston Red Sox}$$

   is true, while

$$\text{Jones} \in S$$

   is false.

Exercises

1. Consider the set $A = \{0, \emptyset, 1\}$.
   (a) How many elements does A have? Explain briefly.
   (b) List the subsets of A.

2. Can a set be a proper subset of itself? Explain.

3. Let $A = \{1, 2, -1, 5\}$, $B = \mathbb{N}$, $C = \{-2, -1, 0, 1, 2\}$, $D = \{1, \{1, 2\}, 2\}$, $E = \{-2, -1, 0, 2, 7\}$. In each case find a set S satisfying all the stated conditions and having the property that no set properly containing S also satisfies the conditions. (The set S is then said to be *maximal* with respect to the conditions.)

   (a) $S \subseteq A$ and $S \subseteq E$
   (b) $S \subseteq A$ and $S \subseteq D$
   (c) $S \subseteq B$ and $S \subseteq E$
   (d) $S \subseteq E$ and $S \not\subseteq C$
   (e) The statement

   $$x \in C \text{ or } x \in D$$

   is true for each $x \in S$.
   (f) The statement

   $$x \in C \text{ and } x \in D$$

   is true for each $x \in S$.

4. Prove the following implications.
   (a) $A \subset B \implies B \not\subset A$
   (b) $A \subseteq \emptyset \implies A = \emptyset$

5. Let S be a set. Prove the statement
   $$a \in S \iff \{a\} \subseteq S.$$

6. Prove that if $A \subset B$ and $B \subseteq C$ then $A \subset C$.

7. Prove the following two propositions.
   (a) $\{x \in \mathbb{N} \mid x^2 \leq 6\} \subseteq \{x \in \mathbb{Z} \mid x = 2^n \text{ for some } n \in \mathbb{Z}\}$
   (b) $(A \subseteq B \text{ and } B \subseteq C \text{ and } C \subseteq A) \implies A = B = C$

8. Let A and B be sets. Show that $A \subseteq B$ if and only if every subset of A is a subset of B.

9. Give sets A and B such that the statements $A \in B$ and $A \subseteq B$ are both true.

10. Prove or disprove: $\{\{\{5\}\}\} = \{\{5\}\}$.

2.5 UNION, INTERSECTION, AND COMPLEMENT

Throughout this section, capital letters $A, B, C, \ldots$ will denote sets. Our goal here is to begin to compare sets with each other and to see how sets can interact to produce offspring: a process informally known as "the joy of sets."

Given sets A and B, it is natural to consider the subset of A that consists of precisely those elements of A that don't belong to B.

2.21 DEFINITION. If A and B are sets, we define their **difference** to be

$$A - B = \{x \in A \mid x \notin B\},$$

also called the **complement of B in A**. In the event that all sets under consideration in a particular discussion are subsets of a given set U (referred to as the **universal set** throughout that discussion), the complement of any set A in U is called the **complement of A**, denoted A'.

2.22 EXAMPLES.

(a) Let $A = \{1, 2, 3\}$ and $B = \{2, 3, 4\}$. Then $A - B = \{1\}$ and $B - A = \{4\}$, so here $A - B \neq B - A$.

(b) $\{\emptyset, \{\emptyset\}\} - \{\emptyset\} = \{\{\emptyset\}\}$

PROOF OF (b). The set $A = \{\emptyset, \{\emptyset\}\}$ has two elements, namely, $\emptyset$ and $\{\emptyset\}$; and $B = \{\emptyset\}$ is a set with $\emptyset$ as its only element. So the only element that belongs to A and not to B is $\{\emptyset\}$. That is, $A - B = \{\{\emptyset\}\}$. □

2.23 EXERCISES. Verify each of the following assertions.

(a) $A - A = \emptyset = \emptyset - A$

(b) $A - \emptyset = A$

(c) $A - B = B - A \Leftrightarrow A = B$

(d) If U is the universal set, then

$$(A')' = A, \qquad \emptyset' = U, \qquad U' = \emptyset.$$

Our next constructions build sets either by uniting two sets into one or by restricting our attention to the elements two sets share.

2.24 DEFINITION. The **union** of A and B is the set

$$A \cup B = \{x \mid x \in A \text{ or } x \in B\}.$$

(Remember, our "or" is *inclusive;* recall Table 1.10.) The **intersection** of A and B is the set

$$A \cap B = \{x \mid x \in A \text{ and } x \in B\}.$$

2.25 EXAMPLE. Let $A = \{1, 2, 3, 4\}$, $B = \{0, 1, 3, 5, 7\}$, and $C = \{2, 4, 6, 8\}$. Then

$$A \cup B = \{0, 1, 2, 3, 4, 5, 7\},$$
$$A \cap B = \{1, 3\}, \qquad B \cap C = \emptyset,$$
$$(A \cup B) \cap C = \{2, 4\} = A \cap C.$$

We sometimes find it useful to represent sets pictorially by disks or blobs, shading appropriate regions to correspond to the subsets of interest. Such pictures are called *Venn diagrams* (after John Venn, who is credited with originating the procedure in the 1880s). For instance, we can represent difference, union, and intersection by

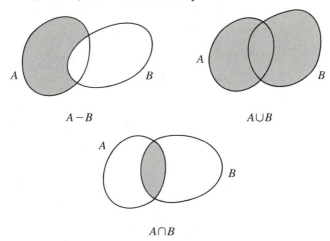

$A - B$ $\qquad\qquad\qquad\qquad$ $A \cup B$

$A \cap B$

and the complement by

A'

Here the universal set U is represented by the entire rectangle.

Venn diagrams are a useful visual aid in the search for theorems about set equality. For example, if in the diagram

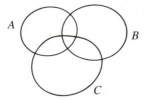

we shade the region corresponding to $A \cap (B \cup C)$, and in another copy of the diagram we shade the region for $(A \cap B) \cup (A \cap C)$, the result in each case looks like

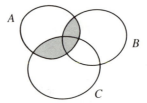

This similarity suggests the equality $A \cap (B \cup C) = (A \cap B) \cup (A \cap C)$. It does not *prove* the equality, it *suggests* it. Pictorial arguments can be misleading, and their worship should be avoided. To illustrate: if $A \cap B = \emptyset$ (in which case we say A and B are **disjoint**), an intersection portrait like

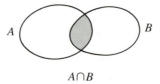

$A \cap B$

suggests that A and B have common elements, contrary to the hypothesis.

The equalities in the following theorem all follow from logical manipulations (justifiable by truth tables) and the definitions of union and intersection. The proof strategy goes back to the definition of set equality in 2.1. Should you feel the urge to draw Venn diagrams to supplement the proofs, go ahead.

2.26 THEOREM.

(a) $A \cap (B \cap C) = (A \cap B) \cap C$ $\left.\right\}$ *Associative laws*
(b) $A \cup (B \cup C) = (A \cup B) \cup C$

(c) $A \cap (B \cup C) = (A \cap B) \cup (A \cap C)$ $\left.\right\}$ *Distributive laws*
(d) $A \cup (B \cap C) = (A \cup B) \cap (A \cup C)$

(e) $A - (B \cup C) = (A - B) \cap (A - C)$
(f) $A - (B \cap C) = (A - B) \cup (A - C)$ } *De Morgan's laws*
(g) $(A \cup B)' = A' \cap B'$
(h) $(A \cap B)' = A' \cup B'$

PARTIAL PROOF.

(a) $x \in A \cap (B \cap C) \Leftrightarrow x \in A$ and $x \in (B \cap C)$
$\Leftrightarrow x \in A$ and $(x \in B$ and $x \in C)$
$\Leftrightarrow (x \in A$ and $x \in B)$ and $x \in C$
$\Leftrightarrow x \in (A \cap B) \cap C$

Therefore the definition of set equality yields the desired conclusion. [Basically the argument is a consequence of the logical equivalence $P \wedge (Q \wedge R) \equiv (P \wedge Q) \wedge R$.]

(b) This proof is essentially the same as the proof of (a), merely replacing "$\cap$" by "$\cup$" and "and" by "or."
(c)–(h) Exercises.   □

2.27 THEOREM.

(a) If $X \subseteq Z$ and $Y \subseteq Z$ then $X \cup Y \subseteq Z$.
(b) If $Z \subseteq X$ and $Z \subseteq Y$ then $Z \subseteq X \cap Y$.

PROOF.

(a) If $w \in X \cup Y$ then either $w \in X$ or $w \in Y$. In either case $w \in Z$, since $X \subseteq Z$ and $Y \subseteq Z$. Therefore $X \cup Y \subseteq Z$.
(b) This is left as an exercise.   □

Exercises

1. Let $A = \{1, 2, 3, 4, 5\}$, $B = \{3, 4, 5, 6\}$, $C = \{x \in \mathbb{Z} \mid x$ is even$\}$. Determine each of the following sets.
   (a) $A - B$ (f) $(A \cap B) \cap C$
   (b) $B - A$ (g) $(A \cap B) \cup C$
   (c) $B \cap A$ (h) $B \cup \emptyset$
   (d) $A \cup B$ (i) $\emptyset \cap C$
   (e) $C - B$

2. Take $\mathbb{Z}$ as the universal set. Using the sets A and C defined in Exercise 1, determine the sets A', C', and $(A \cap C)'$.

3. Verify the following propositions. (Here A is any set.)

   (a) $A \cup \emptyset = A$
   (b) $A \cap \emptyset = \emptyset$
   (c) $(A - A) - A = \emptyset$
   (d) $A - (A - A) = A$
   (e) $A \cup A = A$

4. On one copy of the following diagram shade the region corresponding to $A - (B - C)$; on another copy shade $(A - B) - C$. If the results are the same, prove the proposition $(A - B) - C = A - (B - C)$. If the results are different, give an explicit counterexample to the proposition.

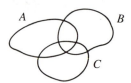

5. Prove each of the following.

   (a) $(A - B) \cap (B - A) = \emptyset$
   (b) $(A \cap B) \cap (A - B) = \emptyset$
   (c) $A = (A \cap B) \cup (A - B)$

6. Prove parts (c), (e), and (g) of Theorem 2.26.

7. Prove part (b) of Theorem 2.27.

8. Let A and B be sets. Define the **symmetric difference** to be

$$A \oplus B = (A - B) \cup (B - A).$$

For example, if $A = \{-3, 0, 1, 2\}$ and $B = \{1, 2, 3, 4\}$, then

$$A \oplus B = \underbrace{\{-3, 0\}}_{A - B} \cup \underbrace{\{3, 4\}}_{B - A} = \{-3, 0, 3, 4\}.$$

The set $A \oplus B$ consists of those elements of $A \cup B$ that belong to exactly one of the sets A, B. Prove the following statements.

   (a) $A \oplus A = \emptyset$
   (b) $A \oplus \emptyset = A$

(c) $(A \oplus B) - (A - B) = B - A$

(d) $A \oplus B = \emptyset \Rightarrow A = B$ (Therefore, combining parts (a) and (d) gives the equivalence $A \oplus B = \emptyset \leftrightarrow A = B$.)

(e) $(A \oplus B) \oplus C = A \oplus (B \oplus C)$ (Start with a Venn diagram.)

9. Let S be a nonempty set. All capital letters in this problem will denote subsets of S; and throughout this problem we will write $A + B$ instead of $A \cup B$, and AB instead of $A \cap B$.

(a) Verify the following three assertions.

$A + B = B + A$ and $AB = BA$

$A(B + C) = AB + AC$

$A + BC = (A + B)(A + C)$ (unfortunate!)

(b) Given A, for which sets B is it true that $A + B = A$? (Find them all.)

(c) Given A, for which sets B is it true that $AB = A$? (Find them all.)

(d) Show that $AX = A$ for all A if and only if $X = S$.

(e) Show that $A + X = A$ for all A if and only if $X = \emptyset$.

(f) Show that $A(B - A) = \emptyset$.

(g) Show that $A + (B - A) = A + B$. (Here "$-$" is the usual set difference.)

10. Let A and B be sets of data stored in a computer memory, with m elements and n elements, respectively. Assume A is stored in locations numbered 1 through m, and B in locations $m + 1$ through $m + n$, with exactly one element in each location.

(a) Write a procedure for listing the elements of $A \cap B$ in consecutively numbered memory locations, with one element per location and no element listed twice.

(b) Do the same for $A - B$.

(c) Do the same for $A \cup B$.

2.6 INDEXED SETS

Imagine that a dog barks four times. How can we denote the set of barks? An easy (and quiet) answer: write $B = \{b_1, b_2, b_3, b_4\}$, where b_1 stands for the first bark, b_2 for the second bark, etc. Here we have used the members of the set $I = \{1, 2, 3, 4\}$ as labels or **indexes** for the members of B; I is therefore called the **index set.** Once the indexes have been assigned we can write $B = \{b_i \mid i \in I\}$. There may be more than one reasonable way to index the members of a set and, if the set is large, choosing an optimal indexing scheme is not an inconsequential matter. For instance, if someone is concerned with the problem of storing large amounts of data in such a way that information can be easily retrieved and manipulated, the way in which the data has been indexed can be crucial.

2.28 EXAMPLES.

(a) Suppose you own many books on a wide variety of subjects, and you want to arrange them on your bookshelves in an optimal way. If you position them alphabetically by author, and books by the same author alphabetically by title, then it will be a triviality for you to locate a book with a given author and title; but, should the need arise to survey all your books that deal with a particular subject, you may have some trouble. On the other hand, if the books are filed by subject, you may have difficulty locating a book with a given author and title if the book's subject area is unknown to you.

(b) Each account file in a savings bank is a set of information about the account holder (name, address, employer, etc.) together with a listing of the amount of money in the account. If electronic access to the money could be had by anyone knowing the name of the account holder, then no depositor's money would be safe. So access is available only to someone knowing a secret number assigned to the account holder. That is, account files are indexed by code numbers and not by peoples' names.

(c) Analytic geometry grew out of René Descartes's insight (in 1637) that the points in a plane Π can be indexed with ordered pairs of real numbers by superimposing a pair of axes on Π and labelling each point according to its position with respect to the axes.

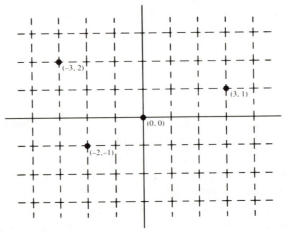

A piece of Π.

When the set $I = \{(a, b)| \ a, b \in \mathbb{R}\}$ is the index set for Π, a typical point $P \in \Pi$ can be completely identified by using its full indexed name $P_{(a,b)}$. In practice we simplify the notation by suppressing the letter P, and we speak of "the point (a, b)." It is not essential that the axes be perpendicular or that the same units of distance be used on both. For example, in the mathematics of crystallography it is often more reasonable to use axes that suggest the geometry of the crystalline structure under consideration.

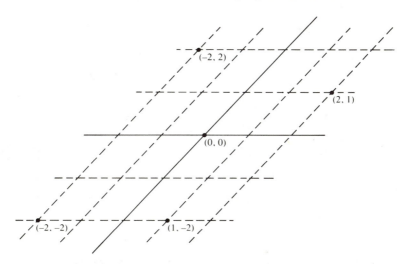

In this case the index set I is the same as it was for Descartes, but the scheme for assigning elements of I to elements of Π is different from his. There are also **polar coordinates** for Π: We choose a point in Π, label it O, and call it the **origin** or **pole**. Then we choose a ray (the **polar axis**) originating at O, and a unit distance. We index a point $P \in \Pi$ by its distance r from O and the angle α that the ray OP makes with the polar axis:

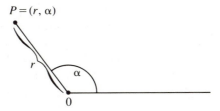

The three indexing procedures just described provide dramatically different descriptions of the elements of Π.

(d) Consider the family F of all of the vertical lines in a given coordinatized plane. Each such line intersects the x-axis at exactly one point $(i, 0)$, and we can use the real number i to index the line

Thus for each $i \in \mathbb{R}$ we let L_i denote the vertical line through $(i, 0)$.

Now let's see another way to index the family F. Each vertical line has exactly one point with y-coordinate 1. The line through the origin and that point makes an angle α (say in radians) with the x-axis, and α can be used to label the line: call the line L_α. In this way the open interval $I_0^\pi = \{\alpha \mid 0 < \alpha < \pi\}$ can serve as the index set for F.

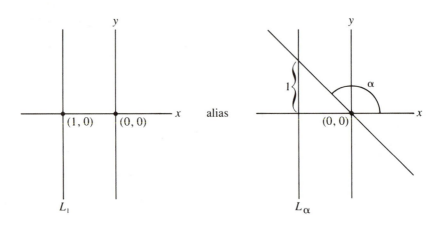

The notation $S = \{x_i \mid i \in I\}$ for an indexed set is often compactified to $S = \{x_i\}_{i \in I}$. Thus in Example 2.28(d), the two indexing procedures for the set F of vertical lines yield these notations for F: $\{L_i\}_{i \in \mathbb{R}}$ and $\{L_\alpha\}_{\alpha \in I_0^\pi}$.

2.29 EXAMPLES.

(a) A collection of sixteen sets

$$\{A, B, C, D, E, F, G, H, I, J, K, L, M, N, O, P\}$$

can be relabelled and written more compactly as the indexed family $\{A_i\}_{i \in I}$ with index set $I = \{i \in \mathbb{Z} \mid 1 \leq i \leq 16\}$.

(b) In the usual xy-plane, we associate the following sets with a real number r:

$$H_r = \{(x, y) \mid x \geq r\}, \quad \text{a } \textit{right half-plane}$$

and, if $r \geq 0$,

$C_r = \{(x, y)| \sqrt{x^2 + y^2} = r\}$, the *circle* with radius r and center $(0, 0)$

$D_r = \{(x, y)| \sqrt{x^2 + y^2} \leqslant r\}$, the *closed disk* with radius r and center $(0, 0)$

This gives three indexed families:

$$\{H_r\}_{r \in \mathbb{R}}, \qquad \{C_r\}_{r \geqslant 0}, \qquad \{D_r\}_{r \geqslant 0},$$

where the index set for C_r and D_r has been indicated by its defining property.

Now we extend our notions of union and intersection, defined earlier for two sets (2.24), to arbitrary indexed families of sets.

2.30 DEFINITION. Let $\{A_i\}_{i \in I}$ be an indexed family of sets. We define the

$$\textbf{union} \quad \bigcup_{i \in I} A_i$$

as the set of elements that belong to *at least one* A_i, and the

$$\textbf{intersection} \quad \bigcap_{i \in I} A_i$$

as the set of elements that belong to *all* A_i. More compactly:

$$\bigcup_{i \in I} A_i = \{x | \ x \in A_i \text{ for some } i \in I\}, \qquad \bigcap_{i \in I} A_i = \{x | \ x \in A_i \text{ for all } i \in I\}.$$

With full formality:

$$\bigcup_{i \in I} A_i = \{x | \ (\exists i \in I)(x \in A_i)\}, \qquad \bigcap_{i \in I} A_i = \{x | \ (\forall i \in I)(x \in A_i)\}.$$

To ease the printer's burden and simpify the notation, if no ambiguity is likely we sometimes write the union as $\bigcup_{i \in I} A_i$ or even $\bigcup A_i$, and we treat intersections similarly.

REMARK. From the definition of $\bigcap_{i \in I} A_i$ it follows that the statement $x \notin \bigcap_{i \in I} A_i$ is equivalent to $\sim (\forall i \in I)(x \in A_i)$, which in turn is equivalent to $(\exists i \in I)(x \notin A_i)$, by our rules for negating quantified statements. In words: if x is not simultaneously in all A_i, then there is some A_i that does not have x as a member. Similar considerations hold for $\cup_{i \in I} A_i$. (Check this.)

2.31 EXAMPLES.

(a) Given sets A and B, we can relabel them as the indexed family $\{A_1, A_2\}$ with index set $I = \{1, 2\}$. Having done that, we have $\cup_{i \in I} A_i = A \cup B$ and $\cap_{i \in I} A_i = A \cap B$. So the concepts defined in 2.30 include those in 2.24 as special cases.

(b) Consider the family $F = \{L_i\}_{i \in \mathbb{R}}$ defined in 2.28(d). We claim that $\cup_{i \in \mathbb{R}} L_i$ is the entire plane, while $\cap_{i \in \mathbb{R}} L_i = \emptyset$. Let's verify the first of these assertions in detail. (The second is left as an exercise.) Denote the plane by E: we must show $E = \cup L_i$. Recall that to do this it suffices to show that $E \subseteq \cup L_i$ and $\cup L_i \subseteq E$. Conversationally: Each point in E is on some vertical line, hence each point is in the union of all the vertical lines, so $E \subseteq \cup L_i$. Conversely, every point in $\cup L_i$ is on some L_i; but L_i is contained in E, so $\cup L_i \subseteq E$. Here is a more detailed proof. If $P \in E$, then $P = (a, b)$ for some $a, b \in \mathbb{R}$. But then $P \in L_a$, since L_a is the vertical line through $(a, 0)$ [that is, $L_a = \{(a, y) \mid y \in \mathbb{R}\}$], and so $P \in \cup L_i$. Conversely, if $P \in \cup L_i$ then $P \in L_i$ for some i, by Definition 2.30. But $L_i \subseteq E$, and therefore $P \in E$.

(c) Let's briefly discuss the unions and intersections of the families defined in Example 2.29(b). As in part (b) of this example, we use E for the plane. We assert that $\cup H_r = E$. Every point in the union is in some H_r, and $H_r \subseteq E$, proving "$\subseteq$." Conversely, every point $(a, b) \in E$ is in the half-plane H_{a-1}, for example, hence in $\cup H_r$. Also, $\cap H_r = \emptyset$ because no point is in every half-plane; for instance, $(a, b) \notin H_{a+1}$. Next, $\cup C_r = E$, essentially because every point in the plane is on a circle centered at the origin, namely, (a, b) is on $C_{\sqrt{a^2+b^2}}$. (Really this only shows "$\supseteq$.") But $\cap C_r = \emptyset$ since no point is on more than one such circle. Finally, $\cup D_r = E$ since every point of E is in some disk (in fact many disks). We claim $\cap D_r = \{(0, 0)\}$. Clearly $(0, 0)$ is an element of every disk centered at $(0, 0)$. Moreover, any point $(a, b) \neq (0, 0)$ is excluded from small enough disks (for example, from $D_{\sqrt{(a^2+b^2)/2}}$) and therefore from the intersection $\cap D_r$.

2.32 THEOREM. Let A be a set and let $\{B_i\}_{i \in I}$ be an indexed family of sets. Then

(a) $A - \bigcap_{i \in I} B_i = \bigcup_{i \in I} (A - B_i),$

(b) $A - \bigcup_{i \in I} B_i = \bigcap_{i \in I} (A - B_i).$

(This theorem extends Theorem 2.26, and is therefore also known as *De Morgan's laws*.)

PROOF OF (a).
$$x \in A - \bigcap_{i \in I} B_i \Leftrightarrow x \in A \quad \text{and} \quad x \notin \bigcap_{i \in I} B_i$$
$$\Leftrightarrow x \in A \quad \text{and} \quad (\exists i \in I)(x \notin B_i)$$
$$\text{(recall 2.30)}$$
$$\Leftrightarrow (\exists i \in I)(x \in A \quad \text{and} \quad x \notin B_i)$$
$$\Leftrightarrow (\exists i \in I)(x \in A - B_i)$$
$$\Leftrightarrow x \in \bigcup_{i \in I} (A - B_i) \quad \square$$

The proof of part (b) is left as an exercise.

Exercises

1. Let $A_1 = \{1, 2, 3, 4\}$, $A_2 = \{0, 1, 2\}$, $A_3 = \{-1, 0, 1\}$, and let $I = \{1, 2, 3\}$. Determine the following sets.

(a) $\bigcup_{i \in I} A_i$ (b) $\bigcap_{i \in I} A_i$

Take the set of integers, $\mathbb{Z}$, as the universal set. Determine the following sets.

(c) $\bigcup_{i \in I} A_i'$ (d) $\bigcap_{i \in I} A_i'$

2. Take the set of all living people as the universal set. Define

$$A_i = \{x \mid x \text{ was born on the } i\text{th day of his/her birth year}\},$$
$$B_j = \{x \mid x \text{ was born during the } j\text{th month of the year}\},$$
$$C_k = \{x \mid x \text{ was born on the } k\text{th day of the week}\}$$

(start counting with Sunday). Use words to describe the elements of each of the following sets.

(a) $B_1 \cap (\bigcup_{1 \leq k \leq 3} C_k)$

(b) $(\bigcap_{1 \leq j \leq 5} B_j) \cup (\bigcup_{12 \leq i \leq 15} A_i)$

(c) $C_6 \cap (\cup_{1 \leqslant j \leqslant 4} B_j)$

(d) $(\cup_{1 \leqslant k \leqslant 5} C_k)'$

(e) $\cup_{1 \leqslant k \leqslant 5} C_k'$

(f) $\cap_{1 \leqslant j \leqslant 4} B_j'$

3. Let n be a positive integer, and suppose a set A_i is given for each $i \in I = \{1, 2, 3, \ldots, n\}$, with $A_i \subseteq A_j$ whenever $i \leqslant j$. Show that $\cup_{i \in I} A_i = A_n$ and $\cap_{i \in I} A_i = A_1$.

4. An indexed family of sets $\{A_i\}_{i \in I}$ is said to be **disjoint** if $\cap_{i \in I} A_i = \emptyset$, and the family is said to be **pairwise disjoint** if $A_i \cap A_j = \emptyset$ whenever $i \neq j$. In words: a family of sets is disjoint if there is no element shared by all sets in the family; a family is pairwise disjoint if no two sets with different indexes share a common element. The following pictures help to clarify these ideas.

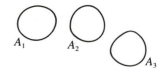

Not disjoint   Disjoint but not pairwise disjoint   Pairwise disjoint

(a) Let $A_1 = \{1, 2, 3, 4\}$, $A_2 = \{3, 4, 5, 6\}$, $A_3 = \{1, 6, 7\}$. Show that the family $\{A_1, A_2, A_3\}$ is disjoint but not pairwise disjoint.

(b) Show that if $\{B_1, B_2, B_3\}$ is a pairwise disjoint family of sets then it is disjoint.

(c) Repeat part (b) for an arbitrary indexed family $\{A_i\}_{i \in I}$.

5. Recall the closed interval notation $[-\alpha, \alpha] = \{x \in \mathbb{R} \mid -\alpha \leqslant x \leqslant \alpha\}$. Determine each of the following sets.

(a) $\cup_{\alpha > 0} [-\alpha, \alpha]$

(b) $\cap_{\alpha > 0} [-\alpha, \alpha]$

(c) $\cap_{\alpha \geqslant 5} [-\alpha, \alpha]$

6. For each integer n, let $M(n)$ be the set of all integer multiples of n. Thus, for example,

$$M_0 = \{0\},$$
$$M_1 = M_{-1} = \mathbb{Z},$$
$$M_2 = M_{-2} = \{0, \pm 2, \pm 4, \pm 6, \ldots\},$$
$$M_3 = M_{-3} = \{0, \pm 3, \pm 6, \pm 9, \ldots\}.$$

Determine each of the following sets.

(a) $\bigcup_{n \in \mathbb{N}} M_n$

(b) $\bigcap_{n \in \mathbb{N}} M_n$

(c) $\bigcup_{p \text{ prime}} M_p$

(d) $\bigcap_{p \text{ prime}} M_p$

(e) $\bigcup_{n \geqslant 6} M_n$

(f) $\bigcup_{n \in M_5} M_n$

7. Prove Theorem 2.32(b).

8. Describe each of the following as an indexed family of sets. Here Π denotes the coordinatized xy-plane.

   (a) The family of closed intervals of length 1 on the real line.
   (b) The family of all circles in Π of radius 1 whose center is on the y-axis.
   (c) The family of all circles in Π of radius 1.
   (d) The family of all lines in Π with slope 6.
   (e) The family of all lines in Π with y-intercept 5.

9. Find $\bigcup_{i \in I} A_i$ and $\bigcap_{i \in I} A_i$ for each family $\{A_i\}_{i \in I}$ in Exercise 8. That is, describe the union and intersection in each case, and briefly justify your claims. A complete proof is not required.

2.7 THE POWER SET

Mathematicians, surgeons, nuclear physicists, and ax murderers achieve many of their objectives by dissecting complicated structures and manipulating their pieces. In this section we begin such a dissection by considering the set of all subsets of a given set.

2.33 DEFINITION. The **power set** $P(A)$ of a set A is the collection of all subsets of A:

$$P(A) = \{X \mid X \subseteq A\}.$$

So for any set X the statements "$X \subseteq A$" and "$X \in P(A)$" are equivalent. (We'll see that many proofs about power sets amount to remembering this equivalence.) Thus a subset of $P(A)$ is a set whose members are subsets of A. If A is small we can describe $P(A)$ by listing the subsets of A between a pair of braces.

2.34 EXAMPLES.

(a) $P(\{3\}) = \{\emptyset, \{3\}\}$
(b) $P(\{1, 2\}) = \{\emptyset, \{1\}, \{2\}, \{1, 2\}\}$
(c) $P(\emptyset) = \{\emptyset\}$

PROOF OF (c). We have $\emptyset \subseteq \emptyset$ (recall Theorem 2.14), so $\emptyset \in P(\emptyset)$, and therefore $\{\emptyset\} \subseteq P(\emptyset)$. Conversely,

$$A \in P(\emptyset) \Rightarrow A \subseteq \emptyset$$
$$\Rightarrow A = \emptyset \qquad \text{(If } A \neq \emptyset \text{ then } A \text{ has a member}$$
$$x, \text{ and the inclusion } A \subseteq \emptyset \text{ would}$$
$$\text{force } x \in \emptyset, \text{ an absurdity.)}$$
$$\Rightarrow A \in \{\emptyset\}$$

Hence $P(\emptyset) \subseteq \{\emptyset\}$. When we combine this with the first part of the proof, we deduce $P(\emptyset) = \{\emptyset\}$, as claimed.   □

(d) $P(\{1, 2, 3\}) = \{\emptyset, \{1\}, \{2\}, \{3\}, \{1, 2\} \{1, 3\}, \{2, 3\}, \{1, 2, 3\}\}$

Examples 2.34 lead us naturally to this conjecture:

If A has n elements then $P(A)$ has 2^n elements.

We'll prove the conjecture toward the end of the chapter, but for now, a proof sketch goes like this: If $A = \{a_1, \ldots, a_n\}$, we can associate each subset X of A with an n-digit binary expression $b_1 b_2 \ldots b_n$ (that is, a string of 0s and 1s) given by

$$b_i = \begin{cases} 1 & \text{if } a_i \in X, \\ 0 & \text{if } a_i \notin X. \end{cases}$$

(For example, $\emptyset$ is associated with $00\ldots0$, and A is associated with $11\ldots1$.) There are 2^n such expressions because there are two possible choices for the first digit on the left and, for each of these possibilities, two choices for the next digit to its right, and so on. Altogether, since there are n decisions in all, there are 2^n possible results.

It is helpful to represent the decision process graphically by a figure called a **decision tree,** which is pictured in 2.35.

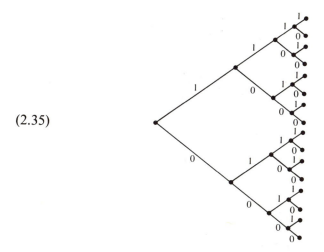

(2.35)

Here the marked points are called **vertices** (singular: **vertex**), and the line segments that connect two vertices are called **edges**. [There is also biological terminology for all this: the edges in a tree are often called **branches**, and a vertex at the end of exactly one branch (in the diagram, one of the far right vertices) is called — you guessed it — a **leaf**.] Each vertex at which a branching takes place (moving from left to right) represents a moment of decision, and an edge emanating to the right from a vertex represents a possible choice at that moment. In our pictured tree, we have labelled the edges to indicate the corresponding choices. Thus a path from the first vertex on the left (called the **root** of this tree) to the fifth vertex up on the far right of the tree corresponds to the four-digit binary string 0100. There are exactly $16 = 2^4$ left-to-right paths, one for each possible four-digit binary string. Similarly, a tree associated with the choice of an n-digit binary string will have 2^n left-to-right paths.

Having counted the n-digit binary expressions in one way or another, we conclude that A has 2^n subsets.

Once we have accepted the concepts of the power set and the empty set, we are forced to admit the existence of very large sets. To see this, watch what happens upon repeatedly taking power sets:

$$P(\emptyset) = \{\emptyset\}, \qquad P(P(\emptyset)) = \{\emptyset, \{\emptyset\}\},$$
$$P(P(P(\emptyset))) = \{\emptyset, \{\emptyset\}, \{\{\emptyset\}\}, \{\emptyset, \{\emptyset\}\}\},$$

and so on. In general,

$$P(P(\cdots P(P(\emptyset))\cdots)) \quad \text{has} \quad 2^{2^{\cdot^{\cdot^{2^0}}}} \Big\} n \text{ 2s} \quad \text{elements.}$$
$$\underbrace{}_{n\ P\text{'s}}$$

2.36 THEOREM. Let A and B be sets. Then

(a) $\{\emptyset, A\} \subseteq P(A)$,

(b) $A \subseteq B \iff P(A) \subseteq P(B)$,

(c) $P(A) \cup P(B) \subseteq P(A \cup B)$,

(d) $P(A) \cap P(B) = P(A \cap B)$.

PROOF. Suggestion: first review proof strategies for inclusions and equalities; see especially 2.12–2.17.

(a) We must show that $\emptyset \in P(A)$ and $A \in P(A)$; equivalently, $\emptyset \subseteq A$ and $A \subseteq A$. But we already know this by 2.14.

(b) We first prove "$\Rightarrow$." So assume $A \subseteq B$, and let $X \in P(A)$. (Here think, "And we want to show $X \in P(B)$, which is the same as showing $X \subseteq B$." Then remember the hypothesis!) Then $X \subseteq A \subseteq B$, and so $X \subseteq B$; that is, $X \in P(B)$. We have therefore verified the implication

$$A \subseteq B \implies P(A) \subseteq P(B).$$

Conversely, assume $P(A) \subseteq P(B)$. We have the implication chain

$$x \in A \implies \{x\} \subseteq A \implies \{x\} \in P(A)$$
$$\implies \{x\} \in P(B) \implies \{x\} \subseteq B \implies x \in B,$$

which gives the result.

(c) Since $A \subseteq A \cup B$ and $B \subseteq A \cup B$, by part (b) we know $P(A) \subseteq P(A \cup B)$ and $P(B) \subseteq P(A \cup B)$. Now apply Theorem 2.27(a).

(d) Since $A \cap B \subseteq A$, we have $P(A \cap B) \subseteq P(A)$ by part (b). Similarly, $P(A \cap B) \subseteq P(B)$, so Theorem 2.27(b) gives $P(A \cap B) \subseteq P(A) \cap P(B)$. Conversely, if $X \in P(A) \cap P(B)$ then $X \subseteq A$ and $X \subseteq B$, so $X \subseteq A \cap B$; that is, $X \in P(A \cap B)$. Therefore $P(A) \cap P(B) \subseteq P(A \cap B)$, and we have proved the desired equality.

2.37 EXAMPLE.   Let $A = \{1, 2\}$ and $B = \{2, 3\}$. Then $\{1, 2, 3\} \in P(A \cup B)$, while $\{1, 2, 3\} \notin P(A) \cup P(B)$. So the inclusion symbol in Theorem 2.36(c) cannot be replaced by equality.

Exercises

1. Let $A = \{1, 2, 3, 4, 5\}$ and $B = \{4, 5, 6\}$. Determine the following sets.
    (a) $P(A \cap B)$
    (b) $P(B - A)$
    (c) $P(A) \cap P(B)$

2. Determine each of the following sets.
    (a) $P(\{2\})$
    (b) $P(P(\{2\}))$
    (c) $P(P(P(\{2\})))$

3. Sam has one of each of the following: a penny, a nickel, a dime, a quarter, a half-dollar, a dollar bill, and a two-dollar bill. He will definitely place a bet on his next poker hand. How many possibilities are there for the amount of his bet? Explain.

4. If each of the sets A and B contains an element not in the other, show that $P(A \cup B) \supset P(A) \cup P(B)$. (A Venn diagram may help here and in some other problems.)

5. Show that if $A \subset B$ then $P(A) \subset P(B)$.

6. (a) Show that if A and B are sets then $\emptyset \notin P(A) - P(B)$.
    (b) Deduce from part (a) that the difference of two power sets can never be a power set.
    (c) Prove that $P(A - B) \subseteq (P(A) - P(B)) \cup \{\emptyset\}$.

7. Under what conditions on sets A and B is it true that $P(A - B) = P(A)$? State and prove a condition that is necessary and sufficient.

8. Let A be a set, and suppose $x \notin A$. Describe $P(A \cup \{x\})$.

2.8   ORDERED PAIRS AND CARTESIAN PRODUCTS

There are many common procedures by which we pair, relate, or associate members of one or more sets with each other. (1) With a person's name we can associate his or her street address, or telephone number, or mother's maiden name, and so on. (2) With a piece of data stored in a computer memory we associate its *location* in memory; if we are to retrieve the data easily, we must know where to find it. (3) A travelling salesman planning to visit twenty cities makes a list that indicates the order in which he will visit the cities: the number n is paired with the nth city on his tour. (4) The formula $y = x^3 + e^x - \sin x$ prescribes a rule for pairing each real number x with a real number y.

Fundamental here is the notion of a set on which an *ordering, arrangement,* or *data structure* has been introduced, leading us to view the set as more than just a collection of separate elements. Examples: (a) Consider a line of people waiting to transact business at a bank. A hatful of paper slips with the names of those in line (one name per slip) fails to capture the essence of the line: who precedes whom? And if there is more than one line at the bank, then people in different lines are in a sense incomparable. (b) The development of large and complicated computational schemes demands careful attention to order: must the program be executed in stepwise fashion from first line to last, or can the program be broken into two or more procedures that can be executed simultaneously (so-called *parallel processing*)?

The pairing process is central to all this organizational work. For example, the order in the bank line just mentioned is completely known to us if, for every two-person subset of the line, we know who precedes whom. How can we impose an order structure on a two-element set? There is no difference in meaning between "shoe and sock" and "sock and shoe," while there *is* a distinction between "first sock, then shoe" and "first shoe, then sock." More generally, since $\{a, b\} = \{b, a\}$, something more is needed if we want to indicate that one element is to be viewed as preceding the other. Here is a start.

2.38 QUASI-DEFINITION. Let a and b be elements. The symbol (a, b) denotes an **ordered pair**, with the following understanding: if (c, d) is also an ordered pair, then

$$(*) \qquad\qquad (a, b) = (c, d) \iff a = c \text{ and } b = d.$$

What's "quasi" about 2.38? It appears that a "symbol" is a new sort of undefined creature. This smacks of buck-passing—defining one unknown in terms of another. This problem will be resolved in the following paragraph, in which we define (a, b) as a kind of set; we then *prove* that the desired property (*) follows from the properties of this set.

2.39 DEFINITION.   Let a and b be elements. Then define

$$(a, b) = \{\{a\}, \{a, b\}\}.$$

Here the elements a and b are called the **first** and **second coordinates**, respectively.

The following theorem and its corollary show that this definition of (a, b) achieves what we want: it attends to the order in which the elements are listed.

2.40 THEOREM. $(a, b) = (c, d) \Leftrightarrow a = c$ and $b = d$.

PARTIAL PROOF.   "$\Leftarrow$": If $a = c$ and $b = d$, then $\{a\} = \{c\}$ and $\{a, b\} = \{c, d\}$, and so $\{\{a\}, \{a, b\}\} = \{\{c\}, \{c, d\}\}$; that is, $(a, b) = (c, d)$.

We leave the converse ("$\Rightarrow$") as a tedious but worthwhile exercise in bracket shuffling. Suggestion: first consider the case $a = b$ (when $\{\{a\}, \{a, b\}\} = \{\{a\}\}$), and then consider the case $a \neq b$.   □

2.41 COROLLARY. $(a, b) = (b, a) \Leftrightarrow a = b$.

Of course, statements 2.40 and 2.41 are well known in the context of analytic geometry, where a point is described by an ordered pair of real numbers according to the point's distances from the axes. But the definition of ordered pair given in 2.39 frees the concept from geometrical constraints, so that we can apply it in other contexts. Later we will extend the notion of order to larger sets. This is important for a variety of purposes, including the practical problems of sorting and arranging large sets of data.

2.42 EXAMPLES.

(a) Suppose a group of numbered tennis players participate in a tournament. Losing a match eliminates an individual from contention. A record of the results can be given as a set S of ordered pairs: define

$$S = \{(m, n)| \text{ player } m \text{ defeated player } n\}.$$

That is, $(m, n) \in S$ is an abbreviation for the statement "player m defeated player n." If no one withdraws midtournament, the winner is the unique person whose number does not appear as the second coordinate of an element of S. The number of matches played by an individual is the number of elements of S in which his or her number appears as a first or second coordinate; and the total number of matches in the tournament is the number of elements of S. Question to ponder: Suppose N players enter the tournament. (Here N is a positive integer but not necessarily a power of 2.) How many matches will be played?

(b) Suppose a collection of numerical data is stored in a computer's memory. If the number m is stored in memory cell n, we make note of the ordered pair (m, n); then the set

$$T = \{(m, n) | \ m \text{ is stored at address } n\}$$

is a complete record of what numbers are where. The fact that the pairs are *ordered* is crucial if we are to avoid confusing a cell's contents with its address.

(Answer to the question in part (a): Each match has a loser, and each loser loses only once. So the number of matches is equal to the number of losers. But an N-player tournament has $N - 1$ losers, hence $N - 1$ matches.)

2.43 DEFINITION.   The **Cartesian product** of sets A and B is the set of all ordered pairs with first coordinate in A and second coordinate in B. Symbolically,

$$A \times B = \{(a, b) | \ a \in A \text{ and } b \in B\}.$$

2.44 EXAMPLES.

(a) Let $A = \{1, 2\}$ and $B = \{2, 3\}$. Then

$$A \times B = \{1, 2), (1, 3), (2, 2), (2, 3)\}$$

and

$$B \times A = \{(2, 1), (2, 2), (3, 1), (3, 2)\}.$$

So here $A \times B \neq B \times A$.

(b) The Cartesian product $\mathbb{R} \times \mathbb{R}$ is the set of all ordered pairs of real numbers. In analytic geometry, we view an ordered pair $(a, b) \in \mathbb{R} \times \mathbb{R}$ as the name of a point in the coordinatized xy-plane. Accordingly, in practice we speak of "the plane $\mathbb{R} \times \mathbb{R}$" and forget the formal definition of ordered pair. (This sort of thinking is familiar: we look at a family photograph album and exclaim "There's Shirley!" and not "There's a picture of the woman named Shirley!")

(c) A **closed interval** in $\mathbb{R}$ is a set of the form

$$[a, b] = \{x \in \mathbb{R} \mid a \leqslant x \leqslant b\}.$$

In particular,

$$[0, 1] = \{x \in \mathbb{R} \mid 0 \leqslant x \leqslant 1\}.$$

Then $[0, 1] \times [0, 1] = \{(x, y) \mid x, y \in [0, 1]\}$ is the **unit square** in $\mathbb{R} \times \mathbb{R}$, pictured as follows.

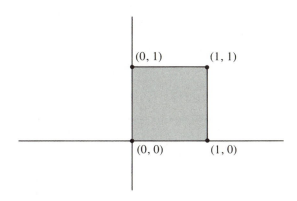

(d) In $\mathbb{R} \times \mathbb{R}$, consider the circle C_1 with radius 1 centered at the origin. (This is the **unit circle**.) Then

$$C_1 \times \mathbb{R} = \{((x, y), z) \mid x, y, z \in \mathbb{R} \text{ and } x^2 + y^2 = 1\}.$$

We associate a typical element $((x, y), z)$ of $C_1 \times \mathbb{R}$ with the point (x, y, z) in coordinatized three-dimensional space. Once we do this we can interpret $C_1 \times \mathbb{R}$ as the circular cylinder of infinite length whose axis is the z-axis and whose radius is 1.

2.45 THEOREM.   Let A, B, and C be sets. Then the following statements hold.

(a) $(A \cup B) \times C = (A \times C) \cup (B \times C)$

(b) $(A \cap B) \times C = (A \times C) \cap (B \times C)$

(c) $(A - B) \times C = (A \times C) - (B \times C)$

(d) If A and B are nonempty sets then

$$A \times B = B \times A \iff A = B.$$

(e) If $A_1 \in P(A)$ and $B_1 \in P(B)$, then $A_1 \times B_1 \in P(A \times B)$.

(f) If A and B each have at least two elements, then not every element of $P(A \times B)$ has the form $A_1 \times B_1$ for some $A_1 \in P(A)$ and $B_1 \in P(B)$.

(g) $\emptyset \times A = \emptyset$

PARTIAL PROOF.   First a general observation. To prove that a Cartesian product $R \times S$ is a subset of a given set T, the definition of set inclusion requires that the proof have the form

$$w \in R \times S \Rightarrow \cdots \Rightarrow w \in T.$$

That is, the proof starts with "Let $w \in R \times S$" and concludes with "then $w \in T$." But we have the equivalence

$$w \in R \times S \iff w = (r, s), \text{ for some } r \in R \text{ and } s \in S.$$

Accordingly, such proofs usually start with something like

$$\text{``}(r, s) \in R \times S \Rightarrow\text{''} \qquad \text{or} \qquad \text{``Let } (r, s) \in R \times S\text{''}$$

with the understanding that $r \in R$ and $s \in S$. Then "$(r, s) \in T$" will be the desired conclusion.

(a) $(x, y) \in (A \cup B) \times C \iff (x \in A$ or $x \in B)$ and $y \in C$
$\iff (x \in A$ and $y \in C)$ or
$(x \in B$ and $y \in C)$
$\iff (x, y) \in A \times C$ or $(x, y) \in B \times C$
$\iff (x, y) \in (A \times C) \cup (B \times C)$

The conclusion now follows from the definition of set equality. [The key to the proof is this logical equivalence from Chapter 1: $(P \vee Q) \wedge R \equiv (P \wedge R) \vee (Q \wedge R)$.]

(b) This part is left as an exercise.

(c) Exercise.

(d) First assume $A = B$. Then

$$(x, y) \in A \times B \Leftrightarrow x \in A \quad \text{and} \quad y \in B$$
$$\Leftrightarrow x \in B \quad \text{and} \quad y \in A \qquad \text{(here we use } A = B\text{)}$$
$$\Leftrightarrow (x, y) \in B \times A$$

Therefore $A \times B = B \times A$, as desired. Conversely, suppose that $A \times B = B \times A$; let's give an indirect proof of the statement $A = B$. If $A \neq B$ then one of the sets A, B contains an element not in the other; say, $a \in A - B$. Pick any $b \in B$. (There *is* such an element b because B is nonempty by hypothesis.) Then $(a, b) \in A \times B$, and so $(a, b) \in B \times A$ (since we have assumed $A \times B = B \times A$). Hence $a \in B$, contradicting the way in which a was chosen. Thus the statement $A \neq B$ must be false.

(e) First note that the definition of the power set notation $P(\ )$ allows us to restate the proposition to be proved:

$$A_1 \subseteq A \quad \text{and} \quad B_1 \subseteq B \implies A_1 \times B_1 \subseteq A \times B.$$

With this in mind, assume $A_1 \subseteq A$ and $B_1 \subseteq B$. Then

$$(x, y) \in A_1 \times B_1 \implies x \in A_1 \quad \text{and} \quad y \in B_1$$
$$\implies x \in A \quad \text{and} \quad y \in B \qquad \text{(from the hypothesis)}$$
$$\implies (x, y) \in A \times B.$$

Therefore $A_1 \times B_1 \subseteq A \times B$, as claimed.

(f) Exercise.

(g) Exercise. □

Exercises

1. Let T be the collection of telephone lines for incoming calls at the Hotel Acme switchboard, and let R be the set of rooms in the hotel. Assume that each room has exactly one telephone.

   (a) Show that $T \times R$ can be interpreted as the set of all possible connections of incoming calls with room telephones.

(b) Suppose four conversations are in progress. To what kind of subset of $T \times R$ does that correspond?

2. Show that if S is any set, then

$$\{1, 2\} \times S = (\{1\} \times S) \cup (\{2\} \times S).$$

3. Is it true that if $\{a, b, c\} = \{a, b, d\}$ then $c = d$?

4. Prove: $A \times B = \emptyset \iff A = \emptyset$ or $B = \emptyset$.

5. Describe the following subsets of $\mathbb{R} \times \mathbb{R}$ (viewed as the xy-plane) in words and pictures.

(a) $[-2, 1] \times [3, 5]$ (c) $\mathbb{Z} \times \mathbb{R}$

(b) $\mathbb{Z} \times \mathbb{Z}$ (d) $\mathbb{R} \times \mathbb{N}$

6. (a) Exhibit a subset S of $\{1, 2, 3\} \times \{1, 2, 3\}$ for which there are no sets $A, B \subseteq S$ such that $A \times B = S$.

(b) Which of the following pictures represent(s) a set that has the form $A \times B$ for some $A, B \subseteq \mathbb{R}$?

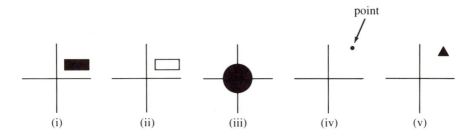

(i)   (ii)   (iii)   (iv)   (v)

7. Prove assertions (b), (c), (f), and (g) of Theorem 2.45.

8. Prove the implication "$\Rightarrow$" in Theorem 2.40.

9. Prove the equality $(\bigcup_{i \in I} A_i) \times S = \bigcup_{i \in I} (A_i \times S)$.

2.9   SET DECOMPOSITION: PARTITIONS AND RELATIONS

Zoologists study mammals, birds, reptiles, insects, and so on. Physiologists are concerned with body types: mesomorph, ectomorph, and endomorph. Mathematicians speak of even and odd integers; of rational and irrational numbers; of numbers that are positive, negative, or zero; and so

on. People tend to classify and, indeed, *need* to classify: it helps us to sort
out the universe. Our main goal in this section is to provide a mathematical framework for the notion of classification.

We begin with the concept of *partitioning* a set, which, according to
standard English usage, means to break it into nonoverlapping pieces.
(After World War II, Germany was partitioned into eastern and western
regions. The office was partitioned in order to reduce the interaction
among staff members.) Before giving a precise definition, here is a useful
example.

2.46 EXAMPLE.   Consider a three-piece jigsaw puzzle. How many
different states of assembly are there? (Call two puzzle states different if
one of them has two pieces interlocked that are not interlocked in the
other.) The answer is five, and here is a pictorial justification:

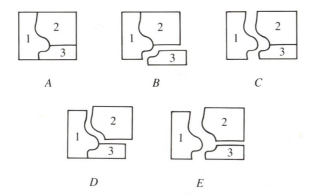

Now represent each cluster of interlocked pieces by the set containing the
corresponding numbers. Then each of the pictures is represented by a
family of such sets, as follows (here the Greek letter Π stands for *partition*):

$$\Pi_A = \{\{1, 2, 3\}\}$$
$$\Pi_B = \{\{1, 2\}, \{3\}\}$$
$$\Pi_C = \{\{2, 3\}, \{1\}\}$$
$$\Pi_D = \{\{1, 3\}, \{2\}\}$$
$$\Pi_E = \{\{1\}, \{2\}, \{3\}\}$$

Each of these is a family of nonempty sets whose union is the set $\{1, 2, 3\}$,
and no two sets in the same family share any elements. We say that each
of Π_A through Π_E is a partition of the set $\{1, 2, 3\}$.

The preceding example leads us to the following definition.

2.47 DEFINITION.   Let S be a nonempty set. A **partition** Π of S is a family $\Pi = \{A_i\}_{i \in I}$ of nonempty subsets of S satisfying these conditions:

$$\bigcup_{i \in I} A_i = S \qquad \text{(every member of } S \text{ belongs to some } A_i\text{)}$$

$$A_i \cap A_j = \emptyset \quad \text{if } i \neq j \qquad \text{(different pieces don't overlap)}$$

The A_i are called the **blocks** of the partition.

2.48 EXAMPLES.

(a) The set $S = \{1, 2, 3, 4\}$ has

$$\{\{1, 2, 3, 4\}\}$$

as its only one-block partition (that is, the only partition that has just one block). It has 7 two-block partitions:

$$\{\{1\}, \{2, 3, 4\}\}, \qquad \{\{2\}, \{1, 3, 4\}\}, \qquad \{\{3\}, \{1, 2, 4\}\},$$
$$\{\{4\}, \{1, 2, 3\}\}, \qquad \{\{1, 2\}, \{3, 4\}\},$$
$$\{\{1, 3\}, \{2, 4\}\}, \qquad \{\{1, 4\}, \{2, 3\}\}.$$

It has 6 three-block partitions (Exercise: list them), and its only partition into four blocks is

$$\{\{1\}, \{2\}, \{3\}, \{4\}\}.$$

(b) A jigsaw puzzle has two extreme states: full assembly and total disassembly. Similarly, a nonempty set S has two extreme partitions: the partition $\{S\}$ with only one block, and the partition into one-element blocks:

$$\{\{x\} \mid x \in S\}.$$

(c) To calculate effectively with a large set S of data, we may need to partition S into smaller blocks that can be treated separately. For instance, if S is a set of real numbers, and our calculations require that we treat numbers differently according to sign, we use the partition $\Pi = \{A_1, A_2, A_3\}$, where

$$A_1 = \{x \in S \mid x > 0\}, \qquad A_2 = \{x \in S \mid x < 0\}, \qquad A_3 = \{0\}.$$

Caution: This is a bona fide partition only if S contains positive numbers, negative numbers, and zero.

(d) Let E be the standard xy-plane, and consider the families $\{L_i\}_{i\in\mathbb{R}}$, $\{H_r\}_{r\in\mathbb{R}}$, $\{C_r\}_{r\geq0}$, and $\{D_r\}_{r\geq0}$ discussed in Examples 2.28 and 2.29; call them F_1, F_2, F_3, and F_4, respectively. The family F_1 is a partition of E; for, the fact that every point is on some vertical line yields $\cup L_i = E$, and two different vertical lines are parallel and therefore don't intersect. But F_2 is not a partition of E, because distinct half-planes intersect. (Indeed, $H_r \cap H_s = H_s$ if $s \geq r$.) It is left to the reader to verify that F_3 is a partition of E and F_4 is not.

We usually decide to partition a set during the course of a procedure when there seems to be no unified method for handling all its elements simultaneously. A useful strategy is to construct a partition such that all the elements in a given block share a common feature that allows a unified treatment.

2.49 EXAMPLES.

(a) The oddness or evenness of an integer is called its **parity**. The relation of "having the same parity" leads to a partition of $\mathbb{Z}$ into two blocks: the set of odd integers and the set of even integers.

(b) Let S be the set of points in the xy-plane that lie on neither axis. That is,

$$S = \{(x, y) \in \mathbb{R} \times \mathbb{R} \mid xy \neq 0\}.$$

For points $P = (p_1, p_2)$ and $Q = (q_1, q_2)$ in S, we say that P and Q "are in the same quadrant" if $p_1q_1 > 0$ and $p_2q_2 > 0$. This partitions S into the usual four quadrants of analytic geometry. If we take as a fifth set the union of the x-axis and y-axis, we achieve a partition of $\mathbb{R} \times \mathbb{R}$ into five blocks:

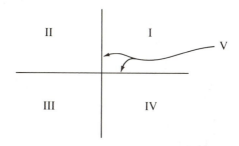

From these examples we see that the consideration of partitions leads us to study relations among elements in a set. But what exactly do we mean by a "relation"? Our objective now is to make this notion precise, to build a vocabulary for discussing the properties of relations, and to examine the interaction between relations and partitions.

Imagine that you are trying to teach the meaning of the words "larger" and "redder" to a child who has not yet learned "large" and "red." You have two apples, A_1 and A_2, to use as teaching materials. You notice that A_1 is larger than A_2, and you point this out to illustrate "larger." Then, when the child seems to understand, you move on to explain "redder" by pointing out that A_1 is redder than A_2. The result: the child comes to believe that "larger" and "redder" are synonyms. In fact, as long as the child works only with the set $S = \{A_1, A_2\}$ there is no harm in this belief, since the relations "is larger than" and "is redder than" on S both apply only to the ordered pair (A_1, A_2). Of course, to do a more effective teaching job you need a third apple, A_3. For instance, if A_3 is larger than A_1 but between A_1 and A_2 in redness, then you can distinguish between "larger" and "redder" by pointing out that on the set $\{A_1, A_2, A_3\}$, the relation "is larger than" is associated with the set of ordered pairs

$$\{(A_1, A_2), (A_3, A_1), (A_3, A_2)\},$$

while "is redder than" corresponds to the set of ordered pairs

$$\{(A_1, A_2), (A_1, A_3), (A_3, A_2)\}.$$

(Of course, if you are working with a standard child, you will use less formal terminology.) After this the child may still not have a clear understanding of the relations being defined (for that, he will need to work with sets with many more elements), but at least he will know that they are not synonyms. To sum up: the only way to distinguish between two relations on a given set is to know an ordered pair that belongs to one of the relations but not to the other. Now we can give a precise definition.

2.50 DEFINITION. A **relation** R on a set S is a collection of ordered pairs of elements of S; that is, a subset $R \subseteq S \times S$. The assertion $(x, y) \in R$ is usually abbreviated xRy, and we say x **is related to** y **by** R. (If $(x, y) \notin R$ we write $x\not\!Ry$.)

2.51 EXAMPLES.

(a) Let S be the set of all people and all dogs. Define a relation R on S by

$$pRd \iff (p \text{ is a person, } d \text{ is a dog, and } p \text{ owns } d).$$

Thus the relation of dog ownership is realized as a set of ordered pairs.

(b) For any set S, the relation $S \times S$ is called the **universal relation** on S, and the empty set $\emptyset$ is called the **empty relation** in this context. Neither is of serious interest. On the one hand, if everything is related to everything, then being related is no mark of distinction; on the other hand, if no two elements are related, why waste our time?

(c) The symbol "$<$" denotes a well-known relation on the set $\mathbb{R}$ of real numbers. In the formal sense of Definition 2.50, $<$ is a set of ordered pairs of real numbers, and the statement $a < b$ abbreviates the more cumbersome expression $(a, b) \in <$. Since it is our custom to picture $\mathbb{R} \times \mathbb{R}$ as the coordinatized xy-plane, we have the following formal portrait of the relation $<$ as a set of ordered pairs, where the dashed diagonal line is not included:

Similar considerations hold for the relation "$>$" on $\mathbb{R}$.

(d) Let L be the set of all living people. Define relations P and C on L as follows:

$$P = \{(x, y) \mid x \text{ is a parent of } y\},$$
$$C = \{(y, x) \mid y \text{ is a child of } x\}.$$

Here $xPy \iff yCx$, and we say that each of P and C is the **inverse** of the other, written $P = C^{-1}$ and $C = P^{-1}$. Similarly, on the set $\mathbb{R}$ of real numbers, the relation $<$ is the inverse of the relation $>$.

Sometimes if we know that certain elements are paired in a given relation, that tells us something about other pairings. For example in $\mathbb{R}$, if $x \leqslant y$ and $y \leqslant z$ then $x \leqslant z$. The following definition gives names for three especially important properties of some relations.

2.52 DEFINITION.   Let R be a relation on the set S. Then R is **reflexive** if xRx for all $x \in S$. The relation R is **symmetric** if for all x, $y \in S$, the implication $xRy \Rightarrow yRx$ is true. Finally, R is **transitive** if for all x, y, $z \in S$, the implication $(xRy$ and $yRz) \Rightarrow xRz$ is true.

2.53 EXAMPLES.

(a) Let P be the set of all people who have ever lived. Let A be the relation "is an ancestor of" (that is, $xAy \Leftrightarrow x$ is an ancestor of y) and let F be the relation "is a first cousin of." Then A is transitive but not reflexive or symmetric; and F is symmetric but not reflexive or transitive.

(b) (From a familiar children's game) Let

$$S = \{\text{rock, scissors, paper}\}.$$

Define a relation B (read "beats") on S by

$$B = \{(\text{rock, scissors}), (\text{scissors, paper}), (\text{paper, rock})\}.$$

The relation B is neither reflexive, symmetric, nor transitive. Incidentally, a similar phenomenon occurs in the ranking of professional tennis players. For instance, from the facts that Jones usually defeats Smith and Smith usually defeats Doe, we cannot conclude that Jones usually defeats Doe.

(c) Let $S = \{1, 2, 3\}$ and consider the following relations on S:

$$R_1 = \{(1, 1), (2, 2), (3, 3), (1, 2), (2, 3)\},$$
$$R_2 = \{(1, 2), (2, 3), (1, 3)\},$$
$$R_3 = \{(1, 2), (2, 1)\}.$$

Then R_1 is reflexive, not symmetric (since $1R_1 2$ and $2\cancel{R}_1 1$), and not transitive (since $1R_1 2$ and $2R_1 3$, but $1\cancel{R}_1 3$). We claim that R_2 is transitive, but neither reflexive ($1\cancel{R}_2 1$) nor symmetric ($1R_2 2$ but $2\cancel{R}_2 1$). To verify transitivity of R_2, we must show that for every x, y, $z \in S$, the following implication is true:

$$(xR_2 y \quad \text{and} \quad yR_2 z) \Rightarrow xR_2 z.$$

Recall from Chapter 1 that an implication is true unless its hypothesis is true and its conclusion is false. But the implication's hypothesis is true only when $x = 1$, $y = 2$, and $z = 3$, and in that case xR_2z is also true. Thus R_2 is transitive, as claimed. Finally, R_3 is symmetric but neither reflexive ($1\not R_3 1$) nor transitive [the implication $(1R_3 2$ and $2R_3 1) \Rightarrow 1R_3 1$ is false].

2.54 EXERCISE.   Let $S = \{1, 2, 3\}$. For each pair of properties from the set

$$\{\text{reflexive, symmetric, transitive}\},$$

write down a nonempty relation on S that satisfies both properties in that pair but not the other property.

2.55 DEFINITION.   A relation is an **equivalence relation** if it is reflexive, symmetric, and transitive. If $\sim$ is an equivalence relation and $x \sim y$, we say x and y are **equivalent** with respect to $\sim$. (The symbol "$\sim$" is often used to denote relations, probably because it is quick to write. We will avoid using it in contexts where it might be mistaken for a negation symbol.)

2.56 EXAMPLES.

(a) Let Z be the set of all animals in a particular zoo, and define $x \sim y$ to mean that x and y are of the same species. The relation $\sim$ is symmetric because each animal is of the same species as itself. It is reflexive because if x and y are of the same species (thus $x \sim y$) then so are y and x (thus $y \sim x$). Finally, if x and y are of the same species, and y and z are, too, then so are x and z, and this verifies transitivity.

(b) On the set $S = \{1, 2, 3, 4\}$, the relation

$$R = \{(1, 2), (2, 1), (1, 1), (2, 2)\}$$

is symmetric and transitive (check this); but it is not reflexive since, for example, $3\not R 3$. We can fortify this relation with reflexivity by adjoining more pairs: the larger relation

$$R_1 = R \cup \{(3, 3), (4, 4)\}$$

is reflexive. The relation R_1 inherits the symmetric and transitive properties of R, and R_1 is therefore an equivalence relation. Observe that R_1 was obtained by supplementing R with exactly what was essential (no more, no less) in order to get a reflexive relation containing R. We call R_1 the **reflexive closure** of R.

(c) Let S be the set of partially exposed stones in a stream that you wish to cross. For $x, y \in S$, define $x \sim y$ to mean that you can step directly from x to y without getting wet. Then $\sim$ is reflexive (once you're on a stone, you can hop on it) and symmetric, but *not* generally transitive: if you can step directly from x to y and from y to z, it does *not* generally follow that you can step directly from x to z. (Under what rock-and-stream conditions will $\sim$ be transitive?)

(d) Let S and $\sim$ be as in part (c). For $x, y \in S$, define $x \approx y$ to mean that you can step from x to y by a sequence of dry steps on members of S. Formally, $x \approx y$ means that there is a collection x_1, $x_2, \ldots, x_n$ of elements of S, with $x = x_1$ and $y = x_n$, such that $x_i \sim x_{i+1}$ for each i satisfying $1 \leqslant i \leqslant n - 1$. The new relation $\approx$ is transitive, since if you get from x to y by stepping on members of S and then from y to z likewise, you have travelled from x to z by stepping on members of S. Like $\sim$, the relation $\approx$ is reflexive and symmetric; therefore $\approx$ is an equivalence relation. Notice that the relation $\approx$ has been obtained by supplementing $\sim$ with precisely those pairs needed to acquire transitivity. Accordingly, $\approx$ is called the **transitive closure** of $\sim$.

(e) On the set T of all triangles in a given plane, the relations of congruence and similarity are both equivalence relations.

2.57 DEFINITION. If $\sim$ is an equivalence relation on a set S, the set of all elements of S that are equivalent (with respect to $\sim$) to a given element x constitute the **equivalence class** of x, denoted $[x]$. Thus

$$[x] = \{s \in S \mid s \sim x\}.$$

For instance, in Example 2.56(a), if the zoo contains an elephant e, then the equivalence class $[e]$ is the set of all elephants in Z. In 2.56(b), the equivalence relation R_1 has the following equivalence classes:

$$[1] = [2] = \{1, 2\}, \qquad [3] = \{3\}, \qquad [4] = \{4\}.$$

In part (e), if $\triangle$ is a triangle in the given plane, then, with respect to similarity, the equivalence class $[\triangle]$ consists of all the triangles in the plane that are similar to $\triangle$.

In each equivalence relation that we have discussed, notice that equivalence classes that are not equal have empty intersection. For instance, in Example 2.56(a) there is an equivalence class $[e]$ consisting of elephants and (probably) an equivalence class $[t]$ of tigers, and clearly $[e] \cap [t] = \emptyset$. In part (b) we have

$$[1] \cap [3] = \{1, 2\} \cap \{3\} = \emptyset.$$

The following result generalizes the preceding observation.

2.58 LEMMA. If $\sim$ is an equivalence relation and $[x] \neq [y]$, then $[x] \cap [y] = \emptyset$.

PROOF. We will prove the contrapositive. That is, we assume $[x] \cap [y] \neq \emptyset$, and we will show that $[x] = [y]$. First we prove the inclusion $[x] \subseteq [y]$. Let $s \in [x]$. Because $[x] \cap [y] \neq \emptyset$, there is an element $z \in [x] \cap [y]$. Thus $z \sim x$ and $z \sim y$ and so

$$
\begin{array}{c}
z \in [x] \\
\downarrow \\
s \underset{s \in [x]}{\underbrace{\sim}} x \sim z \underset{z \in [y]}{\underbrace{\sim}} y.
\end{array}
$$

By the transitive property of $\sim$, it follows that $s \sim y$. (In greater detail: the statements $s \sim x$ and $x \sim z$ give $s \sim z$, by transitivity; this, combined with $z \sim y$, gives $s \sim y$, again by transitivity.) Therefore $s \in [y]$, and we have shown the inclusion $[x] \subseteq [y]$. The converse inclusion $[x] \supseteq [y]$ is shown similarly, and it follows that $[x] = [y]$, as desired. $\square$

The equivalence relation "is of the same species" defined on a zoo Z [recall Example 2.56(a)] leads us to picture Z something like this:

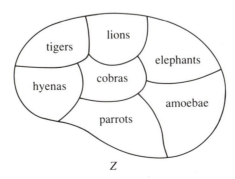

Z

Here we no longer think of Z as an unstructured assortment of animals; instead, the equivalence relation has enabled us to classify the members of Z and to cluster together all the animals of the same species. The resulting family of one-species sets (that is, the collection of "pieces" in the picture) constitutes a partition of Z. Similarly, the relation R_1 in Example 2.56(b) yields the decomposed set

and not just

The procedure for partitioning sets via equivalence relations is made explicit in the following theorem.

2.59 THEOREM. Let $\sim$ be an equivalence relation on a nonempty set S, and let Π be the family of equivalence classes determined by $\sim$. Then Π is a partition of S. (We call Π the partition **induced** by $\sim$.)

PROOF SKETCH. Begin by writing Π as an indexed set: $\Pi = \{C_i\}_{i \in I}$. By Lemma 2.58 we know that $C_i \cap C_j = \emptyset$ whenever $i \neq j$. To prove that Π is a partition it remains to check that $\cup C_i = S$. Since each C_i is a subset of S, so is $\cup C_i$. (Verify this.) Conversely, if $s \in S$ then s belongs to some C_i, namely, to $[s]$ (since $s \sim s$ by reflexivity). Hence $s \in \cup C_i$, so $S \subseteq \cup C_i$, and therefore $S = \cup C_i$. $\square$

Here is another way to design a visual representation for an equivalence relation $\sim$ on a set S of n elements. Draw a point (or **vertex**) corresponding to each member of S, and connect two vertices with an arc or line segment (or **edge**) if the corresponding members of S are related by $\sim$. For the sake of visual economy, it is customary not to draw an arc from each vertex to itself, since reflexivity is understood. The resulting collection of vertices and edges is called the **graph** of the relation $\sim$.

2.60 EXAMPLE. The equivalence relation

$$\sim \ = \ \{(1,1),(2,2),(3,3),(4,4),(5,5),(6,6),(1,5),(5,1),(2,4),(4,2),(2,6),(6,2),(4,6),(6,4)\}$$

on the set $S = \{1, 2, 3, 4, 5, 6\}$ is represented by the graph

Each cluster of connected vertices represents an equivalence class, so the associated partition of S into equivalence classes is

$$\{\{1, 5\}, \{3\}, \{2, 4, 6\}\}.$$

Notice how the graph and the partition display the essence of $\sim$ with an immediacy lacking in the original list of ordered pairs.

We have seen how an equivalence relation on a nonempty set S induces a partition of S into equivalence classes. Conversely, a partition of S can be used to *define* an equivalence relation. The next example is a model for this procedure, which is formalized in the subsequent theorem.

2.61 EXAMPLE.   The principal of Portsmouth Junior High School assigns entering seventh-grade students to one of five homerooms by putting each student's name on a card, shuffling the deck, and dealing the deck into five piles. Once the assignments have been made, an equivalence relation emerges: students x and y are said to be related if they have the same homeroom.

2.62 THEOREM.   Let Π be a partition of the set S. For $x, y \in S$, define $x \sim y$ to mean that x and y belong to the same block of the partition Π. Then $\sim$ is an equivalence relation on S. (This is called the equivalence relation **induced** by the given partition.)

PROOF.   Let $\Pi = \{A_i\}_{i \in I}$. Since Π is a partition, we know (by Definition 2.47) that $A_i \cap A_j = \emptyset$ when $i \neq j$, and also that $\cup A_i = S$. We must prove that the relation $\sim$ is reflexive, symmetric, and transitive.

If $x \in S$ then $x \in A_i$ for some i (since $S = \cup A_i$), and so by definition of $\sim$ we have $x \sim x$, proving reflexivity. For symmetry, if $x \sim y$ then

$$\{y, x\} = \{x, y\} \subseteq A_i, \qquad \text{for some } i \in I,$$

and so $y \sim x$. Finally, transitivity: if $x \sim y$ and $y \sim z$ then $\{x, y\} \subseteq A_i$ and $\{y, z\} \subseteq A_j$ for some $i, j \in I$. But then $y \in A_i \cap A_j$, and therefore $i = j$ (from the definition of partition). Thus $\{x, z\} \subseteq A_i$; that is, $x \sim z$.   □

2.63 EXAMPLES.

(a) Let S be the set of all points in the xy-plane that lie on neither axis. If $p_1, p_2 \in S$, define $p_1 \sim p_2$ to mean that p_1 and p_2 lie in the same quadrant. Then $\sim$ is an equivalence relation, since the four quadrants clearly partition S.

(b) The set $\mathbb{Z}$ of integers has a familiar partition into two blocks: $\mathbb{Z} = A \cup B$, with A the set of even integers and B the set of odd integers. The induced equivalence relation $x \sim y$ means that x and y are both even or both odd. (Another way to say it: $x \sim y$ if and only if $x - y$ is even.)

Exercises

1. Give three different partitions of each of the following sets. Do not use the trivial partition that has only one block. (Caution: Do not confuse the words "partition" and "block.")

    (a) The set $\mathbb{R}$ of real numbers
    (b) The set of all people
    (c) The set $\mathbb{Z}$ of integers

2. List all the partitions of the set $\{1, 2, 3, 4, 5\}$ that have no one-element blocks.

3. Label each of the following statements *true* or *false*.

    (a) $\{\{1, 2\}, \{3, 5\}\}$ is a partition of $\{1, 2, 3, 4, 5\}$.
    (b) $\{\{1, 2\}, \emptyset\}$ is a partition of $\{1, 2\}$.
    (c) $\{\{1, 2, 3, 4\}, \{-1\}\}$ is a partition of $\{-1, 1, 2, 3, 4\}$.
    (d) $\{\{1, 2\}, \{4, 1\}, \{3\}\}$ is partition of $\{1, 2, 3, 4\}$.

4. Prove that if S is a set and A is a nonempty proper subset of S, then $\Pi = \{A, S - A\}$ is a partition of S.

5. If $\Pi_1 = \{A_i\}_{i \in I}$ and $\Pi_2 = \{B_j\}_{j \in J}$ are both partitions of S, we say Π_1 is **finer** than Π_2 if every A_i is contained in some B_j. [Formally, $(\forall i)(\exists j)(A_i \subseteq B_j)$. Intuitively, the blocks in Π_1 are pieces of the blocks in Π_2. Think of a finer grind of coffee bean.] For example, if

$S = \{1, 2, 3, 4, 5\}$, the partition $\Pi_1 = \{\{1, 2\}, \{5\}, \{3, 4\}\}$ is finer than $\Pi_2 = \{\{1, 2, 5\}, \{3, 4\}\}$.

(a) Give four different partitions $\Pi_1, \Pi_2, \Pi_3, \Pi_4$ of the set $\{1, 2, 3, 4, 5, 6\}$, with Π_i finer than Π_{i+1} for $i = 1, 2, 3$.

(b) Do the same for the set $\mathbb{N}$ of natural numbers.

(c) Prove that every set S has a *finest* partition Π. That is, Π has the property that if Π' is any partition of S, then Π is finer than Π'. (Suggestion: First do part (c) when $S = \{1, 2, 3\}$. Think of the smallest "pieces" into which S can be decomposed. Then do the exercise without restricting S.)

6. Suppose $S = \{s_1, \ldots, s_n\}$ is a set of n locations on a tropical island, with $n \geq 1$. Define $s_i R s_j$ to mean that a person can safely get from s_i to s_j without using special equipment.

(a) Can you say for sure that R is reflexive? Symmetric? Transitive?

(b) Suppose location s_n cannot be reached from any of $s_1, \ldots, s_{n-1}$. How does that affect the answer to part (a)?

7. Let R be a relation on S. It can be shown that a unique relation R_1 exists on S with these properties:

(a) $R_1 \supseteq R$

(b) R_1 is reflexive.

(c) No proper subset of R_1 that contains R is reflexive.

The relation R_1 is called the **reflexive closure** of R. Now let $S = \{1, 2, 3, 4\}$, and define

$$R = \{(1, 2), (1, 3), (2, 3), (1, 4), (4, 1)\}.$$

Determine the reflexive closure of R.

8. Repeat Exercise 7 with the word "reflexive" replaced by the word "symmetric" throughout.

9. Repeat Exercise 7 with the word "reflexive" replaced by the word "transitive" throughout.

10. Let C be a set of n cities. For cities $x, y \in C$, define xRy to mean that there is a direct road from x to y that passes through no other cities along the way.

(a) Must R be symmetric? Illustrate any negative answer with a diagram, representing cities by vertices and roads by directed (with arrowheads) arcs or line segments between the cities.

(b) What road construction would have to be undertaken so that the resulting network of roads corresponds to the transitive closure of R? Otherwise put, what's the smallest amount of road construction that must be undertaken in order for the associated road-connection relation to be transitive?

11. We have seen (Example 2.60) how to represent an equivalence relation graphically. More generally, if R is *any* relation on an n-element set S, then R can be represented as follows. Start with a collection of n points in a plane, and label each point with the name of a different element of S; these points are the vertices of the graph. Consider each ordered pair of vertices, and connect vertices p and q with a directed edge from p to q if the ordered pair (p, q) belongs to R.

p $\qquad\qquad\qquad\qquad$ q

After all such edges have been drawn, the resulting configuration is called a **directed graph** associated with R, and we will denote such a graph by $G(R)$.

(a) Draw a directed graph for each relation in Example 2.53(c) and Example 2.56(b).

For each of the following, describe a property of $G(R)$ associated with the given condition on R.

(b) R is symmetric.
(c) R is reflexive.
(d) R is transitive.

12. Let R and $G(R)$ be as in Exercise 11. Describe how to use $G(R)$ to obtain a directed graph for each of the following:

(a) the symmetric closure of R
(b) the reflexive closure of R
(c) the transitive closure of R
(d) the inverse relation R^{-1}

13. Suppose R is a relation on the set $S = \{s_1, s_2, \ldots, s_n\}$. Denote the transitive closure of R by $\hat{R}$. Describe an algorithm (that is, an organized procedure) for determining all the elements $y \in S$ that satisfy the condition $s_1 \hat{R} y$. Your algorithm should involve repeated scanning of S to collect more elements that qualify. [Suggestion: For a start,

notice that every element y that satisfies $s_1 R y$ also satisfies $s_1 \hat{R} y$. This gives an initial batch of elements. Then work with each of these elements in the same way that you worked with s_1, and repeat as often as necessary. Clarifying "in the same way" and "as often as necessary" is the essence of the problem. (Incidentally, the algorithm suggested here is called *breadth-first search*.)]

14. Let R be a relation on the set S. Show that if R is an equivalence relation then so is the inverse relation R^{-1}.

15. A relation R is **antisymmetric** if xRy and yRx together imply that $x = y$. A relation R on S is a **partial ordering** (or simply an **ordering**) if R is reflexive, antisymmetric, and transitive. For example, the relation "$\leq$" on $\mathbb{R}$ is an ordering. Show that each of the following is an ordering.

   (a) The inclusion relation "$\subseteq$" on the power set of a given set A.
   (b) The *divisibility* relation on $\mathbb{N}$, defined as follows: if $a, b \in \mathbb{N}$, define $a \mid b$ (pronounced "a divides b") to mean that $b = ac$ for some $c \in \mathbb{N}$.

16. An ordering R on a set S (see Exercise 15) is a **linear ordering** (or **total ordering**) if every two elements of S are comparable; that is, for each $a, b \in S$ either aRb or bRa.

   (a) Show that neither ordering in Exercise 15 is a total ordering.
   (b) Let W be the set of English words. To use a dictionary one must know the *alphabetical* (or *lexicographic*) ordering of W, which we will denote here by L. Define L precisely. (That is, carefully state the conditions under which one word precedes another in the dictionary.)

17. Let $\sim$ be an equivalence relation on the set $\{1, 2, 3\}$.

   (a) Suppose $\sim$ has exactly three ordered pairs. List them explicitly.
   (b) Is it possible that $\sim$ has exactly *four* ordered pairs? Explain.
   (c) What is the maximum possible number of ordered pairs in $\sim$? When $\sim$ is chosen with this maximum number of ordered pairs, what is the corresponding partition into equivalence classes?

18. Let S be a nonempty set. Verify that the empty relation $\emptyset$ is a symmetric and transitive relation on S, but not a reflexive relation.

19. Let A and B be sets. Many books define a *relation R from A to B* to be

a subset $R \subseteq A \times B$. Show that such an R is a relation on $A \cup B$ according to Definition 2.50.

20. For each of the following, determine whether the given relation is or is not an equivalence relation. Then, if it *is* an equivalence relation, describe the partition it induces.

   (a) On the set H of human beings, define $x \sim y \Leftrightarrow x$ and y weigh within one pound of each other.

   (b) On the set C of all solid-color cars, define $x \sim y \Leftrightarrow x$ and y have the same color.

   (c) On the set $\mathbb{N}$ of positive integers, consider the *divisibility* relation defined in Exercise 15(b).

   (d) On the set $\mathbb{R}$ of real numbers, define $x \sim y \Leftrightarrow x^2 = y^2$.

   (e) On the set $\mathbb{R}$ of real numbers, define $x \not\sim y \Leftrightarrow xy < 0$.

   (f) On the plane $\mathbb{R} \times \mathbb{R}$, define $P \sim Q$ to mean that P and Q have the same y-coordinate.

21. Do Exercise 2.54.

22. Write out the definitions of the equivalence relations induced by the partitions $\{L_i\}_{i \in \mathbb{R}}$ and $\{C_r\}_{r \geq 0}$ discussed in Examples 2.28 and 2.29.

2.10   MATHEMATICAL INDUCTION AND RECURSION

Mathematical induction is a powerful technique for proving statements about large sets, such as the set $\mathbb{N}$ of natural numbers. If we can prove that every natural number n has a certain property P, then we have in effect proved infinitely many theorems: $P(1)$, $P(2)$, $P(3)$, ... (Here we use $P(n)$ to denote the statement, "The integer n has property P.") The miracle of mathematical induction is that it enables us to carry out such a task in only a finite number of steps. Induction is closely linked to the notion of *recursion,* which is a vital ingredient in the definitions of many complex mathematical structures. Recursion is also at the heart of "looping" procedures in computer programming. We begin with three intuitive examples that give the flavor of the inductive process.

2.64 EXAMPLES.

   (a) Imagine that you are swimming across a crocodile-infested lake, and your strength is fading. Suddenly a rescue helicopter arrives and drops down one end of a rope ladder. To be rescued, here is what you must do: (1) You must somehow grab the bottom step of the ladder; and (2) you must be able to get from each step to the next. The second statement needs amplification. Some steps may be slippery or otherwise special, and for your plan to be complete it must accommodate all of these difficulties. For example, getting from the first step to the second may require different strategy from the problem of getting from the eighth step to the ninth. Your plan must either handle individual differences separately or (better) be general enough to treat all cases with one set of instructions. However you do it, once you are assured that you can grab the bottom step and, having done that, that you can always get from where you are to the next level, then you know you will be rescued.

(2.65)

Mathematical induction.

(b) Imagine a massive line of people, beginning in front of you and extending out over the horizon, with everyone's feet embedded in cement. You believe that everyone in the line will soon catch a particular disease, and here is the outline of your proof to government officials. (1) You conduct a test showing that the first person in line has the disease. (2) You prove that the disease is contagious, which means that when one person catches the disease the adjacent person will also catch it. In the manner of part (a), your proof of contagion must accommodate any special characteristics of the individuals in the line.

(c) Suppose you are programming a computer to calculate and print the number $n! = 1 \cdot 2 \cdot 3 \cdots (n-1) \cdot n$ (pronounced "n factorial") for each integer n from 1 to 1000. A first attempt at a program might have 2001 lines as follows (we ignore details of programming format):

```
LET 1! = 1
PRINT 1!
LET 2! = 2·1
PRINT 2!
LET 3! = 3·2·1
PRINT 3!
    ⋮
LET 1000! = 1000·999·998 · · · 3·2·1
PRINT 1000!
END
```

with the dots filled in by actual numbers. Fortunately we can achieve the same results in only eight lines:

```
LET F = 1
LET N = 1
DO WHILE N ≤ 1000
  ┌→ COMPUTE  N·F  AND LABEL THE RESULT  F
  │  PRINT F
  │  ADD 1 TO N AND LABEL THE RESULT  N
  └─ LOOP
END
```

Here, as in the previous examples, rather than treat each step individually we start with the first step of interest ($N = 1$) and then give a general procedure for moving from each step to its successor (from $N = K$ to $N = K + 1$).

Our main goal in this section is to state and apply several versions of the principle of mathematical induction, which is just a formalization of the examples we have been discussing. Though we state the principle without proof, we note that it can be deduced (in all its versions) from the following intuitively reasonable statement, called the *well-ordering principle* for ℕ:

Every set of natural numbers has a least element.

This principle was probably well known during the Stone Age in this form:

If everyone has a rock collection, then someone has a smallest collection.

Notice here that an agreement that a certain minimal number exists is *not* an agreement on how to *compute* it.

In what follows, we continue the convention of using $P(n)$ for the statement, "The number n has property P."

2.66 PRINCIPLE OF MATHEMATICAL INDUCTION (FIRST VERSION). Let $S \subseteq \mathbb{N}$, and suppose the following:

(a) $1 \in S$.

(b) For each $k \in \mathbb{N}$, the following implication is true:

$$k \in S \implies k + 1 \in S.$$

Then $S = \mathbb{N}$.

If for some property P we have $S = \{x \in \mathbb{N} \mid P(x)\}$, then 2.66 can be reformulated as follows.

2.67 PRINCIPLE OF MATHEMATICAL INDUCTION (SECOND VERSION). Suppose P is a property for which the following statements hold:

(a) $P(1)$.

(b) For each $k \in \mathbb{N}$, the following implication is true:

$$P(k) \implies P(k + 1).$$

Then $P(n)$ holds for every $n \in \mathbb{N}$.

The act of verifying $P(1)$ is the **basis step** of the induction proof, and verifying statement (b) is the **induction step**. The statement $P(k)$ in part (b) is called the **induction hypothesis**. In the imagery of Figure 2.65, the basis step fastens our grip onto the bottom step of the ladder, and the induction step guarantees that once we reach any step, we can get to the next step. The result is that we can reach every step of the ladder. (In 2.66 and 2.67, the "ladder" has a step corresponding to each natural number.)

A technique we often use in proving the induction step is to express one or more parts of statement $P(k + 1)$ in terms of expressions involved in $P(k)$; then we can use the information supplied by $P(k)$. Incidentally, a proof by induction of a statement of the form "$P(n)$ is true for every $n \in \mathbb{N}$" is called a proof "by induction on n."

2.68 EXAMPLE. We claim that for every $n \in \mathbb{N}$,

$$1 + 2 + \cdots + n = \frac{n(n + 1)}{2}.$$

This proposition is suggested by the following observation:

$$1 + 2 + 3 + \cdots + 100 = \underbrace{(1 + 100)}_{101} + \underbrace{(2 + 99)}_{101} + \underbrace{(3 + 98)}_{101}$$

$$+ \cdots + \underbrace{(50 + 51)}_{101}$$

$$= 50 \cdot 101 = \frac{100}{2} \cdot 101.$$

More generally, if n is any even natural number, it is easy to believe the equation

$$(*) \quad 1 + 2 + \cdots + n = \underbrace{1 + n}_{n + 1} + \underbrace{2 + (n - 1)}_{n + 1} + \underbrace{3 + (n - 2)}_{n + 1} + \cdots,$$

a sum of $n/2$ terms each equal to $n + 1$; hence the claim. A similar argument can be made if n is odd. (Try this.) Geometrically inclined individuals may be more convinced by a picture:

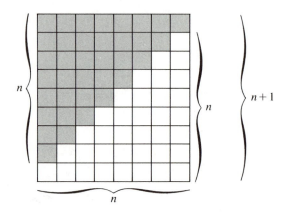

The sum $1 + 2 + \cdots + n$ is half of the rectangle's area. (Just count the white squares.)

Each of these arguments provides some insight, but each has a weak moment. The dots at the right end of (*) are unsatisfying. Also, the picture actually shows an 8×9 rectangle, and perhaps we have been seduced by a special property of 8×9 rectangles that doesn't carry over to $n \times (n + 1)$ rectangles for all n. Now let's actually prove the result using mathematical induction.

DETAILED PROOF WITH SUBCONSCIOUS REMARKS. We verify statements (a) and (b) in 2.67. Here the basis $P(1)$ is the statement $1 = 1 \cdot (1 + 1)/2$, which is clearly true. Now assume the induction hypothesis $P(k)$; that is, assume the statement is true in the case $n = k$: $1 + 2 + \cdots + k = k(k + 1)/2$. We must deduce that $P(k + 1)$ holds; in other words, that $1 + 2 + \cdots + (k + 1) = (k + 1)[(k + 1) + 1]/2$. We now have

$$
\begin{aligned}
1 + 2 + \cdots + (k + 1) &= (1 + 2 + \cdots + k) && \text{(this sets us up to} \\
& \quad + (k + 1) && \text{apply } P(k)) \\[2mm]
&= \frac{k(k + 1)}{2} + (k + 1) && \text{(by the induction} \\
& && \text{hypothesis } P(k)) \\[2mm]
&= \frac{k(k + 1) + 2(k + 1)}{2} \\[2mm]
&= \frac{k^2 + 3k + 2}{2} \\[2mm]
&= \frac{(k + 1)(k + 2)}{2} \\[2mm]
&= \frac{(k + 1)[(k + 1) + 1]}{2}. && \text{(by arithmetic)}
\end{aligned}
$$

So $P(k + 1)$ holds, and we're done. $\square$

MORE STANDARD, CONCENTRATED PROOF
(WITHOUT THE MUMBLING).

$$
P(1): \quad 1 = \frac{1 + 1}{2}
$$

Assume $P(k)$: $\quad 1 + \cdots + k = k(k + 1)/2$. Then

$$
\begin{aligned}
1 + 2 + \cdots + (k + 1) &= (1 + 2 + \cdots + k) + (k + 1) \\[2mm]
&= \frac{k(k + 1)}{2} + (k + 1) && \text{(by } P(k)) \\[2mm]
&= \frac{(k + 1)(k + 2)}{2},
\end{aligned}
$$

which proves $P(k + 1)$. $\square$

Remember, proving that we can get from one step of a ladder to the next does not prove that we can get *on* the ladder! In a proof by induction, the induction step of the proof tells us how to move up the ladder; but the basis step is needed to get us *onto* the ladder in the first place. Similarly, a proof of the conditional statement "$P(k) \Rightarrow P(k+1)$" is not a proof of $P(k)$. In fact, for values of k for which $P(k)$ is false, the statement "$P(k) \Rightarrow P(k+1)$" is *automatically* true; that is why we focus on those cases in which $P(k)$ is true. Thus the line "assume $P(k)$" is legitimate and does not represent a gap in the logic.

2.69 EXAMPLE. Look for a pattern in the following data.

Arithmetic   *Geometry*

n

n

Add the lengths of the bent strips.
(There are n of them, and the sum is n^2.)

Based on this data, it is natural to make the following conjecture:

The sum of the first n odd natural numbers is n^2.

Notice the equalities

$$1 = 2(1) - 1, \quad 3 = 2(2) - 1, \quad 5 = 2(3) - 1, \quad 7 = 2(4) - 1;$$

so the ith odd natural number is the number $2(i) - 1$. The conjecture can now be restated like this:

$$\text{For all } n \in \mathbb{N}, \quad 1 + 3 + 5 + \cdots + (2n - 1) = n^2.$$

Now for the proof by induction. Let $P(n)$ denote the equation to be proved for each n. Then, as desired,

$$P(1): \quad 2(1) - 1 = 1 = 1^2.$$

Now assume $P(k)$: $1 + \cdots + (2k - 1) = k^2$. [We must prove $P(k + 1)$: $1 + \cdots + (2(k + 1) - 1) = (k + 1)^2$.] We have

$$1 + \cdots + (2(k + 1) - 1) = \underbrace{1 + \cdots + (2k - 1)}_{k^2} + \underbrace{(2(k + 1) - 1)}_{2k + 1}$$

$$\text{(by } P(k)\text{)}$$

$$= k^2 + 2k + 1 = (k + 1)^2.$$

The proof of the implication $P(k) \Rightarrow P(k + 1)$ is now complete. □

Sometimes we want to prove a property to be true for every integer greater than or equal to some given integer N, and for this purpose the following variant of the induction principle is needed.

2.70 PRINCIPLE OF MATHEMATICAL INDUCTION (THIRD VERSION). Let N be a fixed integer. Suppose P is a property for which the following two statements hold:

(a) $P(N)$

(b) For all integers $k \geq N$, the implication $P(k) \Rightarrow P(k + 1)$ is true.

Then $P(n)$ is true for all integers $n \geq N$.

2.71 THEOREM. If S is a set with n elements then the power set $P(S)$ has 2^n elements.

PROOF OUTLINE. We will use induction in the format of 2.70 with $N = 0$. Here a set with m elements will be called an *m-set;* and in order to avoid confusion about the meaning of $P(\ )$, in this proof we use $P(\ )$ only for the power set notation.

First consider the case $n = 0$. The only 0-set is $\emptyset$, and $P(\emptyset) = \{\emptyset\}$ is a 1-set ($1 = 2^0$). Now assume the result is true for $n = k$: the power set of a k-set is a 2^k-set. Let S be a $(k + 1)$-set. Fix an element $s_0 \in S$. Then $S = \{s_0\} \cup T$, where T is a k-set. With each subset $A \subseteq T$ we associate two subsets of S, namely, A and $A \cup \{s_0\}$. This accounts for every subset of S. Thus S has twice as many subsets as T, so by the induction hypothesis for k-sets, S has $2 \cdot 2^k = 2^{k+1}$ subsets. □

We will extend this result on the comparative sizes of a set and its power set in Chapter 4 (see Cantor's theorem).

In constructing an induction proof, it is not always an easy matter to recognize how to use the induction hypothesis $P(k)$ in proving $P(k + 1)$. The following example is a typical result in elementary number theory; here the key to the arithmetic juggling is the addition of zero to an expression in such a way as to force the quantity in $P(k)$ to appear. Recall that an integer x is a **multiple** of an integer y if $x = yt$ for some integer t.

2.72 EXAMPLE. Prove that for every integer $n \geqslant 0$, the number $4^{2n+1} + 3^{n+2}$ is a multiple of 13.

PROOF. We use induction on n.

$$P(0): \quad 4^{2(0)+1} + 3^{0+2} = 4 + 9 = 13 = 13 \cdot 1$$

Assume $P(k)$: $\quad 4^{2k+1} + 3^{k+2} = 13t$ for some integer t. We must prove

$$P(k + 1): \quad 4^{2(k+1)+1} + 3^{(k+1)+2} \quad \text{is a multiple of 13.}$$

We have

$$
\begin{aligned}
4^{2(k+1)+1} + 3^{(k+1)+2} &= 4^{(2k+1)+2} + 3^{(k+2)+1} \\
&= 4^2(4^{2k+1}) + 4^2\underbrace{(3^{k+2} - 3^{k+2})}_{0} + 3 \cdot 3^{k+2} \\
&= 4^2\underbrace{(4^{2k+1} + 3^{k+2})}_{\substack{13t \\ \text{(by } P(k))}} + 3^{k+2}\underbrace{(-4^2 + 3)}_{-13} \\
&= 13(16t - 3^{k+2}), \quad \text{a multiple of 13.} \qquad \square
\end{aligned}
$$

Sometimes we want to carry out an induction argument, but we need more than $P(k)$ to deduce $P(k + 1)$. For this purpose the following formulation of the induction principle is useful.

2.73 PRINCIPLE OF MATHEMATICAL INDUCTION (FOURTH VERSION). Let N be a fixed integer. Suppose P is a property for which the following two statements are true:

(a) $P(N)$
(b) For every integer $k \geqslant N$, if $P(t)$ is true for all integers t satisfying $N \leqslant t \leqslant k$, then also $P(k + 1)$ is true.

Then $P(n)$ is true for every integer $n \geqslant N$.

Once again we illustrate with a result from number theory. Recall that a **prime number** is an integer $p > 1$ that has no integer factorization $p = ab$ in which $a > 1$ and $b > 1$.

2.74 THEOREM. Every integer $n \geqslant 2$ is a product of prime numbers. (Here a prime number is itself to be viewed as a one-factor product of prime numbers.)

PROOF. We argue by induction on n, using 2.73. The number 2 is clearly prime, so the result holds when $n = 2$.

Now suppose for some $k \geqslant 2$ that every integer t satisfying $2 \leqslant t \leqslant k$ is a product of primes; we must now show $k + 1$ to be a product of primes. If $k + 1$ is prime we are done. Otherwise write $k + 1 = ab$, with $a > 1$ and $b > 1$. Then in fact $2 \leqslant a \leqslant k$ and $2 \leqslant b \leqslant k$, so by the induction hypothesis, a and b are both products of primes, say, $a = p_1 \cdots p_r$ and $b = p_{r+1} \cdots p_s$, where the p's are primes. Then $k + 1 = ab = p_1 \cdots p_r \cdot p_{r+1} \cdots p_s$, and the proof is complete. □

Notice in the proof of Theorem 2.74 that $P(k)$ by itself would not have enabled us to prove $P(k + 1)$, since there is no obvious link between the factorizations of k and $k + 1$.

We conclude this chapter with a brief discussion of the use of an induction procedure in defining families of objects. Imagine that you are planning a hike that will take you many days. The weather in the region is unpredictable, you are unfamiliar with the details of the terrain, and your physical condition has not been recently evaluated. It is therefore impossible for you to plan the details of timing and route before the actual trip. But conditions do allow a detailed plan for an hour's hike from any point you have reached. So you proceed as follows. You decide when and where to start, and you decide your direction for the first hour's hike. At the end of each hour you will stop and evaluate your situation, and you have an algorithm (a recipe, an organized procedure) that will determine your plans for the next hour according to several clearly defined criteria (such as the slope of the land, the weather, and so on). Thus if h_k denotes your activity in hour k of the hike, your algorithm determines h_{k+1} based on the results of $h_1, h_2, \ldots, h_k$. We say your hike is being defined **by induction (inductively)** or **by recursion (recursively, by recurrence relation)**.

To consider recursion in a mathematical context, we will first return to the statement of Example 2.68:

$$(*) \qquad \text{For every } n \in \mathbb{N}, \quad 1 + 2 + \cdots + n = \frac{n(n+1)}{2}.$$

What we understand by $1 + 2 + \cdots + n$ is this: take 1, then add 2, then add 3 to that, then add 4 to that, and continue until the last term, n, has been added. We can be more precise in our specifications; in fact we can express the left and right sides of the equation in (*) by the following two programs:

```
Left side                          Right side
INPUT N                            INPUT N
LET K = 1                          LET NEWSUM = N(N + 1)/2
LET SUM = 0                        END
DO WHILE K≤N
   ADD K TO SUM AND
      LABEL THE RESULT SUM
   ADD 1 TO K AND LABEL
      THE RESULT K
LOOP
END
```

The meaning of (*) is this: for every input n, SUM = NEWSUM. Thus mathematical induction has enabled us to prove, in a few lines, that two quite different programs compute the same number for all possible inputs!

The stepwise construction of SUM in the left-hand program is an example of recursion in action. Now suppose a number a_i is given for each $i \in \mathbb{N}$. The intuitive expression $a_1 + a_2 + \cdots + a_n$ is usually abbreviated by the

$$\text{summation symbol} \quad \sum_{i=1}^{n} a_i$$

The following recursive scheme formally defines this symbol for each $n \in \mathbb{N}$. In effect, the definition is a program for evaluating the symbol.

$$\sum_{i=1}^{1} a_i = a_1$$
$$\sum_{i=1}^{k+1} a_i = \left(\sum_{i=1}^{k} a_i \right) + a_{k+1}$$

Likewise, we recursively define the

$$\text{product symbol} \quad \prod_{i=1}^{n} a_i \quad \text{and} \quad \text{factorial symbol} \quad n!$$

respectively by

$$\prod_{i=1}^{1} a_i = a_1 \qquad \left.\begin{array}{c}\\[0.5em]\\\end{array}\right\} \quad \text{and} \quad \left\{\begin{array}{l} 0! = 1 \\[1em] (k+1)! = k!(k+1)\end{array}\right.$$

$$\prod_{i=1}^{k+1} a_i = \left(\prod_{i=1}^{k} a_i\right) \cdot a_{k+1}$$

Notice that these definitions have no dots (except for a multiplication symbol), and they are easily translated into computer programs. In a line of text, the sum and product symbols will usually appear as $\Sigma_{i=1}^{n} a_i$ and $\Pi_{i=1}^{n} a_i$, respectively.

Similarly, we can define families of sets recursively. For example,

$$A_1 = \{1, 5\}$$
$$A_{k+1} = P(A_k) \qquad (P(\ ) \text{ is the power set notation})$$

defines a set A_n for each $n \in \mathbb{N}$.

We will return to recursion when we discuss *sequences* in Chapter 3.

Exercises

1. Use mathematical induction to prove the following statement for every $n \in \mathbb{N}$:

$$1^2 + 2^2 + \cdots + n^2 = \frac{n(n+1)(2n+1)}{6}.$$

2. Prove for every integer $n \geq 1$:

$$1^3 + 2^3 + \cdots + n^3 = \left(\frac{n(n+1)}{2}\right)^2.$$

3. Prove that the inequality $n^2 \geq n$ holds *for every integer*.

4. Prove that for every integer $n \geq 0$ the number $n^4 - 4n^2$ is divisible by 3.

5. Assume that $S \subseteq \mathbb{N}$ and $3 \in S$, and assume further that the implication

$$x \in S \Rightarrow x + 3 \in S$$

is true. Deduce that

$$\{3n \mid n \in \mathbb{N}\} \subseteq S.$$

6. Prove for every integer $n \geq 2$:

$$\frac{1}{1 \cdot 2} + \frac{1}{2 \cdot 3} + \cdots + \frac{1}{(n-1)n} = 1 - \frac{1}{n}.$$

7. Prove that $2^n > n^3$ for every integer $n \geq 10$.

8. What is wrong (if anything) with the following induction argument?
   Proposition. All horses are the same color.
   "Proof" outline. It suffices to show for every $n \in \mathbb{N}$ that in any collection of n horses, all n of them are the same color. If $n = 1$ the result is trivial. Now assume that in any collection of k horses, all k horses are the same color. Consider a set of $k + 1$ horses. By the induction hypothesis, the first k horses are the same color and the last k horses are the same color. But the first and last are the same color as the middle horses.

   Therefore all the horses are the same color.

9. It is claimed that every positive real number α has the property $\alpha^{n-1} = 1$ for all $n \in \mathbb{N}$, and here is a "proof" using the principle of mathematical induction (fourth version). If $n = 1$ we have $\alpha^{n-1} = \alpha^{1-1} = \alpha^0 = 1$, as desired. Now suppose the result has been checked for all values of n satisfying $1 \leq n \leq k$. Then

$$\alpha^{(k+1)-1} = \alpha^k = \frac{\alpha^{k-1} \cdot \alpha^{k-1}}{\alpha^{k-2}}$$

$$= \frac{1 \cdot 1}{1} \qquad \text{(by the induction hypothesis)}$$

$$= 1. \quad \square$$

   Is the claim now proved? Explain.

10. There is ambiguity to the exponential expression

(*)

$$a_n^{a_{n-1}^{\cdot^{\cdot^{\cdot^{a_2^{a_1}}}}}}$$

(For example, does 2^{3^4} mean $2^{(3^4)} = 2^{81}$ or $(2^3)^4 = 2^{12}$?) Write a recursive procedure for computing (*), with the interpretation that the exponent of each a_i is everything above it.

CHAPTER 3

Functions

A group of . . . aristocrats lost their way hiking in the Alps. One of them, it is said, took out a map, and after studying it for a long time, exclaimed: "Now I know where we are!" "Where?" asked the others. "See that big mountain over there? We are right on top of it."

GEORGE GAMOW
One Two Three . . . Infinity

3.1 DEFINITIONS AND EXAMPLES

In Chapter 2 we introduced sets: what they are, how to build new ones from old ones, and how to decompose them into subsets. In this chapter we will begin to consider more dynamic set interactions, called *functions,* which link the elements of one set to elements of another in a rather special way.

The theory of functions provides the foundation for what is called *mathematical modelling,* whereby one mathematical object represents certain features of another or of some nonmathematical object, perhaps with the elimination of some distracting complexities. Moreover, functions give a mathematically precise framework for the intuitive idea of *transformation.* Ordered pairs are fundamental for everything that we will be discussing here.

3.1 EXAMPLES.

(a) Each member of a television studio audience takes a numbered raffle ticket from a well-mixed bowl; this associates a unique number with each member of the audience. A complete record of this data is a list of ordered pairs of the form {(John Doe, 467), (Molly Brown, 93), . . . }, with one pair for each audience member.

(b) Pair each real number with its square; this process produces the set of all ordered pairs of the form (x, x^2) for $x \in \mathbb{R}$.

Notice that in Example 3.1(a), no two people are paired with the same number, while in 3.1(b) a nonzero number x and its negative are both paired with x^2. Both examples in 3.1 are functions, according to the definition about to be given, but we will find it useful to remain aware of this distinction between them.

Now we are ready for a precise definition of function. First recall that the Cartesian product $A \times B$ of sets A and B is the set of all ordered pairs of the form (a, b), with $a \in A$ and $b \in B$.

3.2 DEFINITION.
Let A and B be sets. A **function** f **from** A **to** B is a set of ordered pairs

$$f \subseteq A \times B$$

with the property that for each element x in A there is exactly one element y in B such that $(x, y) \in f$. The statement "f is a function from A to B" is usually represented symbolically by

$$f : A \to B \qquad \text{or} \qquad A \xrightarrow{f} B.$$

In Example 3.1(a), the given set of ordered pairs is a function from the audience to the set of ticket numbers; and in 3.1(b) the given set of pairs is a function from $\mathbb{R}$ to $\mathbb{R}$.

Suppose f is a function from A to B. Then A is called the **domain** of f, denoted dom f; its members are the first coordinates of the ordered pairs belonging to f. If $(x, y) \in f$ we usually write $y = f(x)$ and call y the **image** of x; and x is a **pre-image** of y. In Example 3.1(a) the domain is the audience, and f(Molly Brown) $= 93$. In Example 3.1(b) the domain is the set $\mathbb{R}$ of real numbers, and $f(-.2) = f(.2) = .04$; so $-.2$ and $.2$ are both pre-images of .04.

Notice that in 3.1(a) there is no abbreviated way to present the function: all the pairs must be listed if we are to know what the number assignments are. On the other hand, in 3.1(b) an explicit listing of all the pairs is impossible, but we can completely describe the function by giving the rule

$$f(x) = x^2 \qquad \text{for all } x \in \mathbb{R}.$$

Thus it is sometimes convenient to view a function as a set of ordered pairs, as in the definition; and at other times we will view a function f from A to B as *a rule or formula for pairing each member of A with a unique element of B*. Also, we may think of f as a kind of programmed machine that emits a unique element of B whenever an element of A is fed in: input x always yields output $f(x)$.

If $y = f(x)$, we sometimes say that f **transforms** or **takes** x to y, denoted by $x \overset{f}{\mapsto} t$ or $x \mapsto y$. We also say that f **acts** or **operates** on the members of its domain. An unnamed function from A to B is often indicated by $A \to B$. Functions are also called **mappings, maps,** or **transformations**.

3.3 EXAMPLES.

(a) Suppose $S \subseteq T$. Define a function i from S to T by

$$i(x) = x \qquad \text{for all } x \in S.$$

This is called the **inclusion mapping**. If, in this setting, $S = T$, we write $i = i_S$ and call it the **identity mapping** on S.

(b) Given nonempty sets A and B, fix an element $b_0 \in B$ and define a function f from A to B by

$$f(a) = b_0 \qquad \text{for all } a \in A.$$

As a set of ordered pairs, $f = \{(a, b_0) \mid a \in A\}$. This is called the **constant function** with value b_0.

(c) Let F be the set of all the living fathers of living daughters, and let D be the set of all the living daughters of living fathers. Define

$$g = \{(x, y) \mid x \in F \text{ and } y \text{ is the daughter of } x\}.$$

Is g a function from F to D? (Pause and think, *then* read on.) We have $g \subseteq F \times D$, and for each $x \in F$ there is a $y \in D$ such that

$(x, y) \in g$. So far so good. But if a father x has two living daughters, say with the delightful names y_1 and y_2, then $(x, y_1) \in g$ *and* $(x, y_2) \in g$, contradicting the "exactly one" condition in the definition of function. Therefore g is *not* a function. On the other hand, if

$$h = \{(y, x) \mid y \in D \text{ and } x \text{ is the father of } y\},$$

then h is a function from D to F, since every daughter has exactly one father.

(d) What happens when we depress a key on a computer keyboard? Clearly the answer depends on two things: the state of the machine (for example, the model, the data that has already been fed in) and our choice of key. That is, the new state is determined by the ordered pair (q, σ), where q is the former state and σ is the input symbol. Thus if S denotes the set of all possible states of the machine, and Σ denotes the set of keyboard symbols, there is a function

$$\delta: S \times \Sigma \to S$$

defined as follows: $\delta((q, \sigma)) = q'$ means that if the symbol σ is received as input when the machine is in state q, then the machine will go to state q'. The function δ is called the **transition function** for the machine. Notice a new wrinkle in this example: the members of the domain of δ are themselves ordered pairs. In this situation it is customary to suppress one pair of parentheses and write $\delta(q, \sigma)$ instead of $\delta((q, \sigma))$. We will follow this practice from now on.

(e) Let S be the set of all propositions, that is, the set of all truth-valued statements. Define the **truth-value function**

$$v: S \to \{T, F\}$$

by

$$v(P) = \begin{cases} T & \text{if } P \text{ is a true proposition,} \\ F & \text{if } P \text{ is a false proposition.} \end{cases}$$

As we saw in Chapter 1, if we know the values taken by v on some propositions, we can deduce its values on others. For instance, if $t(P) = T$ and $t(Q) = F$, then $t(P \wedge Q) = F$ and $t(P \vee Q) = T$.

(f) A function $f: \mathbb{R} \to \mathbb{R}$ is said to be a **polynomial function of degree** n if there are real numbers $a_0, a_1, \ldots, a_n$, with $a_n \neq 0$, such that

(*) $$f(x) = a_n x^n + a_{n-1} x^{n-1} + \cdots + a_1 x + a_0$$

for all $x \in \mathbb{R}$. The numbers $a_0, a_1, \ldots, a_n$ are called the **coefficients** of f. (It is a fact, but not an *obvious* one, that the coefficients of f are uniquely determined by f. That is, the formula for f cannot be rewritten using other coefficients in place of the given ones without producing a different function. We will omit the proof here, since it requires the use of results from Chapter 6.)

Given a polynomial function f as in (*) and a real number x, how many multiplications are required to compute $f(x)$? (For the purposes of this discussion, we will assume that the coefficients of f are all nonzero.) Our mastery of high school algebra yields an instant answer: computing each term $a_i x^i$ requires exactly i multiplications, so evaluating the whole expression requires $1 + 2 + \cdots + n = n(n + 1)/2$ multiplications. (Here we have used the summation formula obtained in Example 2.68.)

Now imagine that each multiplication has a cost, and we want to minimize the overall expense. (This is a fundamental concern in numerical programming.) Once the quantity x^i has been computed, if we commit it to memory, then the computation of $a_{i+1} x^{i+1}$ will require just two more multiplications:

$$a_{i+1} x^{i+1} = a_{i+1} \cdot x^i \cdot x.$$

Thus the computation of $a_1 x$ requires one multiplication, and then the computation of all the terms $a_i x^i$ for $2 \leqslant i \leqslant n$ requires a total of $2(n - 1)$ more, for a grand total of $2n - 1$ multiplications. This represents an improvement over the $n(n + 1)/2$ multiplications required by the first method, as long as $n > 1$. (Exercise: check this.)

We can do even better by reformulating f as follows:

$$f(x) = (\cdots ((a_n x + a_{n-1}) x + a_{n-2}) x + \cdots + a_1) x + a_0.$$

Let's recite the corresponding algorithm in words. Start with a_n, and then go through the following loop (that is, cycle of activity) until all the coefficients have been exhausted (watch for dizziness and irregular breathing). Multiply the number in hand by x, then add the coefficient whose subscript is one less than the subscript of the previous coefficient used. This algorithm, known as **Horner's rule**, requires only n multiplications; it is therefore the cheapest of the three procedures we have discussed, according to our multiplicative measure of cost.

From this work with polynomial functions we see that having a *formula* for a function is not the same as having an efficient algorithm for its evaluation.

Many practical problems amount to asking whether a function exists that satisfies certain constraints. Among these problems are *assignment problems,* and we will now briefly consider the *marriage problem* as an example. Let $X = \{x_1, \ldots, x_n\}$ and $Y = \{y_1, \ldots, y_m\}$ be the sets of unmarried women and men, respectively, in a given region. Suppose that each woman x_i determines a set $M_i \subseteq Y$ of all men she considers acceptable for marriage. The problem is this: Is every woman able to marry an acceptable man, assuming that bigamy is illegal? Rephrased in terms of functions: Is there a function $f: X \rightarrow Y$ satisfying

$f(x_i) \in M_i,$ (each woman marries an acceptable man)
$f(x_i) \neq f(x_j)$ whenever $i \neq j$. (different women marry different men)

There are some conditions that must be avoided in order to achieve a happy solution. For instance, if $M_1 = M_2 = M_3 = \{y_1, y_2\}$ then no solution is possible, for x_1, x_2, and x_3 cannot all find husbands from the two-element set $\{y_1, y_2\}$. So it is clearly necessary that the set $M_1 \cup M_2 \cup M_3$ contain at least three elements. More generally, for each k satisfying $1 \leq k \leq n$, each union of k of the M_i's must contain at least k elements. In 1935, the mathematician Philip Hall showed that this condition on unions is in fact sufficient to guarantee the existence of the desired function f that solves the marriage problem. The proof, however, is beyond the scope of our present treatment.

Readers whose present concerns are more with jobs than with marriage can rephrase the preceding problem as follows. Let $X = \{x_i\}_{1 \leq i \leq n}$ be the set of applicants for jobs with a major corporation, let $Y = \{y_i\}_{1 \leq i \leq m}$ be the set of available jobs, and let M_i denote the set of jobs for which x_i is qualified. Can everyone be hired?

Here is a more general statement of the marriage problem. Given a family $\{M_i\}_{1 \leq i \leq n}$ of nonempty sets, find a set of n distinct elements $\{y_i\}_{1 \leq i \leq n}$ such that $y_i \in M_i$ for each i. (The reader should check that this really is the same problem as before.) Such a set $\{y_i\}_{1 \leq i \leq n}$ is called a **system of distinct representatives** for the given family $\{M_i\}_{1 \leq i \leq n}$

3.4 EXAMPLE.   Find a system of distinct representatives for the following family of sets: $\{1\}, \{1, 2, 3\}, \{3, 4\}, \{2, 4, 5\}, \{3, 6\}, \{1, 4, 7\}, \{6\}$.

SOLUTION. We start by moving through the sets in the order in which they are listed, and in each case we choose as representative the smallest number compatible with the choices already made. This results in the respective choices

$$1, 2, 3, 4, 6, 7$$

for the first six sets. But now we are stuck: we can't make the required choice of 6 for the representative of the final set, since 6 was already chosen. Therefore we must amend our selection to

$$1, 2, 3, 4, \cancel{6}, 7,$$
$$3$$

which requires the change

$$1, 2, \cancel{3}, 4, \cancel{6}, 7,$$
$$4 \quad 3$$

which in turn demands

$$1, 2, \cancel{3}, \cancel{4}, \cancel{6}, 7.$$
$$4 \ 5 \ 3$$

Now we are free to select 6 as the representative of the final set, giving us the list

$$1, 2, 4, 5, 3, 7, 6$$

as our system of distinct representatives of the respective sets.

A function $f: A \rightarrow B$ can be informally represented by a picture, as follows:

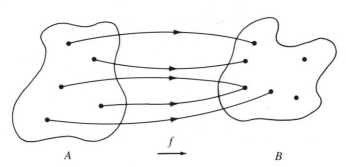

with the understanding that each point in A is taken by f to the point in B to which it is connected in the picture. If A and B are sets of real numbers, then $f \subseteq A \times B \subseteq \mathbb{R} \times \mathbb{R}$, and the pairs $(x, y) \in f$ can be represented as points in the usual coordinatized plane. The resulting diagram is the **graph** of f. For example, the graph of the function $f(x) = x^2$ is the following familiar parabola.

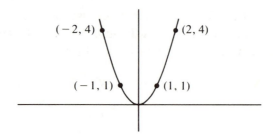

So the graph of a function $\mathbb{R} \to \mathbb{R}$ is just a pictorial representation of the ordered pairs belonging to it.

We have defined a function as a set of ordered pairs satisfying certain conditions; accordingly, functions f and g are **equal** if they are the same sets of ordered pairs. Let's reformulate this statement. Notice that dom f consists of all the first-coordinate elements x in the pairs (x, y) belonging to f, and similarly for g. Also, the equation $f = g$ implies that each element $x \in$ dom f is taken to (that is, paired with) the same element by f as it is by g. It follows that

$$(3.5) \qquad f = g \quad \Leftrightarrow \quad \begin{cases} \text{dom } f = \text{dom } g \\ \qquad \text{and} \\ f(x) = g(x) \quad \forall x \in \text{dom } f. \end{cases}$$

In computer jargon: functions f and g are equal if they have the same sets of acceptable input and if f and g produce the same output for each piece of input.

3.6 EXAMPLE.   Consider functions f, g, h defined by

$$\begin{aligned} f(x) &= x^2 - 1, &\forall x \in \mathbb{Z}, \\ g(x) &= x^2 - 1, &\forall x \in \mathbb{R}, \\ h(x) &= (x + 1)(x - 1), &\forall x \in \mathbb{R}. \end{aligned}$$

Here $g = h$: both functions have domain $\mathbb{R}$, and $g(x) = h(x)$ for all $x \in \mathbb{R}$. Thus the truth of "$g = h$" follows from 3.5, but *not* from the less precise

view of a function as a "rule or formula." (Indeed, the given formulas for
g and h are different.) But $f \neq g$ since dom $f = \mathbb{Z} \neq \mathbb{R} =$ dom g. (That is, f
and g have different sets of acceptable input.)

The functions f and g in Example 3.6 operate via the same formula,
but f applies that formula to a more restricted domain; accordingly, f is
called the **restriction** of g to $\mathbb{Z}$, denoted $f = g|_{\mathbb{Z}}$. More generally:

3.7 DEFINITION. Let f and g be functions. Suppose that dom $f =$
$A \subseteq$ dom $g = B$, and $f(x) = g(x)$ for all $x \in A$. Then f is called the **restric-
tion of g to A**, written $f = g|_A$; and g is called an **extension of f to B**.

3.8 EXAMPLES.

(a) Let $f = \{(1, 3), (3, 2)\}$ and $g = \{(1, 3), (2, 2), (3, 2)\}$. Then f and g
are functions, with dom $f = \{1, 3\}$ and dom $g = \{1, 2, 3\}$. More-
over, $f(1) = g(1) = 3$ and $f(3) = g(3) = 2$. Therefore $f = g|_{\{1, 3\}}$.

(b) For any real number x, let $[x]$ denote the greatest integer less than
or equal to x. Thus $[\pi] = 3$, $[\sqrt{2}] = 1$, $[-2.4] = -3$, and
$[-5] = -5$. The function $f: \mathbb{R} \to \mathbb{R}$ given by $x \mapsto [x]$ is called the
greatest integer function. Notice that $[x] = x$ for all $x \in \mathbb{Z}$; thus
$f|_{\mathbb{Z}} = i_{\mathbb{Z}}$, the identity function on $\mathbb{Z}$. (See Example 3.3(a).) It follows that the greatest integer function and the identity function $i_{\mathbb{R}}$
are both extensions of $i_{\mathbb{Z}}$ to $\mathbb{R}$.

Exercises

1. Let $A = \{1, 2, 3, 4\}$ and $B = \{2, \pi, e\}$. Which of the following sets of
ordered pairs are functions from A to B? Explain each case in which
your answer is negative.

(a) $f = \{(1, \pi), (3, \pi), (4, \pi), (2, e)\}$
(b) $g = \{(1, e), (2, e), (3, \pi), (2, e), (4, 2)\}$
(c) $h = \{(1, 2), (2, 2), (3, \pi), (4, 2), (1, \pi)\}$
(d) $j = \{(2, \pi), (3, e)\}$

2. (a) Give sets A and B such that the following set f is a function from
A to B: $f = \{(3, 2), (1, 1), (8, 5), (9, -4), (\pi, 1)\}$.

(b) With f as in part (a), $f(8) = $ _____ and the pre-image(s) of 2
is (are) _____.

3. Let X and Y be nonempty sets. Show that the Cartesian product $X \times Y$ is a function if and only if Y has exactly one element.

4. For functions whose domains are sets of real numbers, it is common practice to use a formula to describe a function's pairing rule, with the understanding that the domain of the function is the set of all real numbers for which the formula gives a unique real number, unless further restrictions are imposed. For example, the function f given by $f(x) = \sqrt{x-3}$ has domain $\{x \in \mathbb{R} \mid x \geqslant 3\}$, for if $x \geqslant 3$ the formula gives a unique real number, while if $x < 3$ then $x - 3 < 0$ and no square root exists. (Recall that $\sqrt{w}$ is the nonnegative real number whose square is w.) In each of the following, determine the domains of f and g, and then use statement 3.5 in the text to determine whether f and g are equal.

   (a) $f(x) = 1$, $\quad g(x) = \dfrac{x-5}{x-5}$

   (b) $f(x) = \sqrt{x}$, $\quad g(x) = \sqrt{|x|}$

   (c) $f(x) = |x|$, $\quad g(x) = \sqrt{x^2}$

   (d) $f(x) = x^2 - x - 6$, $\quad g(x) = (x-4)(x+3) + 6$

   (e) $f(x) = x^2$, $\quad g(x) = \begin{cases} x^2 & \text{if } x \text{ is rational} \\ 0 & \text{if } x \text{ is irrational} \end{cases}$

5. Let S and T be sets of three elements. How many functions are there from S to T? Justify briefly.

6. Suppose a function $f: A \to B$ is given. Define a relation $\sim$ on A as follows:
$$a_1 \sim a_2 \quad \Leftrightarrow \quad f(a_1) = f(a_2).$$
   (a) Prove that $\sim$ is an *equivalence relation* on A.

   Since $\sim$ is an equivalence relation, it induces a *partition* of A into equivalence classes. Describe these equivalence classes in each of the following cases.

   (b) A is the set of all solid-color cars, B is the set of all colors, $f(x)$ is the color of x.

   (c) $A = B = \mathbb{R}$, $f(x) = x^2$

   (d) $A = B = \mathbb{R}$, $f(x) = |x|$

   (e) $A = \mathbb{R} \times \mathbb{R}$, $B = P(\mathbb{R})$ (the *power set* of $\mathbb{R}$), $f(x, y) = \{x, y\}$

   (f) $A = \mathbb{R} \times \mathbb{R}$, $B = \mathbb{R}$, $f(x, y) = x + y$

   (g) $A = \mathbb{R} \times \mathbb{R}$, $B = \mathbb{R}$, $f(x, y) = x^2 + y^2$

7. Let S be a set. For each subset $A \subseteq S$ define a function $\chi_A \colon S \to \{0, 1\}$, called the **characteristic function** of A, as follows:

$$\chi_A(x) = \begin{cases} 1 & \text{if } x \in A, \\ 0 & \text{if } x \notin A. \end{cases}$$

For example, if $S = \{1, 2, 3, 4\}$ and $A = \{1, 2, 4\}$, then

$$\chi_A(1) = \chi_A(2) = \chi_A(4) = 1 \quad \text{and} \quad \chi_A(3) = 0.$$

(a) Give the formulas for the characteristic functions $\chi_\emptyset$ and χ_S.

(b) Let A' denote the complement of A in S. Compare the formulas for χ_A and $\chi_{A'}$.

(c) Show that if A and B are subsets of S then

$$A = B \quad \Leftrightarrow \quad \chi_A = \chi_B.$$

(The proof will involve the careful consideration of *set equality*, which appears in the left-hand statement, and *function equality*, which appears in the right-hand statement.)

(d) Show that if A and B are subsets of S then

$$\chi_{A \cap B}(x) = \chi_A(x) \cdot \chi_B(x), \qquad \forall x \in S,$$
$$\chi_{A \cup B}(x) = \chi_A(x) + \chi_B(x) - \chi_A(x)\chi_B(x), \qquad \forall x \in S.$$

(Suggestion: For each verification, draw a Venn diagram that represents A and B sitting appropriately inside S. Then compare the values obtained on both sides of the alleged equation as x is considered in different sections of the diagram.)

8. This exercise uses the notation of Example 3.3(d).

(a) Fix $q_0 \in S$. What is the intuitive meaning of the following set?

$$\{\delta(q_0, \sigma) \mid \sigma \in \Sigma\}$$

(b) Fix $\sigma_0 \in \Sigma$. What is the intuitive meaning of the following set?

$$\{\delta(q, \sigma_0) \mid q \in S\}$$

(c) Suppose that with the machine initially in state q_0, the symbol σ_1 is typed on the keyboard, and then the symbol σ_2 is typed. (That is, the two-letter "word" $\sigma_1\sigma_2$ is typed.) Write an expression that represents the resulting state of the machine.

(d) Generalize part (c) by writing an expression that represents the resulting state when the n-letter word $\sigma_1\sigma_2\cdots\sigma_n$ is input. Assume that the initial state is q_0. If you wish (though this is not required) you may express your answer recursively.

9. Show that the system of distinct representatives produced by the solution to Example 3.4 is the only acceptable answer.

10. For each family of sets, either find a system of distinct representatives or show that none exists.

(a) $\{2, 5\}, \{2, 3, 4, 5, 6\}, \{1, 3, 5\}, \{1, 4, 5\}, \{2\}$
(b) $\{1, 4\}, \{2, 3, 4, 5, 6\}, \{2, 4, 5, 6\}, \{4, 6\}, \{1, 4, 6\}, \{1, 4, 6\}$

11. This exercise deals with properties of the **greatest integer function**, $f(x) = [x]$. (See Example 3.8(b).)

(a) Write a simple formula for $[\alpha] + [-\alpha]$, where $\alpha \in \mathbb{R}$.
(b) Show that if n is an integer then

$$n + [\alpha] = [n + \alpha], \qquad \text{for all } \alpha \in \mathbb{R}.$$

(Suggestion: Consider the cases $\alpha \in \mathbb{Z}$ and $\alpha \notin \mathbb{Z}$ separately.)
(c) Suppose α is a real number that is not an integer, and suppose further that $\alpha > 2$. Show that there is a real number β in the interval $(1, 2)$ such that

$$[\alpha][\beta] < [[\alpha]\beta].$$

(d) Let n and m be positive integers. Show (without proving all the details) that there are exactly $[n/m]$ integers x that are multiples of m and satisfy the inequality $1 \leqslant x \leqslant n$.
(e) Part (d) can be used to show that the highest power of 7 that is a divisor of 1000! is

$$\left[\frac{1000}{7}\right] + \left[\frac{1000}{49}\right] + \left[\frac{1000}{343}\right] = 142 + 20 + 2 = 164.$$

(That is, 7^{164} is a divisor of 1000!, but 7^{165} is not.) Give a convincing explanation of why this is so.

12. This exercise concerns two functions that, like the greatest integer function, assign a nearby integer to each real number. Give a formula for each of them in terms of the greatest integer function, and choose the symbols for your formulas from the following:

$$], \times, +, [, -, \tfrac{1}{2}$$

(a) The **nearest integer function:** $f(x)$ is the nearest integer to x; if x is midway between two integers, $f(x)$ is the larger of the two.

(b) The **ceiling function:** $f(x)$ is the smallest integer that is greater than or equal to x.

[The greatest integer function is also called the **floor function**. The notations $\lfloor x \rfloor$ and $\lceil x \rceil$ are sometimes used for the floor and ceiling of x, respectively. (The zealous reader may wish to call $\lfloor x \rfloor - 1$ and $\lceil x \rceil + 1$ the "basement" and "attic" of x, respectively, but this is not standard terminology.)]

13. (*Challenging*) If A and B are sets, let B^A denote the set of all functions from A to B.

(a) Determine the set $\{1, 2\}^{\{1,2\}}$. That is, list its members explicitly.

(b) Show that if A, B, C are sets, and $A \subseteq B$, then $A^C \subseteq B^C$.

(c) Show that the sets $\{1, 2\}^{\{1,2\}}$ and $\{1, 2\}^{\{1,2,3\}}$ are disjoint.

(d) Generalize part (c) by showing that if $B \neq C$ then $A^B \cap A^C = \emptyset$.

(e) Show that if A is a nonempty set then $\emptyset^A = \emptyset$.

(f) Show that if B is any set then $B^\emptyset = \{\emptyset\}$.

3.2   SURJECTIONS, INJECTIONS, BIJECTIONS, SEQUENCES

A function $f: A \to B$ might not use every element of B as an image of some element of A. The subset of B consisting of all those elements of B that *are* images is called the **range** or **image** of f, usually denoted $f(A)$ or im f. Thus

$$f(A) = \text{im } f = \{f(x)|\ x \in A\}.$$

More generally, if $S \subseteq A$ we write

(3.9) $$f(S) = \{f(x)|\ x \in S\}.$$

In a picture:

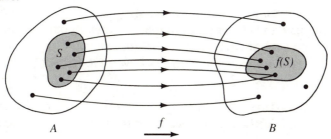

For example, for the trigonometric function $\sin: \mathbb{R} \to \mathbb{R}$, we have $\text{dom}(\sin) = \mathbb{R}$, with $\sin(\mathbb{R}) = [-1, 1]$, and $\sin([0, \pi/2]) = [0, 1]$.

A function $f: A \to B$ is said to be **surjective with respect to B** or **onto B** if $f(A) = B$; equivalently, for each $b \in B$ there exists an element $a \in A$ such that $f(a) = b$. If the set B is understood from the context, we usually say more briefly that such a function f is **surjective** or **onto**, and f is said to be a **surjection**. Notice that the definition dictates a method for *proving* that a function f is surjective: we pick an element $b \in B$, and then we show that its being in B implies the existence of an element $a \in A$ such that $f(a) = b$.

3.10 EXAMPLES.

(a) The function $f: \mathbb{R} \to \mathbb{R}$ given by the formula $f(x) = x^2$ is not surjective, because if $y < 0$ no element $x \in \mathbb{R}$ can be found such that $f(x) = y$.

(b) The function $g: \mathbb{R} \to \mathbb{R}$ given by $g(x) = x + 1$ is surjective, because for each $y \in \mathbb{R}$ we have $g(y - 1) = (y - 1) + 1 = y$. On the other hand, the function $h: \mathbb{N} \to \mathbb{N}$ defined by $h(x) = x + 1$ is *not* surjective, since there is no $x \in \mathbb{N}$ such that $h(x) = 1$.

(c) Let A be an n-element set, with $n \in \mathbb{N}$, and let B be a two-element set. How many nonsurjective functions are there from A to B?

SOLUTION.   Suppose $B = \{b_1, b_2\}$. If a function $f: A \to B$ is not surjective, then either $f(a) = b_1$ for all $a \in A$ or $f(a) = b_2$ for all $a \in A$. (Otherwise, there would exist elements a_1 and a_2 such that $f(a_1) = b_1$ and $f(a_2) = b_2$, and this would imply that f is surjective, contrary to the hypothesis.) Thus there are exactly *two* nonsurjective functions from A to B, namely the two constant functions. (See Example 3.3(b).)

(d) In Chapter 2 we discussed indexed sets without giving the notion a precise definition. Now that we have functions available for use, we can remedy the situation. If S and I are sets and $f: I \to S$ is a surjective function, we say that f has **indexed** the set S by I. Once such an f has been given, we usually write s_i instead of $f(i)$. Return to Examples 2.28 and 2.29 for an abundant supply of indexed sets. For instance, in 2.29(b) we used C_r for the circle with radius r and center $(0, 0)$ in the xy-plane. In our present notation, S is the set of circles centered at $(0, 0)$ and I is the set of nonnegative real numbers; for each $r \in I$, we have $f(r) = C_r$.

3.11 DEFINITION.   A function $f: A \to B$ is said to be **one-to-one** (abbreviated 1-1), **injective**, or an **injection** if the following implication is true for every $a_1, a_2 \in A$:

$$f(a_1) = f(a_2) \implies a_1 = a_2.$$

In words: elements of A with the same image must be equal. The contrapositive is equally useful:

$$a_1 \neq a_2 \implies f(a_1) \neq f(a_2).$$

That is, different elements in the domain yield different images.

3.12 EXAMPLES

(a) The function $f: \mathbb{R} \to \mathbb{R}$ given by $f(x) = x^3$ is injective.

PROOF. We will verify the first implication in Definition 3.11 for $f(x) = x^3$. We have

$$\begin{aligned} f(x_1) = f(x_2) \implies & x_1^3 = x_2^3 \\ \implies & x_1^3 - x_2^3 = 0 \\ \implies & (x_1 - x_2)(x_1^2 + x_1 x_2 + x_2^2) = 0 \\ \implies & x_1 - x_2 = 0 \quad \text{or} \quad x_1^2 + x_1 x_2 + x_2^2 = 0. \end{aligned}$$

To conclude that $x_1 = x_2$, it remains to prove the implication

(*)
$$x_1^2 + x_1 x_2 + x_2^2 = 0 \implies x_1 = x_2.$$

The hypothesis of (*) yields

(**)
$$\begin{aligned} 0 &= x_1^2 + x_1 x_2 + x_2^2 \\ &= (x_1^2 + x_1 x_2 \quad) + x_2^2 \\ &= (x_1^2 + x_1 x_2 + \tfrac{1}{4}x_2^2) + (x_2^2 - \tfrac{1}{4}x_2^2) \\ &= (x_1 + \tfrac{1}{2}x_2)^2 + \tfrac{3}{4}x_2^2. \end{aligned}$$

Because squares of nonzero real numbers are positive, we conclude that $x_2 = 0$, and then that $x_1 = 0$. □

Argument (**) is an instance of a useful algebraic trick called *completing the square*. Readers who find the method unappealing have other options. For instance, the proof of (*) can be broken into separate cases as follows. If exactly one of x_1, x_2 is nonzero the hypothesis of (*) is false, so (*) is automatically true. If neither x_1 nor x_2 is zero and $x_1 \leqslant x_2$, then $x_2^2 + x_1 x_2 \geqslant 0$, hence $x_1^2 + x_1 x_2 + x_2^2 > 0$, a contradiction. The case

$x_2 \leq x_1$ is similar. Alternatively, readers who have studied calculus can use the derivative $f'(x)$ to show that $f(x) = x^3$ is strictly increasing and then observe that a strictly increasing function is injective (why?).

(b) Let Σ stand for the usual English alphabet. A *four-letter word over* Σ is a function $w: \{1, 2, 3, 4\} \rightarrow \Sigma$. If $w(i) = a_i$ for $1 \leq i \leq 4$, we write

$$w = a_1 a_2 a_3 a_4.$$

For example, the word w given by $1 \mapsto s$, $2 \mapsto p$, $3 \mapsto a$, $4 \mapsto r$ is written

$$\text{spar}$$

Now we let W_4 denote the set of four-letter words over Σ, and we let $P(\Sigma)$ denote the power set of Σ (that is, the set of all subsets of Σ). We define a function $f: W_4 \rightarrow P(\Sigma)$ by

$$f(a_1 a_2 a_3 a_4) = \{a_1, a_2, a_3, a_4\}.$$

That is, the image of a word under f is its set of letters. Then f is not injective, because two different words can have the same letters. (For instance, $f(\text{bozo}) = f(\text{zoob}) = \{b, z, o\}$.) And f is not surjective, because no set of five or more letters is in the image of f.

Return for a moment to the discussion of indexing in Example 3.10(d). Notice that there is no requirement that an indexing function be injective. For example, let S be the set of all lines through the origin in the xy-plane and suppose, for each $\alpha \in \mathbb{R}$, that $f(\alpha)$ is the line through $(0, 0)$ at angle α (radians) with the positive x-axis. Then S is indexed by $\mathbb{R}$ by means of f, and $s_\alpha = s_{\alpha + n\pi}$ for all integers n.

3.13 DEFINITION. A function $f: A \rightarrow B$ is said to be **bijective**, a **bijection**, or a **one-to-one correspondence** if it is both injective and surjective.

It is a useful exercise to check that when the definitions of injection and surjection are stated in terms of ordered pairs, the following characterization of bijections emerges:

(3.14)   A set $f \subseteq A \times B$ is a bijection from A to B if and only if each $a \in A$ is the first coordinate of exactly one pair belonging to f and each $b \in B$ is the second coordinate of exactly one pair belonging to f.

3.15 EXAMPLE.   Let $A = \{1, 2, 3\}$ and $B = \{r, s, t, u\}$. Consider

$$f = \{(1, r), (2, u), (3, s)\}, \qquad g = \{(1, r), (2, r), (3, u)\},$$
$$h = \{(1, r), (1, s), (2, u), (3, t)\}, \quad j = \{(1, r), (2, s)\}.$$

Here f and g are functions from A to B, and f is injective but g is not, since $g(1) = g(2) = r$. Neither f nor g is surjective, since $t \notin f(A) \cup g(A)$. The set h is not a function, since 1 is paired by h with two different elements of B. Finally, j is not a function from A to B (since dom $j \neq A$), but j *is* an injective function from $\{1, 2\}$ into B.

None of the given sets of pairs is a bijection from A to B, but f is a bijection from A to $\{r, u, s\}$. More generally, *any* injective function is a bijection from its domain to its range. (Check this.)

At the end of Chapter 2 we discussed the use of recursion to define certain families of mathematical objects; we will now employ recursion in the definition of functions.

NOTATION.   If n is a natural number, we write

$$\mathbb{N}_n = \{x \in \mathbb{N} \mid 1 \leqslant x \leqslant n\} = \{1, 2, \ldots, n\}.$$

3.16 DEFINITION.   Let A be a set. A **finite sequence** in A is a function $f \colon \mathbb{N}_n \to A$ for some $n \in \mathbb{N}$. (This is also called a **sequence of length n**.) An **infinite sequence** in A is a function $f \colon \mathbb{N} \to A$. We will follow the standard practice of using the word **sequence** for both varieties.

If f is a sequence in A, we usually write $f(1) = a_1$, $f(2) = a_2$, and so on. If f is an infinite sequence we visualize it as a never-ending list:

$$a_1, a_2, a_3, \ldots.$$

If f is a sequence of length n, we usually denote f by the symbol

$$(a_1, a_2, \ldots, a_n)$$

and call it an **n-tuple**, with a_i its ith **coordinate**. By our criteria for function equality (see 3.5), it follows that

$$(a_1, \ldots, a_n) = (b_1, \ldots, b_n) \leftrightarrow a_i = b_i \text{ for each } i, \ 1 \leq i \leq n.$$

We sometimes denote a sequence of length n by $\{a_i\}_{1 \leq i \leq n}$ and an infinite sequence by $\{a_n\}_{n \in \mathbb{N}}$ or $\{a_n\}_{n \geq 1}$. (The letters i and n are the world's most popular subscripts.) More informally and less precisely, both kinds of sequences are often denoted $\{a_n\}$. The element a_n is also called the **nth term** of the sequence.

A recursive definition of a sequence $\{a_n\}$ begins by specifying one or more initial terms $a_1, \ldots, a_N$ explicitly; it then gives a formula that defines each succeeding term $a_{N+1}, a_{N+2}, \ldots$ via its predecessors.

3.17 EXAMPLE. We will now consider a recursive procedure for computing the square root of a positive number m. We start by guessing a number r satisfying $0 < r \leq m$. If the guess is correct (that is, $r = \sqrt{m}$) then $m/r = r$. If the guess is wrong, then either m/r is too small and r is too large, or vice versa. (Why?) The correct answer must therefore lie between r and m/r, so we use the average, $\dfrac{1}{2}\left(r + \dfrac{m}{r}\right)$, as the next guess, and continue in this fashion. This is called the **divide-and-average** method for computing $\sqrt{m}$. More compactly, we define a sequence $\{a_n\}$ recursively as follows:

$$a_1 = 1 \qquad \text{(a thought-free initial guess)}$$

$$a_{k+1} = \frac{1}{2}\left(a_k + \frac{m}{a_k}\right).$$

For instance, if $m = 3$ the first few terms of the sequence $\{a_n\}$ are

$$a_1 = 1,$$

$$a_2 = \frac{1}{2}\left(1 + \frac{3}{1}\right) = 2,$$

$$a_3 = \frac{1}{2}\left(2 + \frac{3}{2}\right) = \frac{7}{4},$$

$$a_4 = \frac{1}{2}\left(\frac{7}{4} + \frac{3}{\frac{7}{4}}\right) = \frac{97}{56},$$

$$a_5 = \frac{1}{2}\left(\frac{97}{56} + \frac{3}{\frac{97}{56}}\right) = \frac{18,817}{10,864}.$$

Compare the last number with the value of $\sqrt{3}$ on a calculator.

3.18 EXAMPLE.   Consider the set $\mathbb{N} \times \mathbb{N}$, which we picture as the collection of all points with integer coordinates in the first quadrant.

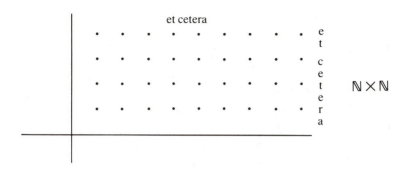

Can we list all these points in some sort of order so that they are the terms in an infinite sequence? That is, is there a surjective function $f: \mathbb{N} \to \mathbb{N} \times \mathbb{N}$? As a first attempt, we might start by listing the points in the first column: $(1, 1)$, $(1, 2)$, $(1, 3)$, ... But this is doomed to failure: we never escape from the first column! Similarly, a horizontal excursion through the bottom row, $(1, 1)$, $(2, 1)$, $(3, 1)$, . . . , doesn't do the job. Success comes by moving along diagonals as pictured here:

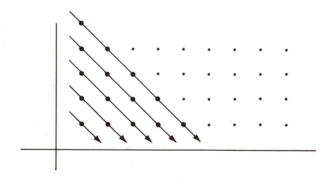

We start on the little diagonal at the lower left and then move down the next higher diagonal, then down the next, and so on. Notice that points (a, b) and (c, d) are on the same diagonal if and only if $a + b = c + d$; in that event, (a, b) precedes (c, d) on their common diagonal if and only if $a < c$. The point (a, b) is on an *earlier* diagonal than (c, d) if and only if $a + b < c + d$. So the list we are constructing satisfies this rule: (a, b) precedes (c, d) on the list if and only if

$$a + b < c + d \quad \text{or} \quad a + b = c + d \text{ and } a < c.$$

Now we will give a recursive definition of the desired function $f: \mathbb{N} \to \mathbb{N} \times \mathbb{N}$ to generate the list we just described. Define $f(1) = (1, 1)$; and if $f(k) = (a, b)$, define

$$(*) \qquad\qquad f(k+1) = \begin{cases} (a+1, b-1) & \text{if } b > 1, \\ (1, a+1) & \text{if } b = 1. \end{cases}$$

Thus $f(2) = (1, 2), f(3) = (2, 1), f(4) = (1, 3)$, and so on. Formula (*) tells us that $f(k+1)$ follows $f(k)$ as we move down $f(k)$'s diagonal, unless $f(k)$ is already at the bottom, in which case $f(k+1)$ is at the top of the next diagonal. We assert that f is surjective, and we will prove this by contradiction. If f is not surjective, move along the diagonals in succession until the first (a, b) is reached that is *not* in $f(\mathbb{N})$.† If $a \neq 1$ then $(a-1, b+1)$ precedes (a, b) on the same diagonal, so $(a-1, b+1) = f(k)$ for some $k \in \mathbb{N}$. But then an application of (*) gives $f(k+1) = (a, b)$, contradicting the assumption that $(a, b) \notin f(\mathbb{N})$. If $a = 1$ then $(b-1, 1)$ is at the bottom of the preceding diagonal, so $(b-1, 1) = f(k)$ for some k, and then (*) yields $f(k+1) = (1, b) = (a, b)$, again a contradiction. Thus our assumption that f is not surjective is false, and we are done. It can also be shown that f is injective, but the details are technical and we omit them. (The argument uses (*) to show that if $k_1 < k_2$ then either $f(k_2)$ is below $f(k_1)$ on the same diagonal or $f(k_2)$ is on some subsequent diagonal.)

Many problems in mathematics and computer science boil down to determining a procedure for listing a given set of objects as the terms in a sequence. It is therefore important to know that certain sets *cannot* be so listed; the following example provides an illustration.

3.19 EXAMPLE. Let F denote the set of all functions from $\mathbb{N}$ to $\mathbb{N}$. Can the members of F be listed as a sequence? More precisely, is there a surjective function $g: \mathbb{N} \to F$? *Suppose there is such a function g;* then for each $n \in \mathbb{N}$ let's write g_n instead of $g(n)$. So we can list the functions from $\mathbb{N}$ to $\mathbb{N}$: $g_1, g_2, g_3, \ldots$. Now define a function $h: \mathbb{N} \to \mathbb{N}$ by the formula

$$h(n) = g_n(n) + 1, \qquad \text{for all } n \in \mathbb{N}.$$

Then $h \in F$, and therefore $h = g_k$ for some $k \in \mathbb{N}$. But

$$h(k) = g_k(k) + 1 \neq g_k(k),$$

---

† That is, among all those points (x, y) not in $f(\mathbb{N})$, consider those whose coordinate sum $x + y$ is minimal, and among those choose the point (a, b) with smallest first coordinate.

and therefore $h \neq g_k$, a contradiction! Conclusion: no surjection $g: \mathbb{N} \to F$ exists, and therefore *no* procedure exists for listing the functions from $\mathbb{N}$ to $\mathbb{N}$.

The main idea underlying Example 3.19 is called **diagonalization**, a method first used by Georg Cantor in the nineteenth century. Cantor's idea is an adaptation of this fact: every row of a checkerboard contains a square on the main diagonal.

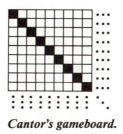

Cantor's gameboard.

Each of the functions $g_n: \mathbb{N} \to \mathbb{N}$ in Example 3.19 is represented by a row of numbers:

$$g_n(1) \quad g_n(2) \quad g_n(3) \quad g_n(4) \quad \cdots$$

That is, g_n is represented by the *n*th row of the following array:

$$\begin{array}{cccccc}
g_1(1) & g_1(2) & g_1(3) & g_1(4) & g_1(5) & \cdots \\
g_2(1) & g_2(2) & g_2(3) & g_2(4) & g_2(5) & \cdots \\
g_3(1) & g_3(2) & g_3(3) & g_3(4) & g_3(5) & \cdots \\
\vdots & \vdots & \vdots & \vdots & \vdots
\end{array}$$

An important slant on sequences.

The function $h: \mathbb{N} \to \mathbb{N}$ is represented by the row

$$h(1) \quad h(2) \quad h(3) \quad h(4) \quad h(5) \quad \cdots$$

Then if h were one of the g_n, one of the terms $h(1), h(2), h(3), \ldots$ would be on the diagonal of the array; but that possibility has been eliminated by the manner in which we defined h.

We will hear more from Cantor in Chapter 4.

Exercises

1. Determine the image of each function in Exercise 4 of Section 3.1. (No proofs required.)

2. Let N be the set of all living nephews of living aunts, and let A be the set of all living aunts of living nephews.

   (a) Define a function from N to A.
   For each of the following questions, if the answer is yes, construct a function with the indicated properties; if the answer is no, explain why.
   (b) Is there an *injective* function from N to A?
   (c) Is there a *surjective* function from N to A?

3. In each case, state without proof whether the given function is injective, surjective, bijective.

   (a) $f: \mathbb{N} \to \mathbb{N}$ given by $f(x) = x^3$.
   (b) $g: \{1, 2, 3\} \to \{-2, 5, 6\}$ given by $g = \{(1, 5), (2, -2), (3, 6)\}$.
   (c) $h: \{1, 2, 3\} \to \{-2, 5, 6\}$ given by $h = \{(3, 5), (2, -2), (1, 5)\}$.

4. Let S and T be sets with three elements and two elements, respectively. In each case state the answer and justify briefly.

   (a) How many functions are there from S to T?
   (b) How many injections are there from S to T?
   (c) How many injections are there from T to S?
   (d) How many surjections are there from S to T?
   (e) How many surjections are there from T to S?
   (f) Guess the answers to (a) and (b) if S has m elements and t has n elements.

5. A function $f: A \to B$ is a subset of $A \times B$, hence a subset of $(A \cup B) \times (A \cup B)$. In the terminology of Chapter 2, f is a **relation** on $A \cup B$. Therefore f has an **inverse relation** (see Example 2.51(d)):

   $$f^{-1} = \{(b, a) \mid (a, b) \in f\}.$$

   (a) Write down the inverses g^{-1} and h^{-1} of the functions g and h in Exercise 3.
   (b) Show that if f is not injective then f^{-1} is not a function.
   (c) Show that if f is injective but not surjective then f^{-1} is a function whose domain is a proper subset of B.

6. For each part of this exercise, give sets A, B, and C, with $C \subset A$, and a function $f: A \to B$ satisfying the given conditions. Or, if no such function exists, *show* that none exists. (There is no need to get fancy here. In each case where such an f exists, an example can be constructed in which each of the sets A, B, C has at most two elements.)

    (a) f is surjective and the restriction $f|_C$ is surjective.
    (b) f is surjective but $f|_C$ is not surjective.
    (c) f is injective and $f|_C$ is injective.
    (d) f is injective but $f|_C$ is not injective.

7. Let a and b be real numbers. Consider a function $f: \mathbb{R} \to \mathbb{R}$ given by the formula $f(x) = ax + b$.

    (a) Under what conditions on a and b is f a bijection from $\mathbb{R}$ to $\mathbb{R}$? (For instance, what if $a = 0$?)

    (b) Under what conditions on a and b is the restriction $f|_{\mathbb{Z}}$ a bijection from $\mathbb{Z}$ to $\mathbb{Z}$?

    (c) Under what conditions on a and b is the restriction $f|_{\mathbb{N}}$ a bijection from $\mathbb{N}$ to $\mathbb{N}$?

8. Let Π denote the xy-plane, let O denote the origin, let $\Pi^* = \Pi - \{O\}$ (the "punctured plane"), and let L denote the set of all lines through O. For each point $P \in \Pi^*$, let $f(P)$ be the line containing the segment OP. Show that the function $f: \Pi^* \to L$ is not a bijection, and then find a set $S \subset \Pi^*$ such that the restriction $f|_S$ *is* a bijection from S to L.

9. (This exercise requires the definition of an *ordering* on a set S. This was given in Exercise 15 of Section 2.9.) First an example: Suppose that each of five parachute jumpers is assigned a different number from 1 to 5. Jumper number 1 will jump first, then jumper number 2, and so on. The standard ordering on the set $\{1, 2, 3, 4, 5\}$ has *induced* an ordering on the set of jumpers. This is the idea underlying this exercise.

    Show that if $\leqslant$ is an ordering on S and $f: S \to T$ is a bijection, then f can be used to define an ordering $<$ on T satisfying this property:

$$s_1 \leqslant s_2 \iff f(s_1) < f(s_2), \qquad \forall s_1, s_2 \in S.$$

(Your answer should start with something like this: "If $t_1, t_2 \in T$, define $t_1 < t_2$ as follows: . . . .")

10. The definition of n-tuple given in this section may produce confusion when $n = 2$, because our definition of the 2-tuple (a_1, a_2) differs from our definition of the ordered pair (a_1, a_2) in Chapter 2. This can be remedied with a different definition of n-tuple as follows: Define the 1-tuple $(a_1) = a_1$. Define the 2-tuple (a_1, a_2) to be the ordered pair (a_1, a_2). In general, if the k-tuple $(a_1, a_2, \ldots, a_k)$ has been defined,

define the $(k + 1)$-tuple $(a_1, \ldots, a_k, a_{k+1})$ to be the ordered pair

$$((a_1, \ldots, a_k), a_{k+1}).$$

Prove by induction on n that this recursive definition of n-tuple retains the following essential property:

$$(a_1, \ldots, a_n) = (b_1, \ldots, b_n) \iff a_i = b_i \ \forall i, \ 1 \leqslant i \leqslant n.$$

11. Stored information in a computer memory is usually coded in the form of n-tuples of 0s and 1s, often called **n-bit strings.** For brevity, commas and parentheses are omitted; thus 01101011 is an 8-bit string. To aid in the detection and correction of errors, it is useful to have different code strings differ in more than one coordinate. (The same is true in ordinary English: a one-letter typing error in "house" can yield "horse" or "mouse" and lead to confusion, whereas a one-letter error in "Mississippi" will probably be detected and corrected.) Define the **distance** between two n-bit strings to be the number of coordinates in which they differ. More formally, let $\mathbb{Z}_2^n$ denote the set of n-bit strings, and define a function $d: \mathbb{Z}_2^n \times \mathbb{Z}_2^n \to \mathbb{Z}$ by

$$d(a_1 a_2 \ldots a_n, b_1 b_2 \ldots b_n) = \sum_{i=1}^{n} |a_i - b_i|,$$

where $a_i, b_i \in \{0, 1\}$ for each i. (Example: $d(1101, 1011) = 2$.) Prove that the function d has the following properties:

(a) $d(s_1, s_2) \geqslant 0$ (here the s's are strings)
(b) $d(s_1, s_2) = 0 \iff s_1 = s_2$
(c) $d(s_1, s_2) = d(s_2, s_1)$
(d) For all $s_1, s_2, s_3 \in \mathbb{Z}_2^n$,

$$d(s_1, s_3) \leqslant d(s_1, s_2) + d(s_2, s_3). \qquad \text{(the } \textit{triangle inequality}\text{)}$$

(Suggestion: Use induction on n, starting with $n = 1$.)

12. Use the divide-and-average method to compute $\sqrt{5}$, correct to five decimal places. (No calculators allowed!)

13. Suppose you are using the divide-and-average method to compute a square root, and you want your answer to be correct to n decimal places. (That is, you want all the digits through the first n digits to the right of the decimal point to be correct.) How can you test your current estimate to see whether it is good enough? Explain.

14. Give a recursive definition of each of the following sequences.

(a) 1, 3, 5, 7, 9, 11, 13, 15, 17, . . .
(b) 1, 1, 3, 3, 5, 5, 7, 7, 9, 9, . . .
(c) 2, 3, 5, 8, 12, 17, 23, 30, 38, . . .
(d) 1, 1, 2, 3, 5, 8, 13, 21, 34, 55, . . .   (the sequence of *Fibonacci numbers*)

15. Suppose $f: \mathbb{N} \to A$ and $g: \mathbb{N} \to B$ are surjections. Prove that there is a surjection $h: \mathbb{N} \to A \cup B$. (Suggestion: Consider the list $f(1)$, $g(1)$, $f(2)$, $g(2)$, . . .)

16. Show that there is a one-to-one correspondence between $\mathbb{N}$ and the following set of integers, written here in decimal form:

$$\{2, 11, 101, 1001, 10001, 100001, . . .\}.$$

17. In the first quadrant, draw a connected path starting at (1, 1) that, if continued, would pass through all of $\mathbb{N} \times \mathbb{N}$. (That is, any prechosen point in $\mathbb{N} \times \mathbb{N}$ will eventually be reached.) Your path should have *only horizontal and vertical segments,* and no point should be covered more than once. No proof is required; but be sure to draw enough so that the method for continuing the path is clear to the reader.

18. Write a recursive definition of a sequence whose range is the set of points with integer coordinates in the fourth quadrant. Your formula should correspond to moving along diagonals as pictured here. (Suggestion: review formula (∗) in Example 3.18.)

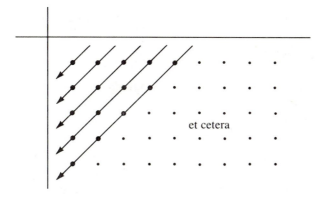

et cetera

19. Draw a connected path starting at (0, 0) that, if continued, would pass through all of $\mathbb{Z} \times \mathbb{Z}$.

3.3   COMPOSITION OF FUNCTIONS

Suppose we are given functions $f: A \to B$ and $g: B \to C$; then each element $x \in A$ is paired by f with a unique element $y \in B$, and y is in turn paired by g with a unique element $z \in C$. It is natural to view x as paired with z by this two-stage process, and the result is a new function called the **composition** of f and g, denoted $g \circ f$. More formally, we define $g \circ f: A \to C$ by the rule

$$(3.20) \qquad (g \circ f)(x) = g(f(x)), \qquad \forall x \in A.$$

The intuitive picture:

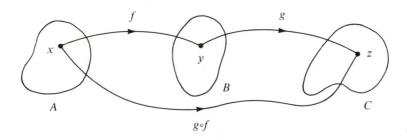

A consequence of our definition is that $\text{dom}(g \circ f) = \text{dom } f$.

3.21 EXAMPLES.

(a) Suppose $f: \mathbb{R} \to \mathbb{R}$ and $g: \mathbb{R} \to \mathbb{R}$ are defined by the formulas

$$f(x) = x + 3 \quad \forall x \in \mathbb{R} \qquad \text{and} \qquad g(x) = x^2 \quad \forall x \in \mathbb{R}.$$

Then for each $x \in \mathbb{R}$ we have

$$(f \circ g)(x) = f(g(x)) = f(x^2) = x^2 + 3$$

and

$$(g \circ f)(x) = g(f(x)) = g(x + 3) = (x + 3)^2 = x^2 + 6x + 9.$$

Note that here $g \circ f \neq f \circ g$ since, for example, $(f \circ g)(1) = 4$, while $(g \circ f)(1) = 16$. (We remark that $(f \circ g)(-1) = (g \circ f)(-1) = 4$; so unequal functions with a common domain need not disagree at every point of that domain.)

(b) Consider functions f and g from $\mathbb{R} \times \mathbb{R}$ to $\mathbb{R} \times \mathbb{R}$ defined by $f(x, y) = (x, 0)$ and $g(x, y) = (0, y)$ for all $(x, y) \in \mathbb{R} \times \mathbb{R}$. (The functions f and g are called the **projections** onto the horizontal and vertical axes, respectively.) Here we have

$$(f \circ g)(x, y) = f(g(x, y)) = f(0, y) = (0, 0),$$

and similarly $(g \circ f)(x, y) = (0, 0)$ for all $(x, y) \in \mathbb{R} \times \mathbb{R}$. Therefore in this example we have $f \circ g = g \circ f$.

In elementary arithmetic we have the dual notions of multiplication and factorization. We can multiply two integers and produce a single product, and we can factor some integers into products of smaller integers. Similarly, we can compose certain functions to get one function, and we can express some functions as compositions of simpler ones. A complicated procedure is sometimes more manageable if it is decomposed into simpler steps.

3.22 EXAMPLE. Consider the function $f: \mathbb{R} \to \mathbb{R}$ given by $f(x) = (x^2 + 1)^3$. Then we can express f in a variety of ways as a composition of functions from $\mathbb{R}$ to $\mathbb{R}$, and here are some samples:

$$f = h \circ u = s \circ r = (s \circ h) \circ g = s \circ (h \circ g),$$

where $g(x) = x^2$, $h(x) = x + 1$, $r(x) = x^2 + 1$, $s(x) = x^3$, and $u(x) = x^6 + 3x^4 + 3x^2$. Let's check in detail that $f = (s \circ h) \circ g$ to illustrate the method; the other equalities are left for the reader. For each $x \in \mathbb{R}$ we have

$$
\begin{aligned}
((s \circ h) \circ g)(x) &= (s \circ h)(g(x)) && \left.\begin{array}{l}\text{(from the definition of}\\ \text{function composition)}\end{array}\right. \\
&= s(h(g(x))) \\
&= s(h(x^2)) && \text{(since } g \text{ acts by squaring)} \\
&= s(x^2 + 1) && \text{(since } h \text{ acts by adding 1)} \\
&= (x^2 + 1)^3 && \text{(since } s \text{ acts by cubing)} \\
&= f(x). && \text{(by definition of } f)
\end{aligned}
$$

This shows that $(s \circ h) \circ g$ and f have the same action on each element of their common domain $\mathbb{R}$. Therefore $(s \circ h) \circ g = f$ by 3.5.

We verbalize the existence of functions f and g such that $f \circ g \neq g \circ f$ by saying that composition of functions is not **commutative**. Mathematicians use this word in a variety of settings, with varying degrees of precision. In general, objects or procedures are said to **commute** with one another if the order in which they appear in a statement makes no crucial difference. For example, 3 and 5 commute with each other additively but not with respect to division: $3 + 5 = 5 + 3$, but $5 \div 3 \neq 3 \div 5$. The procedures of putting on socks and shoes do not commute with each other. We will give a more precise definition of commutative operations in Chapter 6.

The computation displayed in Example 3.22 is an exercise in parenthesis juggling (a favorite sport (of (many) mathematicians)), and the key part comes in the first two lines, in which $((s \circ h) \circ g)(x)$ is transformed into $s(h(g(x)))$. A similar computation will show that $(s \circ (h \circ g))(x)$ is also equal to $s(h(g(x)))$, from which it follows that $(s \circ h) \circ g = s \circ (h \circ g)$. Because the formulas that define s, h, and g are nowhere involved in this deduction, we sense that our example is an instance of a more general law, and the following theorem makes that law explicit.

3.23 THEOREM (ASSOCIATIVE LAW OF FUNCTION COMPOSITION). Given functions $f: A \to B$, $g: B \to C$, and $h: C \to D$; then

$$h \circ (g \circ f) = (h \circ g) \circ f.$$

PROOF. We have

$$\mathrm{dom}(h \circ (g \circ f)) = \mathrm{dom}((h \circ g) \circ f) = A.$$

(Why?) So to complete the proof we must verify that $(h \circ (g \circ f))(x) = ((h \circ g) \circ f)(x)$ for all $x \in A$. But the definition of composition gives

$$(h \circ (g \circ f))(x) = h((g \circ f)(x)) = h(g(f(x)))$$
$$= (h \circ g)(f(x)) = ((h \circ g) \circ f)(x),$$

and this storm of parentheses completes the proof.　　$\square$

Thanks to Theorem 3.23, the composite functions $h \circ (g \circ f)$ and $(h \circ g) \circ f$ can now be written $h \circ g \circ f$ without fear of confusion.

The following theorem shows what happens when one injection or surjection is composed with another. We will apply these results again and again in the subsequent chapters.

3.24 THEOREM. Let $f: A \to B$ and $g: B \to C$ be functions.

(a) If f and g are injections then $g \circ f: A \to C$ is an injection.
(b) If f and g are surjections then so is $g \circ f$.
(c) If f and g are bijections then so is $g \circ f$.

PROOF.

(a) Let $a_1, a_2 \in A$. We have the following series of implications:

$$(g \circ f)(a_1) = (g \circ f)(a_2) \Rightarrow g(f(a_1)) = g(f(a_2)) \quad \text{(by definition of composition)}$$
$$\Rightarrow f(a_1) = f(a_2) \quad \text{(since } g \text{ is injective)}$$
$$\Rightarrow a_1 = a_2. \quad \text{(since } f \text{ is injective)}$$

Therefore $g \circ f$ is injective.

(b) Let $c \in C$. We must find an $a \in A$ such that $(g \circ f)(a) = c$. Since g is onto C, there is an element $b \in B$ such that $g(b) = c$. Since f is onto B, there exists $a \in A$ such that $f(a) = b$. Then

$$(g \circ f)(a) = g(f(a)) = g(b) = c.$$

(c) This part follows immediately from the preceding parts. □

3.25 EXAMPLE. Let $\mathbb{Q}^+$ denote the set of positive rational numbers. In Example 3.18 we exhibited a surjection $f: \mathbb{N} \to \mathbb{N} \times \mathbb{N}$. Now define a function $g: \mathbb{N} \times \mathbb{N} \to \mathbb{Q}^+$ by the rule $(a, b) \mapsto a/b$. Check that g is surjective. (But g is not injective since, for example, $g(2, 3) = g(4, 6) = 2/3$.) By Theorem 3.24(b), it follows that the function $g \circ f: \mathbb{N} \to \mathbb{Q}^+$ is surjective. Imagine that each point representing a natural number n on the real line is physically uprooted and replanted on top of the point representing the image $(g \circ f)(n)$. Then surjectivity of $g \circ f$ shows the remarkable fact that there are enough integer points on the positive real axis to completely cover all the rational points. Let's say it again:

N $\mathbb{Q}^+$

1   2   3   4   ···   0

These   can cover   these.

Suppose A and B are sets and $f: A \to B$ is a bijection. We now define a new function, the **inverse** of f, denoted f^{-1}, that transforms each element of B back to the element of A from which it came. There *is* such an element of A since f is surjective; and it is unique because f is injective. More formally, define $f^{-1}: B \rightarrow A$ by

$$(3.26) \qquad\qquad f^{-1} = \{(b, a) \in B \times A \mid (a, b) \in f\}.$$

Now we can see that f^{-1} has the desired action:

$$f^{-1}(b) = a \quad \Leftrightarrow \quad (b, a) \in f^{-1} \quad \Leftrightarrow \quad (a, b) \in f \quad \Leftrightarrow \quad f(a) = b.$$

It follows from 3.14 that f^{-1}, like f, is a bijection.†

The next theorem shows that the actions of f and f^{-1} cancel each other, and that f^{-1} is the *only* function having that behavior with respect to f. Recall that i_A denotes the identity function on A.

3.27 THEOREM. Let $f: A \rightarrow B$ and $g: B \rightarrow A$. Then the following two statements are equivalent:

(a) f is a bijection and $g = f^{-1}$.
(b) $g \circ f = i_A$ and $f \circ g = i_B$.

PROOF. Assume statement (a) is true. We will prove the equation $g \circ f = i_A$. (The proof of $f \circ g = i_B$ is similar, and it is left to the reader.) Pick any element $a \in A$; say, $f(a) = b \in B$. Then

$$(g \circ f)(a) = (f^{-1} \circ f)(a) = f^{-1}(f(a)) = f^{-1}(b) = a = i_A(a).$$

Therefore $g \circ f$ and i_A agree on all elements of their common domain A, and so $g \circ f = i_A$, as desired.

Now suppose statement (b) is true. Then for all $a_1, a_2 \in A$ we have the implications

$$f(a_1) = f(a_2) \implies g(f(a_1)) = g(f(a_2))$$
$$\implies a_1 = a_2. \qquad \text{(since } g \circ f = i_A)$$

Therefore f is injective. Moreover, if $b \in B$ then $g(b) \in A$ and $f(g(b)) = b$ (since $f \circ g = i_B$), so f is surjective. Thus f is a bijection, so we know that the function f^{-1} exists; it remains to prove that $g = f^{-1}$.

Observe that both $g \circ i_B$ and g have domain B; and for each $b \in B$ we

---

† Injectivity of f is essential in order for f^{-1} to be a function; surjectivity of f is required in order to have dom $f^{-1} = B$. See Exercise 5 in Section 3.2 for further discussion.

have $(g \circ i_B)(b) = g(i_B(b)) = g(b)$. Therefore $g \circ i_B = g$; similarly, check that $i_A \circ f^{-1} = f^{-1}$. The conclusion now follows from a chain of equalities:

$$g = g \circ i_B = g \circ (f \circ f^{-1}) = (g \circ f) \circ f^{-1} = i_A \circ f^{-1} = f^{-1}. \qquad \square$$

3.28 COROLLARY (CANCELLATION LAWS). Let f be a bijection. Then

$$f \circ g = f \circ h \Rightarrow g = h, \qquad \text{(left cancellation)}$$
$$r \circ f = s \circ f \Rightarrow r = s. \qquad \text{(right cancellation)}$$

PROOF. Exercise. (Major hint: Compose appropriately with f^{-1}.)
$$\square$$

If a bijection f is given by a formula of the form $f(x) = y$, then computing f^{-1} amounts to determining a formula that tells us how to transform y back to x. For this we reverse the steps leading from x to y, starting with the last step and backing up from there. A mnemonic: to invert "put on socks, put on shoes, walk from A to B," we "walk from B to A, take off shoes, take off socks." For example, consider

$$\mathbb{R}^+ = \{x \in \mathbb{R} \mid x \geqslant 0\} \qquad \text{and} \qquad B = \{x \in \mathbb{R} \mid x \geqslant 5\}$$

and define $f\colon \mathbb{R}^+ \to B$ by $y = f(x) = x^2 + 5$. This formula says "square, add 5" so f^{-1} tells us "subtract 5, take square root." That is,

$$y = x^2 + 5 \Rightarrow y - 5 = x^2 \Rightarrow \sqrt{y - 5} = x.$$

(The assumption $x \geqslant 0$ is crucial in the last step.) Thus $x = f^{-1}(y) = \sqrt{y - 5}$. Although the letters we use to represent points in the domain and range of a function are arbitrary, it is a common (but not universal) custom to use x and y, respectively, for these. With this convention our formula becomes $y = f^{-1}(x) = \sqrt{x - 5}$.

In summary, the procedure for determining inverses can be stated briefly like this: *If a bijection f is given by $y = f(x)$, then the formula for f^{-1} is obtained by solving for x in terms of y. If so desired, the variables can then be relabeled by using the letter y in place of x and x in place of y.*

The following theorem reinforces our understanding that we undo the results of "put on sock, then put on shoe" with "take off shoe, *then* take off sock."

3.29 THEOREM. Let $f: A \to B$ and $g: B \to C$ be bijections. Then $(g \circ f)^{-1} = f^{-1} \circ g^{-1}$.

PROOF. We make repeated use of the associative law for composition of functions:

$$(f^{-1} \circ g^{-1}) \circ (g \circ f) = f^{-1} \circ (g^{-1} \circ (g \circ f)) = f^{-1} \circ ((g^{-1} \circ g) \circ f)$$
$$= f^{-1} \circ (i_B \circ f) = f^{-1} \circ f = i_A.$$

Similarly, check that $(g \circ f) \circ (f^{-1} \circ g^{-1}) = i_C$. Now use Theorem 3.27.

□

Exercises

1. Because $(g \circ f)(x) = g(f(x))$ for all $x \in \text{dom } f$, the strict definition of composition requires that $f(x) \in \text{dom } g$ for all $x \in \text{dom } f$. It is common practice to be sloppy and *not* require this, with the understanding that the domain of $g \circ f$ is the set of all $x \in \text{dom } f$ for which $g(f(x))$ makes sense. For example, if f and g act on real numbers by the formulas $f(x) = x - 5$ and $g(x) = \sqrt{x - 4}$, then

$$(g \circ f)(x) = g(f(x)) = \sqrt{(x - 5) - 4} = \sqrt{x - 9},$$

hence dom $g \circ f = \{x \mid x \geqslant 9\}$. In each of the following cases, write the formula for $g \circ f$ and compute its domain. (The domain of each given function is understood to be the set of all real numbers for which the formula produces a value.)

(a) $f(x) = x^2$, $g(x) = \sqrt{x}$

(b) $f(x) = \sqrt{x}$, $g(x) = x^2$

(c) $f(x) = \sin x$, $g(x) = \dfrac{1}{x^2}$

(d) $f(x) = \begin{cases} \dfrac{\pi}{2} & \text{if } x \in \mathbb{Q}, \\[2mm] \dfrac{\pi}{4} & \text{if } x \notin \mathbb{Q}, \end{cases}$ $g(x) = \tan x$

2. Express each of the following functions as a composition of two functions, neither of which is the identity function.

(a) $f(x) = (x^2 + \cos x)^3$

(b) $f(x) = x^5 - 7x + 1$

(c) $f(x) = x^2(10)^{2x}$

(d) $f(x) = \dfrac{5}{x\sqrt{x-1}}$

3. (a) Give examples of functions f and g from $\mathbb{R}$ to $\mathbb{R}$ that do not commute with respect to composition (not the example in the text, please).

   (b) Give examples of functions f and g from $\mathbb{R}$ to $\mathbb{R}$ (with $f \neq g$) that commute with respect to composition.

4. Suppose that f and g are different constant functions from S to S, where S is a nonempty set. Show that $f \circ g \neq g \circ f$.

5. If $a, b \in \mathbb{R}$, define $a * b$ to be the result of rounding the product ab to the nearest tenth. For example, $3.7 * 12.9 = 47.7$. Verify that

$$(8.2 * 9.2) * 10.2 \neq 8.2 * (9.2 * 10.2).$$

(This exercise is here as a moral lesson: it warns us that this section's associative law actually requires proof. There are instances in which we might wish for such a law but not have our wish granted.)

6. Just for fun, forget for a moment that you know the associative law for composition of functions, and verify that $f \circ (g \circ h) = (f \circ g) \circ h$ when

$$f(x) = \sin x^3, \qquad g(x) = \frac{x^2 - 5}{x + 1}, \qquad h(x) = \sqrt[4]{3x}.$$

7. Suppose $f: A \to B$ and $g: B \to C$ are functions.

   (a) Show that if $g \circ f$ is injective then f is injective.

   (b) Show that if $g \circ f$ is surjective then g is surjective.

8. In each part of this exercise, give examples of sets A, B, C and functions $f: A \to B$ and $g: B \to C$ satisfying the indicated properties.

   (a) g is not injective but $g \circ f$ is injective.

   (b) f is not surjective but $g \circ f$ is surjective.

   (Suggestions: Work with sets having at most three elements; draw pictures.)

9. Define functions f and g from $\mathbb{Z}$ to $\mathbb{Z}$ such that f is not surjective and yet $g \circ f$ is surjective.

10. In our statement of the cancellation laws (3.28), the function f was taken to be a bijection. Now we ask whether this condition is essential.

   (a) Are the cancellation laws valid when f is surjective but not injective? Consider both laws, and in each case either prove that the law holds for such a function f or give a counterexample.

   (b) Repeat part (a) with f injective but not surjective.

11. Let $A = B = \{1, 2, 3, 4\}$, and let $f: A \rightarrow B$ be the function

$$f = \{(1, 3), (2, 4), (3, 2), (4, 1)\}.$$

   If f is a bijection, determine f^{-1}. If f is not a bijection, *show* that it isn't.

12. The function f acts on real numbers according to the formula

$$f(x) = x^{1/4} + 3.$$

   Determine the domain and range of this function, and give a formula for f^{-1}.

13. Refer to Theorem 3.29. Instead of the rule given there, a natural dream would be to have the rule $(g \circ f)^{-1} = g^{-1} \circ f^{-1}$.

   (a) Show that this statement can be made to fail solely on account of incompatibility of domains and ranges.

   (b) Give an example of a set S and bijections f and g from S to S, with $f \neq g$, such that the dream comes true:

$$(f \circ g)^{-1} = f^{-1} \circ g^{-1}.$$

   (c) Regain your senses by giving bijections from S to S such that

$$(f \circ g)^{-1} \neq f^{-1} \circ g^{-1}.$$

14. A function is a set of ordered pairs (satisfying certain conditions), so the **union** of two functions f and g is also a set of ordered pairs. It is therefore reasonable to ask whether $f \cup g$ is also a function and, if it is, to consider how its properties relate to the properties of f and g. This exercise considers these issues.

   Assume $f: A \rightarrow B$ and $g: C \rightarrow D$. All the examples requested in this exercise can be produced using sets A, B, C, D that have at most two elements each.

(a) Show that if $A \cap C = \emptyset$, then $(f \cup g): A \cup C \to B \cup D$.

(b) Give an example in which $A \cap C \neq \emptyset$ and $f \cup g$ is not a function.

(c) Give an example in which $A \cap C \neq \emptyset$ and $(f \cup g): A \cup C \to B \cup D$.

For the remaining parts, assume that $f \cup g$ *is* a function from $A \cup C$ to $B \cup D$.

(d) Show that if f and g are surjective then $f \cup g$ is surjective.

(e) Show that if $f \cup g$ is injective then f and g are injective.

(f) Show that if $f \cup g$ is a bijection then $(f \cup g)^{-1} = f^{-1} \cup g^{-1}$.

(g) Give an example in which $f \cup g$ is injective.

(h) Give an example in which f and g are injective but $f \cup g$ is not injective.

15. Prove the left and right cancellation laws (Corollary 3.28).

CHAPTER 4

Finite and Infinite Sets

4.1 CARDINALITY; FUNDAMENTAL COUNTING PRINCIPLES

Learning to count is one of our first great intellectual triumphs. But what exactly are we doing when we count? Can the process be generalized in a way that lets us compare sets that are larger than numbers can describe? What is meant by an infinite set? Are all infinite sets essentially the same size? Exploring these mysteries is our central goal in this chapter. As we proceed we will examine many of the fundamental principles of counting that we have viewed as intuitively obvious since childhood. (Example: You have n objects, and you give some of them to a friend; then you have given your friend at most n objects.) This will lay the foundation for the many practical counting procedures that we will develop here and in Chapter 5. In Section 4.5 we will use our work on infinite sets to initiate the study of languages and automata, fundamental topics in computer science.

Two primitive shepherds want to determine whose flock is larger, but counting is unknown to them. How can they proceed? Here is a way. They could guide the flocks (call them A and B) into adjoining pens, then select one sheep from *A* and one from *B* and link them with a cord. They would repeat this pairing procedure until one of the flocks has been used up. If one of the flocks has remaining unpaired sheep, that flock is larger; if no unpaired sheep remain in either flock, then *A* and *B* are the same size.

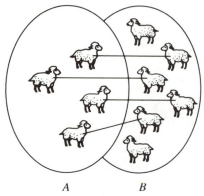

Flock B is larger.

While we're on this subject we can all agree on one more assertion: If flocks *A* and *B* are the same size and someone adds a sheep to *A* but not to *B*, then *A* will have more sheep than *B*.

Now imagine that there is a hotel with infinitely many rooms in a line: a first room, a second room, and so on. Imagine further that there is a person occupying each room; so we can reasonably say that there are exactly as many guests as there are rooms. You arrive at the hotel seeking a room, and the manager tells you, "At the moment the rooms are all occupied, but we always have room for one more." Whereupon she speaks into an intercom connected to all the rooms, saying, "At the sound of the bell, each guest should step out into the corridor and move into the next room in line." This clears out the first room, and you have a place to stay. Notice that although you thought you enlarged the collection of guests by your presence, there are still as many hotel rooms as there are guests. (Contrast this remark with the closing assertion on the flocks of sheep example.)

Always room for one more.

The two examples just discussed (particularly the second one) show that the matter of set comparison—saying that one set has more, fewer, or the same number of elements as another—involves some subtleties, and we will now give a careful treatment of these ideas.

4.1 DEFINITION. Sets A and B are said to have the same **cardinality** or **cardinal number** if there is a bijection $f: A \rightarrow B$. In this situation we also say that A and B are **equipotent** sets, abbreviated $A \approx B$, or that they are **in one-to-one correspondence**. The negation of this statement is written $A \not\approx B$.

The following theorem provides three useful properties of equipotence.

4.2 THEOREM. Let A, B, C be sets. Then

(a) $A \approx A$,

(b) $A \approx B \implies B \approx A$,

(c) $A \approx B$ and $B \approx C \implies A \approx C$.

Consequently, if S is any collection of sets, then equipotence is an equivalence relation on S.

PROOF.

(a) The mapping $i_A: A \rightarrow A$ is a bijection.

(b) If $f: A \rightarrow B$ is a bijection, then so is $f^{-1}: B \rightarrow A$.

(c) If $f: A \rightarrow B$ and $g: B \rightarrow C$ are bijections, then so is $g \circ f: A \rightarrow C$.

The final assertion of the theorem is an immediate consequence of (a), (b), and (c). $\square$

4.3 EXAMPLE. Let S be the set of seats at the Acme Theater, and let P be the set of people waiting to be seated for the next show. The assertion $P \approx S$ means that there is a seat for everyone, with no seats to spare. If $P \not\approx S$ then either there are too many people (no function $P \rightarrow S$ is injective) or, no matter how the people distribute themselves in the seats, there will be empty seats remaining (no function $P \rightarrow S$ is surjective).

4.4 EXAMPLE. Let X be a set with ten elements, let S be the set of all seven-element subsets of X, and let T be the set of all three-element subsets of X. Then $S \approx T$.

PROOF OUTLINE.   For each $A \in S$, let A' denote the complement of A in X. Then $A' \in T$, and the function $S \to T$ given by $A \mapsto A'$ is a bijection from S to T.   □

Intuitively, we usually view two equipotent sets as having the same size or the same number of elements. But there are weaknesses in this idea, as the next examples show.

4.5 EXAMPLES.

(a) Let E denote the set of even integers, and define $f: \mathbb{Z} \to E$ by $f(n) = 2n$. Then f is a bijection (check this), and therefore $\mathbb{Z} \approx E$. Notice the miracle here: We have taken $\mathbb{Z}$, thrown out half of it (namely all the odd integers), and we are left with a subset E that is still in one-to-one correspondence with $\mathbb{Z}$. Therefore $\mathbb{Z}$ is an example of a set that, in a sense, is no larger than half of itself!

(b) If $a, b \in \mathbb{R}$, with $a < b$, consider the **open interval**

$$I_a^b = \{x \in \mathbb{R} \mid a < x < b\}.$$

Define $f: I_a^b \to I_0^1$ by

$$x \xrightarrow{\quad f \quad} \frac{x - a}{b - a}.$$

If $x_1 < x_2$ then clearly $f(x_1) < f(x_2)$, and so f is injective. Also f is surjective. For suppose $t \in I_0^1$; then define $x = a + t(b - a)$ and check that $x \in I_a^b$ and $f(x) = t$. Therefore $I_a^b \approx I_0^1$. (Intuitively, f stretches or contracts I_a^b to cover I_0^1 precisely.)

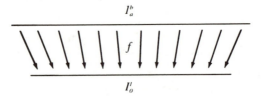

A visit to the shrink.

In particular, this result gives $I_{-\pi/2}^{\pi/2} \approx I_0^1$; so by Theorem 4.2 we obtain the equipotence $I_a^b \approx I_{-\pi/2}^{\pi/2}$. But we know from trigonometry that the mapping $I_{-\pi/2}^{\pi/2} \to \mathbb{R}$ given by $x \mapsto \tan x$ is a bijection, and hence $I_{-\pi/2}^{\pi/2} \approx \mathbb{R}$. Therefore, again by Theorem 4.2, we have $I_a^b \approx \mathbb{R}$. Our conclusion:

(4.6) Every open interval, no matter how short, is in one-to-one correspondence with the set $\mathbb{R}$ of real numbers. Therefore any two open intervals have the same cardinality (even if one interval is properly contained in the other).

For each integer $n \geqslant 0$, define

$$\mathbb{N}_n = \{x \in \mathbb{N} \mid 1 \leqslant x \leqslant n\}.$$

Then $\mathbb{N}_0 = \emptyset$, and clearly $\mathbb{N}_m \subseteq \mathbb{N}_n$ if $m \leqslant n$.

4.7 TERMINOLOGY. A set S is said to be **finite** if $S \approx \mathbb{N}_n$ for some $n \geqslant 0$, and in this case we write $\#S = n$. (The expression $\#S$ is verbalized "the number of elements in S.") A set is **infinite** if it is not finite. To **count** a finite set S is to establish a bijection $f: \mathbb{N}_n \to S$ for some $n \in \mathbb{N}$ or to recognize that S is empty. (This is a precise mathematical formulation of the counting process used by every elementary school child.)

The next few theorems on counting provide a useful exercise in humility. In each case the intuitive formulation of the result is a statement that seems obvious, yet the proof is challenging.

4.8 THEOREM. Let n and m be nonnegative integers, with $n > m$.

(a) There is no injection from $\mathbb{N}_n$ to $\mathbb{N}_m$, and hence $\mathbb{N}_n \not\approx \mathbb{N}_m$.

(b) If A is a set and $\#A = n$, then $\#A \neq m$. (Intuitively: *If a finite set is correctly counted twice, the results will be the same.*)

PROOF.

(a) We use induction on n, starting with $n = 1$. If $n = 1$ then $m = 0$, and there is no injection (indeed, no *function*) from $\mathbb{N}_1$ to $\mathbb{N}_0$, for such a function would have to pair the element 1 in $\mathbb{N}_1$ with an element of $\mathbb{N}_0$; but since $\mathbb{N}_0 = \emptyset$, this is impossible.

Assume the result to be true when $n = k$; that is, if $0 \leqslant m < k$ there is no injection from $\mathbb{N}_k$ to $\mathbb{N}_m$. Suppose there *is* an injection $f: \mathbb{N}_{k+1} \to \mathbb{N}_m$ for some $m < k + 1$; we seek a contradiction. First note that we must have $m \geqslant 1$, since (as in the case $n = 1$) no function exists from $\mathbb{N}_{k+1}$ to $\mathbb{N}_0 = \emptyset$. Now let $g: \mathbb{N}_m \to \mathbb{N}_m$ be the bijection that interchanges m with $f(k + 1)$ and fixes everything else; that is, define g by

$$g(x) = \begin{cases} f(k+1) & \text{if } x = m, \\ m & \text{if } x = f(k+1), \\ x & \text{otherwise.} \end{cases}$$

(So g is just the identity mapping on $\mathbb{N}_m$ if $f(k+1) = m$.) Then $g \circ f \colon \mathbb{N}_{k+1} \to \mathbb{N}_m$ is an injection, since f and g are injections, and $(g \circ f)(k+1) = m$. Therefore the restriction $(g \circ f)|_{\mathbb{N}_k}$ is an injection from $\mathbb{N}_k$ to $\mathbb{N}_{m-1}$. But $m - 1 < k$, so the existence of such a function contradicts the induction hypothesis. Thus an injection such as f cannot exist, and this completes the induction argument.

(b) If we had $\#A = n$ and $\#A = m$, then we would also have $\mathbb{N}_n \approx A \approx \mathbb{N}_m$, an impossibility by part (a).   □

4.9 COROLLARY (PIGEONHOLE PRINCIPLE).   Let A and B be nonempty finite sets, with $\#A > \#B$. Then there is no injection from A to B. Thus for any function $A \to B$, some element in B has at least two preimages. (Intuitively: *If n pigeons fly into m pigeonholes, and $n > m$, then some pigeonhole receives at least two pigeons.* Here the "pigeons" are the elements of A, the "pigeonholes" are the elements of B, and "flying" is the pairing action of the function.)

PROOF.   Suppose $\#A = n$ and $\#B = m$, with $n > m$. Then there are bijections $f \colon \mathbb{N}_n \to A$ and $g \colon B \to \mathbb{N}_m$. (Why?) If there were an injection $h \colon A \to B$, then the function

$$g \circ h \circ f \colon \mathbb{N}_n \to \mathbb{N}_m$$

would also be an injection, contradicting Theorem 4.8. Therefore no such h exists.   □

4.10 EXAMPLES.

(a) The game of musical chairs is played with n people and $n - 1$ chairs. Each chair can hold only one person. The players march around the chairs to music until the music is suddenly stopped, whereupon each player attempts to sit in a chair. By the pigeonhole principle, someone loses. (Let's go through the argument carefully. The pigeonhole principle says that there is no injection from the set P of n players to the set C of $n - 1$ chairs, so at least one player must remain unseated in any attempt to seat everyone. On the other hand, if p_0 is an unseated player then $\#(P - \{p_0\}) = \#C = n - 1$, so there *is* an injection $P - \{p_0\} \rightarrow C$, hence there is only one loser in a given round of the game.) The game now repeats with $n - 1$ players and $n - 2$ chairs, and the foolishness continues until only one player remains: the winner.

(b) In any set of eleven integers, there are two whose difference is divisible by 10. Proof idea: by the pigeonhole principle, two of the integers have the same right-hand digit, hence their difference is divisible by 10.

(c) A cheese cube with 2-inch edges is finely chopped, and nine mice are each fed one of the resulting crumbs. We claim that two of those crumbs were at most $\sqrt{3}$ inches from each other in the original cube. To see this, imagine the original cube partitioned into eight cubes with 1-inch edges, as shown here:

By the pigeonhole principle, two crumbs are from the same small cube, and those two crumbs are at most as far apart as the little cube's diagonally opposite corners. Invocation of Pythagoras's theorem completes the proof of the claim. (Check this.)

Recall that the symbol $\subset$ denotes *strict* set inclusion, so if $A \subset B$, then B contains at least one element not in A.

4.11 THEOREM. Every subset of $\mathbb{N}_n$ is finite; and if $A \subset \mathbb{N}_n$ then $\#A = t$ for some $t < n$.

Before proving this theorem, we pause to note that our terminology and notation make this theorem seem more obvious than it is, and it is in part the *job* of terminology and notation to do this. (See the quotation from Alfred North Whitehead on page 56.) But the assertion "$\#A = t$ for some $t < n$" has a precise meaning spelled out in 4.7, and there are details we must check before we can call the assertion "obvious" with at least a moderately clear conscience. A less suggestive notation would have made it more evident that there is work to be done in order to prove this theorem. Consider the following comment from Bertrand Russell (and compare it with the earlier statement by his pal Whitehead):

> Symbolism is useful because it makes things difficult. Now in the beginning everything is self-evident, and it is hard to see whether one self-evident proposition follows from another or not. Obviousness is always the enemy to correctness. Hence we must invent a new and difficult symbolism in which nothing is obvious.

PROOF OF THEOREM 4.11.   It suffices to prove the second assertion. (The first is an immediate consequence of it.) We use mathematical induction, starting with $n = 0$.

If $n = 0$ the result is true because the set $\mathbb{N}_0 = \emptyset$ has no proper subsets. Now suppose the result is true when $n = k$, and consider a subset $A \subset \mathbb{N}_{k+1}$. We must show that $\#A = t$ for some $t \leqslant k$.

Case (1).   Suppose $k + 1 \notin A$. Then $A \subseteq \mathbb{N}_k$. If $A = \mathbb{N}_k$ then $\#A = k$; and if $A \subset \mathbb{N}_k$ the induction hypothesis yields $\#A = t$ for some $t < k$, and we are done.

Case (2).   Suppose $k + 1 \in A$. Then $A = \{k + 1\} \cup (A \cap \mathbb{N}_k)$, and $A \cap \mathbb{N}_k \subset \mathbb{N}_k$ (since if $A \cap \mathbb{N}_k = \mathbb{N}_k$ we would have $A = \mathbb{N}_{k+1}$, contradicting the hypothesis that $A \subset \mathbb{N}_{k+1}$). From the induction hypothesis, we know that $\#(A \cap \mathbb{N}_k) = s$ for some $s \leqslant k - 1$; thus there is a bijection $f: A \cap \mathbb{N}_k \to \mathbb{N}_s$. Now define a function $g: A \to \mathbb{N}_{s+1}$ by

$$g(x) = \begin{cases} f(x) & \text{if } x \in A \cap \mathbb{N}_k, \\ s + 1 & \text{if } x = k + 1. \end{cases}$$

Then g is a bijection (why?), and therefore $\#A = s + 1 \leqslant k$.   □

The following result is understood and believed by every first-grade child on the basis of experience: If you have a set of objects and you count some, but not all, of the objects, then your answer will be less than the total number. This is the intuitive meaning of assertion (c) of the following theorem.

4.12 THEOREM.

(a) Every subset of a finite set is finite.

(b) Every set containing an infinite set is infinite.

(c) If $A \subset B$ and B is finite then $\#A < \#B$.

PROOF.

(a) Suppose $A \subseteq B$ and B is finite. Then from the definition of "finite" there is a bijection $f: B \to \mathbb{N}_n$ for some integer n. The restricted function $f|_A$ is injective, and it is a bijection from A to $f(A)$. By Theorem 4.11 we have $f(A) \approx \mathbb{N}_t$ for some integer $t \leqslant n$. Thus we have

$$A \approx f(A) \approx \mathbb{N}_t.$$

Therefore $A \approx \mathbb{N}_t$, and so A is finite.

(b) Let $A \subseteq B$. In part (a) we proved the implication

$$B \text{ finite} \quad \Rightarrow \quad A \text{ finite}.$$

The statement "A infinite $\Rightarrow B$ infinite" is just the contrapositive of this, so we are done.

(c) The proof is the same as the proof of statement (a), except that the symbol "$\subset$" replaces "$\subseteq$", and "$<$" replaces "$\leqslant$".   □

Before proceeding further let's display an infinite set for everyone to admire.

4.13 THEOREM.   The set $\mathbb{N}$ of natural numbers is infinite.

PROOF (by contradiction).   If $\mathbb{N}$ is finite then there is a bijection $f: \mathbb{N} \to \mathbb{N}_m$ for some natural number m. Pick any natural number n such that $n > m$. Then $f|_{\mathbb{N}_n}$ is an injection from $\mathbb{N}_n$ into $\mathbb{N}_m$, and this is an impossibility by Theorem 4.8(a).   □

4.14 THEOREM.   If A and B are disjoint finite sets, then $A \cup B$ is finite and

$$\#(A \cup B) = \#A + \#B.$$

PROOF.   Suppose $\#A = m$ and $\#B = n$. Then there exist bijections

$$f: \mathbb{N}_m \to A \quad \text{and} \quad g: \mathbb{N}_n \to B.$$

Now define a function $h: \mathbb{N}_{m+n} \to A \cup B$ by

$$h(i) = \begin{cases} f(i) & \text{if } 1 \leqslant i \leqslant m, \\ g(i-m) & \text{if } m+1 \leqslant i \leqslant m+n. \end{cases}$$

It is a short exercise to check that h is a bijection. Once this has been done, it follows that $\#(A \cup B) = m + n = \#A + \#B.$ □

Recall that sets $A_1, \ldots, A_m$ are said to be **pairwise disjoint** if $A_i \cap A_j = \emptyset$ whenever $i \neq j$.

4.15 COROLLARY. If $A_1, \ldots, A_m$ are pairwise disjoint finite sets, then $\bigcup_{i=1}^{m} A_i$ is finite and

$$\# \left(\bigcup_{i=1}^{m} A_i \right) = \#A_1 + \cdots + \#A_m.$$

PROOF OUTLINE. When $m = 1$ we have $\bigcup_{i=1}^{1} A_i = A_1$, so here the theorem asserts that $\#A_1 = \#A_1$, which is true (since $A_1 \approx A_1$). The case $m = 2$ is covered by Theorem 4.14, and the general case follows by induction on m.   □

4.16 COROLLARY. If A and B are finite sets (not necessarily disjoint), then $A \cup B$ is finite and

$$\#(A \cup B) = \#A + \#B - \#(A \cap B).$$

PROOF. Express $A \cup B$ as the union of three pairwise disjoint sets:

$$A \cup B = \underbrace{(A - B) \cup \overbrace{(A \cap B)}^{B} \cup (B - A)}_{A}$$

(The Venn diagram for $A \cup B$ leads naturally to this decomposition.) Then

$$\#(A \cup B) = \#(A - B) + \#(A \cap B) + \#(B - A) \qquad \text{(by Corollary 4.15)}$$
$$= \underbrace{[\#(A - B) + \#(A \cap B)]}_{\#A} + \underbrace{[\#(B - A) + \#(A \cap B)]}_{\#B} - \#(A \cap B).$$

This gives the result.   □

A more intuitive argument in support of 4.16 goes as follows. We start our computation of $\#(A \cup B)$ by first adding the number of elements in A to the number of elements in B; this assures that all elements of $A \cup B$ are included in our count. Then, upon noticing that the elements in $A \cap B$ have been counted twice, we subtract #($A \cap B$) from the sum. Corollary 4.16 is an instance of the *inclusion–exclusion principle,* a general formula for the number of elements in the union of a finite number of finite sets. (We'll consider the case of three sets in Exercise 7 at the end of this section.)

4.17 THEOREM. If #$A = m$ and #$B = n$, then #($A \times B$) = mn.

PROOF OUTLINE. If $A = \emptyset$ then also $A \times B = \emptyset$, and the result follows from the fact that #$\emptyset = 0$. Otherwise we have $A = \{a_1, \ldots, a_m\}$, say, and then $A \times B = \cup_{i=1}^{m}(\{a_i\} \times B)$, a union of m pairwise disjoint sets, each with the same cardinality as B. Apply Corollary 4.15. $\square$

Theorem 4.17 provides the following fundamental counting principle:

> A shop offering m ice cream flavors and n flavored sauces can serve mn kinds of ice cream sundaes.

More generally:

(4.18) Suppose two choices are to be made in succession. If there are m possibilities for the first choice and, once the first choice has been made, n possibilities for the second, then there are mn possibilities for the ordered pair of choices.

4.19 EXAMPLES. To estimate the weight of a typical goldfish in a tank of 100 fish, it is decided to sweep a one-fish net through the tank until a fish is caught, to record the weight of the captured fish, and then to repeat the whole process once more. (The average of the two weights will be used as the estimate.)

(a) If the first fish is dropped back into the tank before the next selection is made, then by 4.18 there are $100 \times 100 = 10{,}000$ possibilities for the ordered pair of fish being chosen. (This is an example of **sampling with replacement**: there is the possibility that the same fish will be chosen twice.)

(b) If the first fish is kept out of the tank while the second fish is selected, then there are $100 \times 99 = 9{,}900$ possibilities for the ordered pair. (This is **sampling without replacement**.)

(c) Now suppose there are two tanks of fish, each with 100 fish, and *one* fish is to be chosen. How many possible choices are there? This situation is covered *not* by 4.18 but by Theorem 4.14, which tells us that there are 200 fish in all, and so 200 is our answer.

Now we can extend our discussion to decision procedures of arbitrary length. Recall from Chapter 3 that if n is a positive integer, an **n-tuple** $(a_1, \ldots, a_n)$ is a sequence of length n. (See Exercise 10 in Section 3.2 for an alternative definition.)

4.20 DEFINITION.   Let $A_1, \ldots, A_n$ be sets. The **Cartesian product**

$$A_1 \times \cdots \times A_n$$

is the set of all n-tuples $(a_1, \ldots, a_n)$ such that $a_i \in A_i$ for $1 \leq i \leq n$.

The following generalization of Theorem 4.17 can be proved by induction on n; we omit the details.

4.21 THEOREM.   Suppose $A_1, \ldots, A_n$ are finite sets, with $\#A_i = m_i$ for $1 \leq i \leq n$. Then the Cartesian product $A_1 \times \cdots \times A_n$ is also finite, and

$$\#(A_1 \times \cdots \times A_n) = m_1 m_2 \cdots m_n.$$

Theorem 4.21 leads to a natural extension of 4.18, as follows.

4.22 PRODUCT RULE.   Suppose that n consecutive choices are to be made. Suppose further that there are m_1 possibilities for the first choice, and that for each k satisfying $2 \leq k \leq n$, there are m_k possibilities for the kth choice, no matter how the first $k - 1$ choices were made. Then there are exactly $m_1 m_2 \cdots m_n$ possibilities for the succession of n choices.

4.23 EXAMPLE.   A state motor vehicle bureau has decided that all automobile license plates must have exactly seven characters. The bureau directors agree that the characters will be chosen from the English alphabet or from the numerals 0, 1, . . . , 9, but the directors are still discussing whether the characters should satisfy some further conditions. In each of the following cases we will determine the number of licenses that satisfy the given condition.

(a) Each character can be any letter or numeral.

(b) The first three characters are letters and the last four are numerals.

(c) Three consecutive characters (not necessarily the first three) are letters and the others are numerals.

(d) Exactly three characters are letters and the others are numerals.

(e) Suppose the agreement that each license will have exactly seven characters is abandoned. Instead it is decided that a license must have *at most* seven characters, and each character can be any letter or numeral. (No car will get a blank license.)

SOLUTION.

(a) Since there are 26 letters and 10 numerals, there are 36 possibilities for each character. (This is evident; but the underlying theorem is 4.14.) By 4.22 (with $n = 7$ and $m_i = 36$ for $1 \leq i \leq 7$), we see that there are 36^7 possible configurations.

(b) Here there are 26^3 possible letter configurations and 10^4 numeral configurations, so there are $(26^3)(10^4)$ possibilities in all.

(c) First it must be decided which character (counting from the left, say) will be the first one to be a letter; there are five such possibilities, and this choice dictates the locations of the other two letters. Once this decision has been made, there are $(26^3)(10^4)$ possibilities for the digits, by the same reasoning as in case (b). So altogether there are $5(26^3)(10^4)$ possibilities.

(d) Once it is decided which three places will hold the letters, as before there will be $(26^3)(10^4)$ possible choices. In how many ways can we choose the three positions out of seven in which to insert the letters? In Chapter 5, when we discuss *combinations*, we will see that there are 35 ways to do this. Accepting this for now, our present answer is $(35)(26^3)(10^4)$.

(e) Imagine the one-character plates in one pile, the two-character plates in a second pile, and so on. By the argument in case (a), we see that the kth pile will have 36^k plates. Altogether this gives a grand total of

$$36 + 36^2 + 36^3 + 36^4 + 36^5 + 36^6 + 36^7$$

possibilities.

4.24 EXAMPLE (FOR CARD PLAYERS). A *straight* hand in poker consists of five cards in consecutive numerical order, for example,

$$3, 4, 5, 6, 7 \quad \text{and} \quad 10, J, Q, K, A$$

are straights. (Suits are irrelevant.) How many straights are there in all?

SOLUTION. There are ten possibilities for the lowest numerical value of a card in a straight. (An ace can have numerical value 1 or 14.) Once that lowest number has been determined, there are four cards in the deck corresponding to each of the five relevant numerical values. Therefore there are $(10)(4^5) = 10{,}240$ straights in a poker deck.

Exercises

Throughout this exercise set, the letters A, B, C, D will denote sets. Do not assume them to be finite unless that condition is given explicitly.

1. (a) Write a brief explanation of what it means to *count* a finite set. Your answer should be understandable to a fourth grade child who has never heard of functions.
   (b) Express the meaning of statement 4.2(c) in a form accessible to an elementary school child who has never heard of functions.

2. Let L and M be lines in the xy-plane. Prove that $L \approx M$. (Suggestion: First show that every line is equipotent to $\mathbb{R}$.)

3. (a) Exhibit a bijection from $\mathbb{Z}$ to the set of all integers that are 2 more than a multiple of 5.
   (b) Write a formula for the inverse of the bijection in part (a).

4. Prove the following statements.
   (a) $A \times B \approx B \times A$
   (b) If $A \approx C$ and $B \approx D$ then $A \times B \approx C \times D$.
   (c) $(A \times B) \times C \approx A \times (B \times C)$
   (d) If w is any element then $\{w\} \times A \approx A$.

5. (For students who have had differential calculus) Outline the argument showing that the mapping $I_{-\pi/2}^{\pi/2} \to \mathbb{R}$ given by $x \mapsto \tan x$ is a bijection. (This fact is used in Example 4.5(b).)

6. Suppose that m and n are positive integers, with $m \leqslant n$, and assume $\#A = n$. Prove that the number of m-element subsets of A is equal to

the number of $(n - m)$-element subsets of A. (Example 4.4 and Theorem 4.14 may help you.)

7. If A, B, and C are finite sets, write a formula for the number of elements in $A \cup B \cup C$. Your formula may use addition, subtraction, and intersection, but not *union*. (Suggestion: A double application of 4.16 will do the job. Alternatively, draw a Venn diagram and think about it.)

8. A study of 115 breakfast eaters shows that 85 also eat lunch, 58 use dental floss regularly, and 27 subscribe to a morning newspaper. Among those who also eat lunch, 52 floss regularly and 15 get the morning paper, and 10 lunch eaters both floss and get the paper. Four flossers neither eat lunch nor get the paper.

   (a) How many of those in the study neither eat lunch, nor floss regularly, nor get the morning paper?

   (b) How many of those who use dental floss regularly also get the morning paper?

   (c) How many of those who get the morning paper neither use dental floss regularly nor eat lunch?

9. Prove that if $A_1, A_2, \ldots, A_n$ are finite sets, then $\cup_{i=1}^n A_i$ is finite and $\#(\cup_{i=1}^n A_i) \leqslant \#A_1 + \cdots + \#A_n$. (Use induction on n.)

For Exercises 10, 11, and 12, imagine that you are familiar with the set of nonnegative integers, but that you have never heard of the operations of addition, subtraction, and multiplication.

10. Define the *sum* of nonnegative integers m and n as follows. Take disjoint sets A and B, with $\#A = m$ and $\#B = n$, and define

$$m + n = \#(A \cup B).$$

   (a) Show that if a different disjoint pair of sets A_1 and B_1 are used, with $\#A_1 = m$ and $\#B_1 = n$, then the same value of $m + n$ will result.

   (b) Show that the associative law of addition,

$$(m + n) + p = m + (n + p),$$

   is a consequence of the definition of addition given in this exercise. (Here m, n, and p are understood to be nonnegative integers.)

11. Let m and n be integers with $n > m \geq 0$, and define the *difference* $n - m$ as follows. Take sets A and B such that $\#A = m$, $\#B = n$, and $A \subset B$; then define

$$n - m = \#(B - A).$$

(Here the "minus" sign on the left is the symbol being defined; the "minus" sign on the right denotes set difference.) Show that with "$-$" as defined here and "$+$" as in the previous problem, the following equation holds:

$$m + (n - m) = n.$$

12. Let m and n be nonnegative integers, and define the *product mn* as follows. Suppose sets A and B are such that $\#A = m$ and $\#B = n$; then define

$$mn = \#(A \times B).$$

(a) Show that it follows from this definition that $mn = nm$.
(b) Show that from the definition of product given here and the definition of sum in Exercise 10, it follows that

$$m(n + k) = mn + mk$$

when m, n, and k are nonnegative integers.

13. (a) Show that in any group of eight people, there are two (at least) who were born on the same day of the week.
(b) How large a group of people must we consider in order to be sure that at least n of them were born on the same day of the week? (Here n is a positive integer.) Explain.

14. Define a **codeword** of length three to be a three-digit string of 0s and 1s. For instance, 011 and 101 are codewords.

(a) Show that there are eight codewords of length three.
(b) Define the **distance** between two codewords to be the number of digits in which they differ. For instance, the distance between 011 and 101 is 2. Show that there does not exist a set of five different codewords of length three, each of which has distance of at least 2 from each of the others. (Do *not* attempt to list all the five-element sets of codewords and eliminate them one by one. Instead, proceed by contradiction: suppose there *is* such a set, use the pigeonhole principle to show that at least three of these five code-

words have the same first digit, and then go on to consider the second and third digits of these three codewords to deduce a contradiction.)

15. A camper's afternoon at Camp Sweatless begins with an hour of dabbling in one of seven art media, followed by attendance at one of the camp's nine films, and then consumption of a massive "snack." (The camper chooses one from a menu with seventeen possibilities.)

    (a) How many variations are possible in an afternoon's fun schedule?
    (b) Suppose that a new camp director is appointed who allows each camper the additional freedom to choose the order of his or her activities (for example, perhaps first a movie, then a snack, then art). How many variations are possible under this new arrangement?

16. The well-known mail order retailer M. M. Pease carries men's shirts in seven different sleeve lengths and five different collar sizes; and for women's shirts (these are shaped differently from the men's) there are eight different sizes and four different collar styles. (Don't be concerned with colors here.)

    (a) How many variations in shirt construction are there altogether?
    (b) Suppose the men's shirts can be purchased with two or three initials, or without initials. How many varieties of men's shirts are there in all, if you count two shirts as different if they are identically sized but have different initials? (And count differently sized shirts as different.)
    (c) One quarter of all the varieties of women's shirts can be purchased with a pocket as an extra feature. Counting these as different from the shirts without pockets, how many different women's shirts are there?

17. Ten people are in a room, and each person shakes everyone else's hand. It is argued that altogether 90 handshakes have occurred, because each of ten people shakes nine hands, making 90 in all. Do you agree? Explain.

18. Suppose there are two 100-fish tanks, and one fish is to be chosen from each tank.

    (a) How many ways are there to make the selection, assuming that the order in which the fish are chosen is irrelevant? (That is, all that matters is which two fish are chosen, not which is chosen first.)
    (b) How many ways are there to make the selection if order *is* relevant?

19. Let A^B denote the set of all functions from B to A.

   (a) Suppose $\#A = m$ and $\#B = n$. Then use the counting principles of this section to show that

   $$\#(A^B) = m^n.$$

   (b) Determine the number of injections from B to A, assuming that $\#A = m$ and $\#B = n$. Here you will need to use the fact (to be proved in Chapter 5) that if $n \leqslant m$, the number of n-element subsets of an m-element set is

   $$\frac{m!}{n!(m-n)!}.$$

   (c) (Challenging) Without any restrictions on the sets A, B, C, show that

   $$(A \times B)^C \approx A^C \times B^C.$$

20. (a) Suppose that $\sim$ is an equivalence relation on an infinite set S, and suppose the relation partitions the set into a finite number of equivalence classes. Deduce that some equivalence class must contain an infinite number of elements.

   (b) Suppose $\sim$ is an equivalence relation on an infinite set S, with no further conditions. Must some equivalence class be infinite?

4.2   COMPARING SETS, FINITE OR INFINITE

Our focus in this chapter up to now has been the measurement and comparison of finite sets. The strategy has been to assign to each set a number that represents the set's size and then to compare sets by comparing their associated numbers. But the size of an *infinite* set is not represented by a number. So, while our intuition might lead us to declare that every infinite set is larger than every finite set, our methods thus far will yield only knitted brows when we contemplate comparing one infinite set with another. Our first goal in this section is therefore to reformulate our criteria for comparing set magnitudes so that we can compare *any* two sets, finite *or* infinite; and we want to do this in a way that is consistent with what we have achieved so far in the finite case. We will see that just as finite sets come in different sizes, so do infinite sets; and in fact there are infinitely many different levels of infinity! (Our brief taste of infinite sets in Example 4.5 might have led us to believe otherwise.) At the end of the chapter we will have a look at the impact of some of these remarkable concepts on theoretical computer science, when we briefly consider *languages* and the problem of *language specification.*

We begin with a slightly unnerving but useful lemma that shows the empty set to possess surprising vigor.

4.25 LEMMA. If A is any set, then $\emptyset$ is an injective function from $\emptyset$ to A.

PROOF. First we show that $\emptyset$ is a function from $\emptyset$ to A by checking the conditions of Definition 3.2. We have $\emptyset \subseteq \emptyset \times A$ (in fact, $\emptyset = \emptyset \times A$). Moreover, for every element x, the conditional statement

$$x \in \emptyset \quad \Rightarrow \quad (x, y) \in \emptyset \text{ for exactly one } y \in A$$

has a false hypothesis and is therefore true. Thus $\emptyset$ satisfies Definition 3.2 and is a function from $\emptyset$ to A.

The proof of injectivity has a similar flavor: the statement

$$x_1, x_2 \in \emptyset \text{ and } \emptyset(x_1) = \emptyset(x_2) \quad \Rightarrow \quad x_1 = x_2$$

is true because its hypothesis is false. $\square$

The following theorem provides the key to our broadening of the criteria for set comparison in order to accommodate infinite sets. The proof freely uses the basic properties of bijections and inverse functions. (See especially Theorem 3.24 and the material following 3.25.)

4.26 THEOREM. Let A and B be finite sets. Then

(a) $\#A = \#B \quad \Leftrightarrow \quad A \approx B,$
(b) $\#A \leq \#B \quad \Leftrightarrow \quad$ There is an injection from A to $B,$
(c) $\#A < \#B \quad \Leftrightarrow \quad$ There is an injection but no bijection from A to $B.$

PROOF. Write $\#A = m$ and $\#B = n$. So there are bijections

$$\mathbb{N}_m \xrightarrow{f} A \quad \text{and} \quad \mathbb{N}_n \xrightarrow{g} B.$$

We start with a proof of statement (b). If $m \leq n$ there is an injection $j: \mathbb{N}_m \to \mathbb{N}_n$. (For instance, take j to be the inclusion mapping; see Example 3.3(a). Or use Lemma 4.25 if $m = 0$.) Then the mapping

$$A \xrightarrow{g \circ j \circ f^{-1}} B$$

is an injection, as desired. Conversely, if $h: A \to B$ is an injection, then so is

$$\mathbb{N}_m \xrightarrow{g^{-1} \circ h \circ f} \mathbb{N}_n,$$

and therefore $m \leq n$ by Theorem 4.8.

Now to prove statement (a), repeat the proof of (b) using "=" in place of "$\leq$" and "bijection" in place of "injection."

Finally, we check (c). Assume $m < n$. If there were a bijection $h: A \to B$, then $g^{-1} \circ h \circ f: \mathbb{N}_m \to \mathbb{N}_n$ would also be bijective, and this is impossible by 4.8. The converse is an immediate consequence of statements (a) and (b).   □

Notice that the conditions in Theorem 4.26 on the right side of the "$\Leftrightarrow$" symbols make sense *whether or not the sets A and B are finite.* Accordingly, we can use these conditions to rewrite our criteria for set comparison, with full confidence that they will be consistent with the criteria we have been using for finite sets.

4.27 DEFINITION.   Let A and B be sets (finite or infinite). Define the expressions $\#A = \#B$, $\#A \leq \#B$, and $\#A < \#B$ by statements 4.26(a), (b), and (c), respectively.

The expressions defined in 4.27 are read as if they are relations between numbers (which they are, as we have seen, when the sets are finite). For example, the second expression is read, "the cardinality of A is less than or equal to the cardinality of B," or, more casually, "The number of elements in A is. . . ." Intuitively, the expression $\#A < \#B$ means that B is larger than A, because the existence of an injection $A \to B$ tells us that there are enough elements in B to pair each element of A with a different member of B, while the fact that no injection is surjective means that after any such pairing, there are elements of B left over.

We have given meaning to the statement $\#A < \#B$ (and the other statements in 4.27) without explicitly defining the symbol #A when A is infinite. (To do so would take us deeper into the foundations of mathematics than we intend to go here.) Is this really so strange? After all, it is possible to demonstrate that one object weighs less than another without knowing the weight of either. (See picture.) We have merely generalized the criterion of comparison used by the primitive shepherds discussed at the start of the chapter.

How to compare two walruses.

It is natural to suppose that the standard laws of numerical inequalities extend to expressions such as those in 4.27. Indeed, it is easy to see that for every set A we have #$A \leqslant$ #A, since the identity mapping is an injection from A to A. We also have

$$\#A \leqslant \#B \quad \text{and} \quad \#B \leqslant \#C \quad \Rightarrow \quad \#A \leqslant \#C,$$

for the hypothesis provides injections $A \to B$ and $B \to C$, whose composition is injective by 3.24(a). But to handle the *strict* inequalities we need the following theorem, proved independently by E. Schröder and F. Bernstein in the 1890s. We omit the proof,† but we remark that if the sets involved are finite, the theorem is an immediate consequence of the pigeonhole principle. Incidentally, we will continue to write $A \approx B$ instead of the slightly longer #$A =$ #B; there is no harm in this since the expressions are equivalent.

4.28 SCHRÖDER–BERNSTEIN THEOREM. Let A and B be sets. Then

$$\#A \leqslant \#B \quad \text{and} \quad \#B \leqslant \#A \quad \Rightarrow \quad A \approx B.$$

---

† For a highly readable version see P. R. Halmos, *Naive Set Theory* (New York: Springer-Verlag, 1974).

4.29 EXAMPLE.   Consider the open interval (0, 1) and the closed interval [0, 1]. We claim that

$$[0, 1] \approx (0, 1).$$

(Intuitively, this says that attaching the two endpoints to the open interval has no effect on the "number" of elements present.) The Schröder–Bernstein theorem tells us that to verify the claim, it suffices to construct an injection from each of the given intervals to the other. First notice that since $(0, 1) \subseteq [0, 1]$, the *inclusion mapping* $(0, 1) \rightarrow [0, 1]$ is a readily available injection. In the other direction, we map [0, 1] into (0, 1) according to the following scheme:

More formally, define $f\colon [0, 1] \rightarrow (0, 1)$ by the formula

$$f(x) = \frac{x}{2} + \frac{1}{4}$$

and check that f is injective. It follows that $[0, 1] \approx (0, 1)$, by the Schröder–Bernstein theorem. Notice that the theorem tells us that *there exists* a bijection between the two intervals, but it does not spell out an explicit formula for such a bijection (and neither of the injections we have mentioned is surjective). Suggestion: try to construct a bijection from [0, 1] to (0, 1). (We'll come back to this in Exercise 4 at the end of this section.)

4.30 THEOREM.   Let A, B, and C be sets. Then

(a) $\#A < \#B < \#C \implies \#A < \#C$;

(b) $\#A < \#B \leqslant \#C \implies \#A < \#C$;

(c) $\#A \leqslant \#B < \#C \implies \#A < \#C$.

[If the sets in the theorem are not all finite, then these alleged inequalities are *not* numerical inequalities of the high-school variety. (They just *look* like them.) Instead, they are assertions about the existence and nonexistence of certain functions, as spelled out in Theorem 4.26 and Definition 4.27. So there really is something to prove here.]

PROOF OF (a). The hypothesis and the fact that the composition of two injections is an injection give $\#A \leqslant \#C$. So it remains to prove that $A \not\approx C$, and we do this by contradiction.

If in fact $A \approx C$ then there is a bijection $C \to A$; and the hypothesis $\#B < \#C$ provides an injection $B \to C$. Then the composition $B \to C \to A$ is injective, so $\#B \leqslant \#A$. But the hypothesis $\#A < \#B$ implies the weaker statement $\#A \leqslant \#B$, and therefore the Schröder–Bernstein theorem gives $A \approx B$. This contradicts the hypothesis $\#A < \#B$. So the assumption $A \approx C$ must be discarded, and the inequality is strict: $\#A < \#C$. Proofs of (b) and (c) are similar and are left as exercises. □

The careful comparison of infinite sets is outside the range of our childhood experience, and our intuition is no longer completely reliable. For example, it might seem as though all infinite sets are equipotent, because of our vague sense that there are always enough elements in one infinite set to accommodate the needs of another; but this is *not* the case, as the next theorem shows. This theorem was proved by Georg Cantor (1845–1918), whom we first encountered in Chapter 3 (see Example 3.19). Cantor's results on infinite sets in the last quarter of the nineteenth century startled the mathematical world and infuriated much of it, and led to a reexamination and reconstruction of the foundations of mathematics.† Recall from Chapter 2 that the **power set** $P(S)$ of a set S is the set of all subsets of S; so "member of $P(S)$" and "subset of S" are synonymous expressions.

4.31 CANTOR'S THEOREM. Let S be a set with power set $P(S)$. Then

$$\#S < \#P(S).$$

PROOF. First note that if $S = \emptyset$ then $P(S) = \{\emptyset\}$, so $\#S = 0$ and $\#P(S) = 1$, and the inequality is valid.

Now suppose $S \neq \emptyset$. To prove the theorem, the definition of "<" requires that we show the existence of an injection and the nonexistence of a bijection from S to $P(S)$. Define a function $g: S \to P(S)$ by setting

† David Hilbert (1862–1943), another great mathematician, later called set theory "a paradise created by Cantor from which nobody will ever expel us."

$g(x) = \{x\}$ for each $x \in S$. Then g is injective, since

$$g(x_1) = g(x_2) \quad \Rightarrow \quad \{x_1\} = \{x_2\} \quad \Rightarrow \quad x_1 = x_2.$$

This proves that $\#S \leqslant \#P(S)$.

It remains to prove that there is no bijection from S to $P(S)$. Suppose, to the contrary, that there *is* such a bijection f. Then for each $x \in S$ we have $f(x) \in P(S)$; that is, $f(x)$ is a subset of S. Thus either $x \in f(x)$ or $x \notin f(x)$. Define

$$E = \{x \in S \mid x \notin f(x)\}.$$

Then E is a subset of S; that is, $E \in P(S)$. Since f is onto $P(S)$ (remember, f is a bijection), there is an element $z \in S$ such that $f(z) = E$. Now, is the statement "$z \in E$" true or false? Observe:

$$z \in E \iff z \notin f(z) \qquad \text{(from the definition of } E)$$
$$\iff z \notin E. \qquad \text{(since } f(z) = E)$$

This is madness. That is, the assumption that f is a bijection has led to an impossibility: a statement that is equivalent to its negation. Therefore no such bijection can exist. $\square$

The following astonishing result is a corollary of Cantor's theorem: Not only are there infinite sets (for example $\mathbb{N}$), but *there are higher and higher levels of infinity*. In particular:

4.32 COROLLARY. $\#\mathbb{N} < \#P(\mathbb{N}) < \#P(P(\mathbb{N})) < \#P(P(P(\mathbb{N}))) < \cdots.$

Exercises

1. (a) Supplement Lemma 4.25 by showing that the "empty function"

$$\emptyset \colon \emptyset \to A$$

   is surjective if and only if $A = \emptyset$.
   (b) Show that for any set A, the empty function is the *only* function from $\emptyset$ to A.

2. Someone deduces that the following statement is a consequence of the Schröder–Bernstein theorem: "If there are injections $f \colon A \to B$ and

$g: B \rightarrow A$, then each of these injections is a bijection." Either *show* that this is a consequence of the theorem or give a counterexample to the statement.

3. Prove Theorem 4.30(b).

4. In Example 4.29 we deduced the equipotence $[0, 1] \approx (0, 1)$ by invoking the Schröder–Bernstein theorem, but we didn't actually construct a bijection $[0, 1] \rightarrow (0, 1)$.

    (a) Verify that the function h given by the following scheme is such a bijection:

    $$0 \overset{h}{\mapsto} \tfrac{1}{2}$$

    $$1 \mapsto \tfrac{1}{3}$$

    $$\frac{1}{n} \mapsto \frac{1}{n+2} \quad \text{for each integer } n \geq 2$$

    $$x \mapsto x \qquad \text{otherwise.}$$

    (b) The function h in part (a) acts on $[0, 1]$ in a way that resembles the strategy of the hotel manager at the start of the chapter. How?

5. Let A, B, and C be sets. For each of the following, either prove the statement or give a counterexample.

    (a) $\#(A - B) \leq \#A$
    (b) $\#A \leq \#B \implies \#(A \times C) \leq \#(B \times C)$
    (c) $\#(A - B) \leq \#B$

6. Show that if $\#A \leq \#B$ then $\#P(A) \leq \#P(B)$.

7. Let I denote the closed unit interval $[0, 1]$; then $I \times I$ is the closed unit square. (Here "closed" means that the boundary points are included.)

Show that $I \times I \approx I$. You may use (without proof) the fact that every nonzero real number can be written in a unique way as a nonterminating decimal, that is, a decimal without an infinite string of zeros on its right end. (For example, we can write .38000. . .= .37999. . . .)

Suggestion: In mapping $I \times I$ to I as part of the solution, "shuffle" the decimal representations of the two coordinates.

8. (a) Use Cantor's theorem to deduce that $n < 2^n$ for every nonnegative integer n.
   (b) Prove that $n < 2^n$ for every nonnegative integer n, using mathematical induction instead of Cantor's theorem.

4.3 COUNTABLE AND UNCOUNTABLE SETS

4.33 DEFINITION.   A set S is said to be **countably infinite** if $\mathbb{N} \approx S$; that is, if there is a bijection from $\mathbb{N}$ to S. And a set is said to be **countable** or **denumerable** if it is finite or countably infinite.† A set that is not countable is called **uncountable, uncountably infinite**, or **nondenumerable**.

In the language of Chapter 3, a set S is countable if there is a finite or infinite sequence whose terms are the elements of S. Thus, to show that a set is countable, it suffices to exhibit a procedure for listing its elements, and for this it is sufficient to start the list and to exhibit a method for picking the successor of any element on the list. (See the material on definition of sequences by recursion in Section 3.2.)

4.34 EXAMPLE.   Show that the set $\mathbb{Z}$ of integers is countably infinite.

PARTIAL SOLUTION.   The set $\mathbb{Z}$ is infinite because it contains the infinite set $\mathbb{N}$. (See Theorem 4.13.) To show that $\mathbb{Z}$ is countable, we will first informally list its elements, then determine a function that makes our listing process precise. It's easy to start such a list: $0, 1, -1, 2, -2, 3, -3$, and so on. In more detail:

Position on list	Listed element
1	0
2	1
3	-1
4	2
5	-2
6	3
7	-3

† Terminology varies slightly from book to book. Some authors use "countable" or "denumerable" to mean what we have called "countably infinite."

The relation between the left column and the right column is formally described by the function $f: \mathbb{N} \to \mathbb{Z}$ defined by

$$f(n) = \begin{cases} \dfrac{n}{2} & \text{if } n \text{ is even,} \\[2ex] -\dfrac{n-1}{2} & \text{if } n \text{ is odd.} \end{cases}$$

Checking that f is a bijection will complete the solution.

The following theorem explains the "always room for one more" principle in the hotel-with-infinitely-many-rooms discussion.

4.35 THEOREM.   If A is finite and B is countable then $A \cup B$ is countable.

PROOF.   First notice that if B is finite then $A \cup B$ is finite (by 4.16) and hence countable.

Now suppose B is countably infinite. Observe that $A \cup B = (A - B) \cup B$. A corresponding Venn diagram:

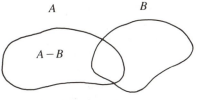

$A \cup B$ *(original version)*.

The set $A - B$ is finite because it is a subset of the finite set A. Moreover, $A - B$ is disjoint from B. If we now relabel by writing A instead of $A - B$, we see that our proof will be complete if we verify the following statement: If A is finite and B is countably infinite and $A \cap B = \emptyset$, then $A \cup B$ is countably infinite.

$A \cup B$ *(revised version)*.

For this it suffices to show that there is a bijection $A \cup B \to \mathbb{N}$. Since A is finite and B is countably infinite, there are bijections

$$f: A \to \mathbb{N}_n \quad \text{(for some } n \geqslant 0) \qquad \text{and} \qquad g: B \to \mathbb{N}.$$

Now define $h: A \cup B \to \mathbb{N}$ by

(*)
$$h(x) = \begin{cases} f(x) & \text{if } x \in A, \\ n + g(x) & \text{if } x \in B. \end{cases}$$

Intuitively we sense that h shifts the image of g to the right on the real line in order to make room for the image of f, just as the guests filling the rooms of the infinite hotel moved down the hall to accommodate a new arrival.

We claim that h is the desired bijection. First note that since A and B are disjoint, there is no ambiguity about which line of definition (*) applies to any given element x. (We say that h is a *well-defined* function.)

Injectivity. For all $x_1, x_2 \in B$, we must verify the implication

$$h(x_1) = h(x_2) \quad \Rightarrow \quad x_1 = x_2.$$

Case (1). Suppose $x_1, x_2 \in A$. Then

$$h(x_1) = h(x_2) \quad \Rightarrow \quad f(x_1) = f(x_2) \qquad \text{(by definition of } h)$$
$$\Rightarrow \quad x_1 = x_2. \qquad \text{(since } f \text{ is injective)}$$

Case (2). Suppose $x_1, x_2 \in B$. Then

$$h(x_1) = h(x_2) \quad \Rightarrow \quad n + g(x_1) = n + g(x_2) \qquad \text{(by definition of } h)$$
$$\Rightarrow \quad g(x_1) = g(x_2)$$
$$\Rightarrow \quad x_1 = x_2. \qquad \text{(since } g \text{ is injective)}$$

Case (3). Suppose $x_1 \in A$ and $x_2 \in B$ (so $x_1 \neq x_2$). Here $h(x_1) = f(x_1) \leqslant n$, while $h(x_2) = n + g(x_2) \geqslant n + 1$, and so $h(x_1) \neq h(x_2)$.

Surjectivity. Let $k \in \mathbb{N}$. If $1 \leqslant k \leqslant n$ then

$$k = f(x) = h(x)$$

for some $x \in A$, since $f: A \to \mathbb{N}_n$ is surjective. If $k \geqslant n + 1$ then $k = n + t$ for some $t \in \mathbb{N}$. Therefore $k = n + g(x) = h(x)$ for some $x \in B$, because $g: B \to \mathbb{N}$ is surjective. $\square$

Proofs of bijectivity are at the heart of most counting arguments, no matter how informal they may seem: injectivity guarantees that nothing is counted more than once, and surjectivity guarantees that everything is counted. Nevertheless, as the preceeding proof shows, the arguments can be long and they tend to have many cases, corresponding to different sections of the underlying Venn diagram. From now on most proofs of bijectivity will be omitted or only informally sketched. In each instance, supplying the details is a worthwhile exercise.

The following two-part theorem is useful in the study of infinite sets. We omit the proof, which is not elementary.

4.36 THEOREM.

(a) Every subset of a countable set is countable.

(b) Every infinite set contains a countably infinite subset.

Now we can prove the countable analogues of some of our earlier results on finite sets.

4.37 THEOREM.   If A and B are countable sets then $A \cup B$ is countable.

PROOF.   As in the proof of 4.35, we can write $A \cup B = (A - B) \cup B$. Notice that $A - B$ is countable, by Theorem 4.36(a). Also, we can assume that $A - B$ and B are both infinite, since otherwise Theorem 4.16 completes the proof for us. Now relabel by writing A instead of $A - B$. We sum up this groundwork as follows: without loss of generality, it suffices to prove that if A and B are disjoint countably infinite sets, then $A \cup B$ is countably infinite.

We know (from 4.5 and 4.34) that the set E of even integers is countably infinite. Similarly, if E' denotes the set of odd integers, the bijectivity of the mapping $\mathbb{Z} \to E'$ given by $n \mapsto 2n - 1$ shows that E' is countably infinite. Our hypothesis that A and B are countably infinite therefore provides bijections $A \xrightarrow{f} E$ and $B \xrightarrow{g} E'$. Now take the union of f and g as sets of ordered pairs.† Because the function f pairs the elements of A with elements of E, and g pairs the elements of B with elements of E', then the function $f \cup g$ pairs the elements of $A \cup B$ with elements of $E \cup E'$ (which is $\mathbb{Z}$). Moreover, since $A \cap B = \emptyset$, there is no ambiguity about the image

† See Exercise 14 in Section 3.3 for further exploration of the properties of functions obtained as unions of given functions.

of each element of $A \cup B$. Explicitly, the function

$$f \cup g: A \cup B \rightarrow E \cup E' = \mathbb{Z}$$

is given by the formula

$$(f \cup g)(x) = \begin{cases} f(x) & \text{if } x \in A, \\ g(x) & \text{if } x \in B, \end{cases}$$

or pictorially by

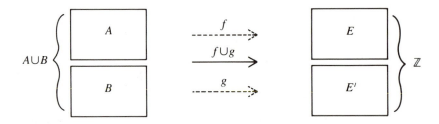

Since f and g are bijections onto E and E', respectively, and $E \cap E' = \emptyset$, it follows that $f \cup g$ is a bijection. Therefore, since $\mathbb{Z}$ is countably infinite, so is $A \cup B$. □

From Theorem 4.37, a short induction argument yields the following.

4.38 COROLLARY. If $A_1, \ldots, A_n$ are countable sets, then $\cup_{i=1}^{n} A_i$ is countable.

Recall that a **prime number** is an integer $p > 1$ having no factorization of the form $p = ab$, where $a > 1$ and $b > 1$ are integers. (Examples: 2, 3, 5, 7, 11, 13, 17, 19, 23 are the first few primes.) The *Fundamental Theorem of Arithmetic* is the following statement: *Every integer $n > 1$ can be written in exactly one way as a product of prime numbers.* (We proved part of this in Theorem 2.74, and we will complete the proof in Chapter 6.) That is, if $n = p_1 p_2 \cdots p_r = q_1 q_2 \cdots q_s$, where the p's and q's are primes, then $r = s$ and the p's and q's are equal (though perhaps listed in different order).

4.39 THEOREM.

(a) $\mathbb{N} \times \mathbb{N}$ is countably infinite.

(b) If A and B are countable then $A \times B$ is countable.

PROOF.

(a) The set $\mathbb{N} \times \mathbb{N}$ is infinite, because it contains the infinite set

$$\{(1, 1), (2, 1), (3, 1), (4, 1), \ldots \}.$$

Now define $f: \mathbb{N} \times \mathbb{N} \to \mathbb{N}$ by

$$f(m, n) = 2^m 3^n.$$

By the Fundamental Theorem of Arithmetic, it follows that f is injective; so, since $f(\mathbb{N} \times \mathbb{N})$ is an infinite subset of the countable set $\mathbb{N}$, it is countably infinite by Theorem 4.36(a). Hence there is a bijection g from $f(\mathbb{N} \times \mathbb{N})$ to $\mathbb{N}$, and the composition $g \circ f$ is a bijection from $\mathbb{N} \times \mathbb{N}$ to $\mathbb{N}$. Therefore $\mathbb{N} \times \mathbb{N}$ is countably infinite. [*Outline of an alternate proof.* Define $g: \mathbb{N} \times \mathbb{N} \to \mathbb{N}$ by

$$g(m, n) = 2^{m-1}(2n - 1),$$

and use the Fundamental Theorem of Arithmetic to show that g is a bijection.]

(b) If A and B are finite, we're through by Theorem 4.17. If $A = \{a_1, \ldots, a_n\}$ and B is countably infinite, then

$$A \times B = (\{a_1\} \times B) \cup \cdots \cup (\{a_n\} \times B).$$

Since $\{a_i\} \times B \approx B$ [via the bijection with action $(a_i, b) \mapsto b$], each set $\{a_i\} \times B$ is countably infinite, so invocation of 4.38 completes the proof. Finally, suppose A and B are both countably infinite; then there are bijections $f: A \to \mathbb{N}$ and $g: B \to \mathbb{N}$. It follows that the mapping $A \times B \to \mathbb{N} \times \mathbb{N}$ given by

$$(a, b) \mapsto (f(a), g(b))$$

is also a bijection (check this), so $A \times B \approx \mathbb{N} \times \mathbb{N}$. But $\mathbb{N} \times \mathbb{N}$ is countable, and therefore $A \times B$ is countable also. $\square$

NOTATION.   The symbol $\aleph_0$ is often used to denote the cardinality of a countably infinite set. ($\aleph$ is *aleph*, the first letter of the Hebrew alphabet; and $\aleph_0$ is read "aleph naught.") That is, the statement "A is a countable set" is symbolically represented by

$$\#A = \aleph_0.$$

With this notation, the infinite cases of Theorems 4.37 and 4.39(b) become this:

$$\text{If } \#A = \#B = \aleph_0, \text{ then } \#(A \cup B) = \aleph_0 \text{ and } \#(A \times B) = \aleph_0.$$

(Compare this with the properties of *finite* sets in 4.16 and 4.17!)

Before you read the next theorem, draw two copies of the real line, mark off the positive integers on one of them and the rational numbers on the other (the rationals will constitute a "fuzz" seeming to fill up most of the line), and consider which set is larger. Meditation on this theme will add to your appreciation of the theorem, which tells us that there is actually a one-to-one correspondence between the two sets. (Also see Example 3.25 for some related discussion.)

4.40 THEOREM.   The set $\mathbb{Q}$ of rational numbers is countable.

PROOF.   First write $\mathbb{Q} = \mathbb{Q}^+ \cup \{0\} \cup \mathbb{Q}^-$, where $\mathbb{Q}^+$ and $\mathbb{Q}^-$ denote the sets of positive and negative rationals, respectively. Clearly $\mathbb{Q}^+ \approx \mathbb{Q}^-$ (the mapping $x \mapsto -x$ does the job), so by 4.38 it will suffice to show that $\mathbb{Q}^+$ is countable.

Every positive rational number can be written in exactly one way in the form a/b, where a and b are positive integers with no common prime factors. (We say the number is "in lowest terms.") With this notation, we define $f: \mathbb{Q}^+ \to \mathbb{N} \times \mathbb{N}$ by $a/b \mapsto (a, b)$. This function is injective, so $\mathbb{Q}^+ \approx f(\mathbb{Q}^+)$. By Theorems 4.39(a) and 4.36(a) we know that $f(\mathbb{Q}^+)$ is countable, and therefore so is $\mathbb{Q}^+$. $\square$

An explicit listing of $\mathbb{Q}$ can be carried out as follows. First imagine $\mathbb{Q}^+$ displayed in a rectangular array as shown here, omitting fractions that are not in lowest terms (so each rational number appears only once).

$$
\begin{array}{cccccccc}
\frac{1}{1} & \frac{1}{2} & \frac{1}{3} & \frac{1}{4} & \frac{1}{5} & \frac{1}{6} & \frac{1}{7} & \frac{1}{8} & \cdots \\
\frac{2}{1} & & \frac{2}{3} & & \frac{2}{5} & & \frac{2}{7} & & \cdots \\
\frac{3}{1} & \frac{3}{2} & & \frac{3}{4} & \frac{3}{5} & & \frac{3}{7} & \frac{3}{8} & \cdots \\
\frac{4}{1} & & \frac{4}{3} & & \frac{4}{5} & & \frac{4}{7} & & \cdots \\
\frac{5}{1} & \frac{5}{2} & \frac{5}{3} & \frac{5}{4} & & \frac{5}{6} & \frac{5}{7} & \frac{5}{8} & \cdots \\
\vdots & \vdots & \vdots & \vdots & \vdots & \vdots & \vdots & \vdots &
\end{array}
$$

Then $\mathbb{Q}^+$ can be listed by moving down the diagonals in succession. Now take a similar list for $\mathbb{Q}^-$ (just negate each entry in the listing of $\mathbb{Q}^+$), merge the two lists by alternately choosing from one list and then from the other, and precede the whole business with zero. This produces a listing of $\mathbb{Q}$:

$$0, 1, -1, \tfrac{1}{2}, -\tfrac{1}{2}, 2, -2, \tfrac{1}{3}, -\tfrac{1}{3}, 3, -3, \tfrac{1}{4}, -\tfrac{1}{4}, \tfrac{2}{3}, -\tfrac{2}{3}, \tfrac{3}{2}, -\tfrac{3}{2}, \ldots$$

We conclude this section with a proof of the uncountability of the real numbers. The proof is very close to the proof of 3.19, in that Cantor's *diagonalization method* again provides the key. We will also need to accept (without proof) the following fact about real numbers:

Every nonzero real number has a unique representation as a nonterminating decimal. (If a number has a decimal representation of finite length, reduce the last nonzero digit by 1 and append an infinite string of 9s. For example, $2.38 = 2.37999. \ldots$)

4.41 THEOREM. The set $\mathbb{R}$ of real numbers is uncountable.

PROOF. It suffices to show that the open unit interval

$$I = \{x \in \mathbb{N} \mid 0 < x < 1\}$$

is uncountable, since we saw earlier (in 4.6) that $\mathbb{R} \approx I$; and we will achieve that by showing that there is no surjection $\mathbb{N} \to I$. (Otherwise stated: every list of numbers from I is incomplete.)

Let $f \colon \mathbb{N} \to I$ be any function. Then f determines a list of nonterminating decimals:

$$
\begin{aligned}
f(1) &= .a_{11} a_{12} a_{13} a_{14} \cdots \\
f(2) &= .a_{21} a_{22} a_{23} a_{24} \cdots \\
f(3) &= .a_{31} a_{32} a_{33} a_{34} \cdots \\
&\ \ \vdots
\end{aligned}
$$

Here a_{ij} is the jth decimal digit of $f(i)$. Define a number $m = .m_1 m_2 m_3. \ldots$ as follows:

$$
m_i = \begin{cases} 2 & \text{if } a_{ii} = 1, \\ 1 & \text{if } a_{ii} \neq 1. \end{cases}
$$

Then $m \in I$; and for each $i \in \mathbb{N}$ we notice that $f(i) \neq m$, since $f(i)$ and m

differ in the ith decimal place. (That is, the decimal expansion of m nowhere intersects the diagonal of the array.) Therefore $m \notin f(\mathbb{N})$; that is, f is not surjective. □

4.42 COROLLARY.   The set of irrational numbers is uncountable.

PROOF.   Let S denote the set of irrational numbers; then $\mathbb{R} = \mathbb{Q} \cup S$. If S were countable then $\mathbb{R}$ would also be countable, in contradiction to Theorem 4.41. □

4.43 EXAMPLE.   Since $\mathbb{Q}$ is countable, the set $\mathbb{Q} \times \mathbb{Q}$, consisting of all the points in the xy-plane that have rational coordinates, is also countable. We will use this fact to sketch a proof of the following remarkable geometric fact:

> If A and B are distinct points in the xy-plane and not in $\mathbb{Q} \times \mathbb{Q}$, then A and B can be connected by a path that contains no points in $\mathbb{Q} \times \mathbb{Q}$.

To see this, let L denote the perpendicular bisector of the line segment AB. It is easy to believe (and prove) that $L \approx \mathbb{R}$. Now for each point $X \in L$ let P_X denote the path $AX \cup XB$ connecting A to B.

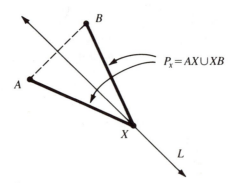

We claim: some path P_X contains no points in $\mathbb{Q} \times \mathbb{Q}$. Suppose, to the contrary, that for each $X \in L$ we have $(\mathbb{Q} \times \mathbb{Q}) \cap P_X \neq \emptyset$. Then we can define a function $f: L \to \mathbb{Q} \times \mathbb{Q}$ by assigning to each $X \in L$ a point in $(\mathbb{Q} \times \mathbb{Q}) \cap P_X$. Notice that if X and Y are different points on L, then A and B are the only points shared by the paths P_X and P_Y. It follows from this that f is injective. But this is impossible, because L is uncountable and $\mathbb{Q} \times \mathbb{Q}$ is countable. □

Exercises

1. Express the set $\mathbb{Z}$ of integers as the union of three countably infinite subsets that are pairwise disjoint.

2. Imagine a hotel with infinitely many rooms. One stormy night a countably infinite crowd of weary travellers arrives seeking shelter. They are also seeking privacy: no two of them will share a room. Upon arriving at the hotel, they learn that the rooms are full. Can the hotel manager accommodate this mass of would-be guests? Explain in detail.

3. This exercise presents a pair of pigeonhole principles for infinite sets.
   (a) Show that if a countably infinite set is partitioned into a finite number of blocks, then at least one of the blocks is countably infinite.
   (b) Repeat (a), but with "countably infinite" replaced by "uncountable."

4. Answer *true* or *false*, and justify: If A is countable and B is a finite subset of A, then $A - B$ is countable.

5. Show that the integers that are 3 more than a multiple of 7 constitute a countable set.

6. Prove Corollary 4.38.

7. (a) Show that the set $\mathbb{N} \times \mathbb{N}$ can be expressed as the union of a countably infinite family of countably infinite sets.
   (b) Use part (a) to prove that a union of countably many pairwise disjoint countably infinite sets is countably infinite.

8. Without attempting to supply all the details, argue convincingly that the union of a countably infinite family of countable sets is countable. Do *not* assume the sets to be pairwise disjoint. (A reasonable solution to this exercise would be to outline a procedure for listing the elements of the union under consideration.)

9. Show that on the real line $\mathbb{R}$, the collection of all closed intervals with rational endpoints is a countable set.

10. Suppose a pairwise disjoint family of intervals on the real line is given. Assume that none of the intervals is just a single point. (For instance, the degenerate interval $[5, 5] = \{5\}$ does not belong to the given family.) A partial picture:

Outline a proof of the fact that this family of intervals is a countable family.

11. (For readers who have had calculus) If you try to mark all the points on the real line that correspond to rational numbers, and thereby draw a "picture" of $\mathbb{Q}$, you will find yourself marking so many points that they will seem to fill up most of the line. (For instance if you were to mark all the points corresponding to rationals with denominator 1,000,000, the line would appear virtually completely marked.) The purpose of this exercise is to show that this impression is dramatically false.

   Since $\mathbb{Q}$ is countable, its members can be listed:

$$\mathbb{Q} = \{q_1, q_2, q_3, \ldots \}.$$

   (a) Surround q_1 by an interval I_1 of length 1/2. (The precise location of I_1 is not important.) Then surround q_2 by an interval I_2 of length $1/2^2 = 1/4$. Continue in this manner, covering each q_n by an interval I_n of length $1/2^n$. (Some of the intervals will overlap, and that is acceptable.) What is the "total" of the lengths of all these intervals, and what does this tell you about the size of $\mathbb{Q}$ as a geometric object?

   (b) Now let ϵ denote a small positive number (the smaller the better), and again cover the rationals by intervals; but this time cover each q_n by an interval of length $\epsilon/2^n$. Again find the "total" of the lengths of all these intervals. *Now* what do you conclude about the size of $\mathbb{Q}$ as a geometric object?

   (c) Why have there been quotation marks around the word "total" in this exercise?

12. (Refer to the result of Exercise 11.) Imagine that you are going to throw a dart at the xy-plane; in this fantasy include the assumption that the tip of the dart will hit exactly one point P. Peer pressure demands that you bet on one of the following three possibilities:

   (a) Neither coordinate of P will be rational.
   (b) Exactly one coordinate of P will be rational.
   (c) Both coordinates of P will be rational.

   What is your best bet? Explain.

13. Prove that in the xy-plane there is a circle centered at the origin that passes through no points whose coordinates are both rational numbers.

4.4 MORE ON INFINITY

The purpose of this brief section is to give another characterization of infinite sets. History tells us that infinite sets were the source of considerable confusion and anxiety among nineteenth century mathematicians. Perhaps the reader has experienced a similar response to the subject. Why should this be? Probably because we grow up counting finite sets, and when we leave that domain we are at a loss for where to begin. Our definition of an infinite set earlier in this chapter reflected that despair by saying, in effect: A set is infinite if no natural number has the "strength" to measure its size.

Now let's seek a more positive description of infinity. We have seen that a finite set cannot be put into one-to-one correspondence with a proper subset. On the other hand, we have exhibited examples of infinite sets for which the opposite is true, and it turns out that this is exactly the property needed to characterize infinite sets. We remark that the following theorem's statement was used as the *definition* of an infinite set by R. Dedekind and C. S. Pierce in the 1880s.

4.44 THEOREM. Let S be a set. Then

$$S \text{ is infinite } \Leftrightarrow S \approx S' \text{ for some } S' \subset S.$$

PROOF. First assume $S \approx S'$ for some $S' \subset S$; we must show S is infinite. If S were finite we would have

$$\mathbb{N}_n \approx S \approx S' \approx \mathbb{N}_m$$

for some $m < n$. But this is impossible, so S must be infinite.

Conversely, assume S is infinite. Then by Theorem 4.36(b), S contains a countably infinite subset $C = \{c_1, c_2, c_3, \ldots \}$. Define $f: S \rightarrow S$ by

$$f(c_i) = c_{2i} \quad \text{for all } c_i \in C,$$
$$f(x) = x \quad \text{for } x \notin C.$$

Then f is a bijection of S onto the proper subset

$$S' = S - \{c_1, c_3, c_5, \ldots \}. \qquad \square$$

4.5   LANGUAGES AND FINITE AUTOMATA

Languages are used by people and computers, and perhaps also by porpoises and other creatures. Computer programs, solutions of mathematical problems, and proofs of theorems are finite lists of statements in a language; and statements are suitably designed finite lists of symbols. If we want to understand what kinds of problems are solvable, and perhaps have a means of measuring the complexity of those problems before seeking a solution, we had better have a clear understanding of the nature and limitations of the language in which we are working. This is a matter of great importance in mathematics and computer science, as well as in psychology and linguistics.

Mathematical modelling of real-world phenomena is often an enormously difficult task. We begin by extracting some essential features of the subject under study and ignoring others. Thus, for example, we may find ourselves studying feathers falling in a vacuum, apples twirling from weightless strings, and balls rolling over frictionless surfaces. Similarly, a simplest possible mathematical definition of "language" must ignore matters of pronounciation, style, grammar, and nearly everything else that makes a language interesting. What's left? Just this: a language is a collection of strings of symbols from some alphabet. The purpose of this section is to formulate this idea in the mathematical framework required by theoretical computer science and to have a brief look at a procedure for specifying languages by means of *finite automata*.

4.45 DEFINITION.   An **alphabet** Σ is a finite nonempty set.

4.46 EXAMPLES. $\{\emptyset\}$, $\{a, b, c, \ldots, z\}$, and $\{0, 1\}$ are alphabets.

4.47 DEFINITION.   Let n be a nonnegative integer. A **word** or **string** w of **length** n over the alphabet Σ is a function $w: \mathbb{N}_n \to \Sigma$.

If w is given by $i \xmapsto{w} a_i$, for $1 \leq i \leq n$, then w is usually written as a string of symbols:

$$w = a_1 \ldots a_n.$$

On the other hand, if $n = 0$ then $w = \emptyset$. (To see this, recall that $\mathbb{N}_0 = \emptyset$ and review Lemma 4.25.) In this case w is called the **empty string**, and we will denote it by ϵ (epsilon). Our intuition is that ϵ denotes a string of *no* symbols. Thus, if we are waiting for printed output from a computer but the printer remains silent, instead of saying "no output" we can say "output ϵ."

4.48 EXAMPLES. Words of length three over the alphabets in 4.46 are, respectively,

$$\text{⚫⚫⚫},\qquad \text{zap},\qquad 001.$$

The empty string is the only string of length zero (no matter what alphabet is in use).

The set of all words over Σ, including the empty word, is denoted Σ^* (pronounced "sigma star"). If $w \in \Sigma^*$, the symbol $|w|$ denotes the length of w.

4.49 THEOREM. If Σ is an alphabet then Σ^* is countably infinite.

PROOF OUTLINE. Let $\Sigma = \{s_1, \ldots, s_t\}$. Then Σ^* contains the countably infinite subset $\{s_1, s_1 s_1, s_1 s_1 s_1, \ldots\}$, and therefore $\#(\Sigma^*) \geq \aleph_0$. Now we can give a method for listing Σ^*. Just as our English alphabet comes equipped with an ordering (*a* precedes *b*, and *b* precedes *c*, and so on), we can impose an ordering on Σ by declaring that s_1 precedes s_2, and s_2 precedes s_3, and so on. Call this the **lexicographic ordering** of Σ. We now list Σ^* as follows: we start with ϵ, then list words of length 1 (think of these as the symbols in Σ) in lexicographic order, then the words of length 2 in lexicographic order, then the words of length 3, and so on. This produces the list

$$\epsilon, \underbrace{s_1, s_2, \ldots, s_t,}_{\text{length 1}} \underbrace{s_1 s_1, s_1 s_2, \ldots, s_1 s_t, \ldots, s_t s_1, s_t s_2, \ldots, s_t s_t,}_{\text{length 2}}$$

$$\underbrace{s_1 s_1 s_1, s_1 s_1 s_2, \ldots, s_t s_t s_t,}_{\text{length 3}} \ldots$$

If the words are read in this order, then any given word in Σ^* will eventually be reached. Therefore Σ^* is countable. $\square$

4.50 DEFINITION. Let Σ be an alphabet. A subset $L \subseteq \Sigma^*$ is called a **language** over Σ.

4.51 THEOREM.   Let Σ be any alphabet. Then there are uncountably many languages over Σ.

PROOF.   The set of all languages over Σ is just the power set $P(\Sigma^*)$. By Cantor's theorem (4.31) we know that $\#(\Sigma^*) < \#P(\Sigma^*)$. But $\#(\Sigma^*) = \aleph_0$ by 4.49, and this completes the proof.   □

How can we describe a language? There is a variety of ways to do this, but each of these ways can be transcribed into a succession of statements made up of symbols from the world's typewriters; and each such succession of statements can be viewed as a word (perhaps a very long word) over this finite set T of typewriter symbols, that is, as an element of T^*. But T^* is countable. Thus there are only countably many possible descriptions of languages and uncountably many languages. The philosophical conclusion: *Not all languages are describable.* Therefore the best we can do is to describe *some* languages, with the hope that we can describe some useful ones. This is a central topic in computer science (and linguistics). For instance, if we are to design a computer that acts in accordance with the instructions we provide it, we must have a precise way of specifying the language of those instructions.

Is the expression

(∗)   *Celui fromage de la parce que maintenant*

a legitimate French sentence? If the reader knows no French, then an approach to this question might start by determining that the alphabet of the expression is the correct one, that the words are listed in a reputable French dictionary, and that they are organized in accordance with the rules of French grammar. (This last step may be no picnic.) There is a simpler way to investigate (∗): take it to someone fluent in French, and ask whether the expression is acceptable.

Similarly, given a language L, suppose that a computer can be constructed that will accept a statement if and only if that statement is in L. Then a symbol string alleged to be in L can be tested by putting it into the machine and seeing whether it is accepted. Conversely, suppose we take a computer and tinker with it (perhaps by modifying its hardware in a haphazard way, or by infecting its memory with a bizarre program), in effect obtaining a new machine, M; then we can *define* a language $L(M)$ to be the set of all strings that M will accept.

Our approach to languages in this section will be that of the previous paragraph, emphasizing the linkage between languages and machines; except instead of an actual computer we will use a primitive mathematical *model* of a computer, designed to "read" strings over a given alphabet Σ and "accept" them or not according to whether they satisfy certain criteria. (The quotation marks reflect our initial discomfort in attributing human activity to a mathematical construction; from now on they will be omitted.)

Let's write down the essential features of a machine designed for the purpose of language recognition. Picture yourself sitting in front of a computer. There is a keyboard filled with symbols; we call that collection of symbols the *alphabet* Σ. The overall condition of the machine's components at a particular time (including, for example, the configuration of data in the memory and the display on the monitor) is called the *state* of the machine at that time; thus the machine has a finite set S of states. When the machine is turned on and prepared to receive its first piece of input, we say that the machine is in its *initial state* $q_0 \in S$. In general, typing in a symbol from Σ may change the state (perhaps modifying what is in the memory, or altering the display on the monitor), depending on what is typed and the state just before typing. This change occurs in a deterministic way; that is, for every state $q \in S$ and symbol $\sigma \in \Sigma$, the ordered pair (q, σ) determines a unique new state, which we will denote by $\delta(q, \sigma)$. (We know the machine is not working properly if an ordered pair (q, σ) determines different states at different times.) For the purpose of language recognition, there must be a subset F of S such that whenever the machine is in a state in F, we can say that the machine has *accepted* the data we have just entered.

Now we can give the formal definition that ties all this together into a single mathematical package. Notice that the definition is unencumbered by any connection with physical reality and depends only on the basic notions of set theory. (We need no electricity, no disks, and no incomprehensible 750-page manual.)

4.52 DEFINITION. A **finite automaton** (or **finite-state machine**) is a five-tuple

$$M = (S, \Sigma, \delta, q_0, F),$$

where

S is a finite nonempty set (the set of **states**);
Σ is a finite nonempty set (the **alphabet**);
δ is a function from $S \times \Sigma$ to S (the **transition function**);
$q_0 \in S$ (the **initial state**);
F is a subset of S (the set of **final states**).

4.53 EXAMPLES.

(a) Consider an automaton M that reads the alphabet $\Sigma = \{a, b\}$, and suppose there are two states: $S = \{q_0, q_1\}$. Let q_0 be the only final state (that is, $F = \{q_0\}$), as well as the initial state. Assume that reading a always causes a state change [more formally: $\forall q \in S$, $\delta(q, a) \neq q$], while reading b has no effect. The transition function δ corresponding to this response pattern can be represented by a table, which follows. The rows are labelled (on the left) by the states of S, and the columns (on top) by the symbols from Σ. The state appearing at the intersection of row q and column σ is $\delta(q, \sigma)$.

σ q	a	b
q_0	q_1	q_0
q_1	q_0	q_1

(b) David wakes up in the morning and checks the weather. If it is sunny he goes fishing; if it is rainy he goes to the movies. He is happy if he sees a good movie or if he finds that the fish are biting; otherwise he is sad. Not every input condition will change David's state; for example, if he has gone to the movies or if he has just awakened then he is unaffected by the activities of the fish. Once David's mood is set, it doesn't change. His criterion for "accepting" the day is that the events make him happy. We represent David's situation by the automaton $M = (S, \Sigma, \delta, q_0, F)$, where

    $S = \{\text{wakeup, fishing, movies, happy, sad}\}$,
    $\Sigma = \{\text{sunny, rainy, biting, not biting, good movie, poor movie}\}$,
    $q_0 = \text{wakeup}$,
    $F = \{\text{happy}\}$,

and the transition function δ is given by the following table:

q \ σ	sunny	rainy	biting	not biting	good movie	poor movie
wakeup	fishing	movies	wakeup	wakeup	wakeup	wakeup
fishing	fishing	fishing	happy	sad	fishing	fishing
movies	movies	movies	movies	movies	happy	sad
happy	happy	happy	happy	happy	happy	happy
sad	sad	sad	sad	sad	sad	sad

Remember: an alphabet can be any finite nonempty set. So, for instance, "biting" is a symbol in our alphabet in this example.

A finite automaton in one of its states.

It is hard to get an intuitive feeling for the operation of an automaton M by looking at the table of its transition function. It is usually more insightful to represent M by a graph called a **transition diagram** or **state graph**. The graph is a collection of points (called **vertices** or **nodes**), one for each state in S, together with a set of *directed* arcs (called **edges**), each labelled by a symbol from Σ. If q_1 and q_2 are states of M and $\sigma \in \Sigma$, then an edge

in the diagram indicates that M goes from state q_1 to state q_2 upon reading σ; that is, $\delta(q_1, \sigma) = q_2$. The vertex of each final state is circled in the diagram. For example, the automaton in 4.53(a) has the following state graph:

(4.54)

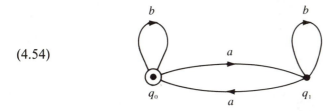

And here is a state graph for 4.53(b):

(4.55)

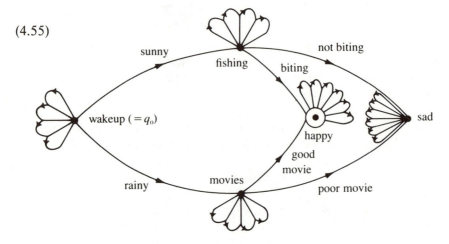

To avoid congestion in this diagram, we have omitted the labels on edges corresponding to inputs that don't cause a state change. (These are the unmarked loops at each vertex.)

Up to now our automata have been able to read only one-symbol words, and it is time to correct this reading disability. (We are viewing the elements of Σ as one-symbol words, so that $\Sigma \subseteq \Sigma^*$.) More formally, we want to extend the transition function δ to a function

$$S \times \Sigma^* \to S.$$

Let $w \in \Sigma^*$; say, $w = \sigma_1 \ldots \sigma_n$. How should we teach M to read w when M is in state q? The answer: First read σ_1, which takes M into state $\delta(q, \sigma_1)$. Then read σ_2, which produces state $\delta(\delta(q, \sigma_1), \sigma_2)$; call the resulting state $\delta(q, \sigma_1 \sigma_2)$. In general, having defined $\delta(q, \sigma_1 \ldots \sigma_k)$, define

(4.56) $\delta(q, \sigma_1 \ldots \sigma_k \sigma_{k+1}) = \delta(\delta(q, \sigma_1 \ldots \sigma_k), \sigma_{k+1}).$

This recursion tells M how to read every string of length 1 or greater, starting in any state q. Finally, define

$$\delta(q, \epsilon) = q, \qquad \forall q \in S.$$

(That is, the state of M doesn't change if we give it no input.) This completes the extension of our original transition function δ to a new function, also denoted δ, from $S \times \Sigma^*$ to S; so *we have taught M to read any* *element of* Σ*. From now on, the symbol δ will always denote the extended transition function that we have defined here.

Now that each of our automata can read any word over its associated alphabet, we can endow them with the ability to accept some of what they read.

4.57 DEFINITION.   Let $M = (S, \Sigma, \delta, q_0, F)$ be a finite automaton, and let $w \in \Sigma^*$. We say that M **accepts** w if $\delta(q_0, w) \in F$. (In words: M accepts w if M goes from its initial state to some final state upon reading w.) Define the **language accepted by** M to be the language

$$L(M) = \{w \in \Sigma^* \mid M \text{ accepts } w\}.$$

4.58 EXAMPLES.

(a) Let M be the automaton represented by diagram 4.54. Then $L(M)$ is the set of all words over the alphabet $\{a, b\}$ that have an even number of a's.

(b) With M as represented by 4.55, an element of $L(M)$ is a string over Σ satisfying one of the following sets of conditions:

1. There is an occurrence of "sunny" that has no occurrence of "rainy" anywhere preceding it; and "biting" occurs somewhere after that "sunny," without "not biting" in between.

2. There is an occurrence of "rainy" that has no occurrence of "sunny" anywhere preceding it; and "good movie" occurs somewhere after that "rainy," without "poor movie" in between.

(c) Some automata are completely tasteless, and others are impossible to please. To be precise, let $M = (S, \Sigma, \delta, q_0, F)$. If $F = S$ then $L(M) = \Sigma^*$; if $F = \emptyset$ then $L(M) = \emptyset$.

(d) Let $\Sigma = \{a, b\}$, and let M have the following platypus-like transition diagram:

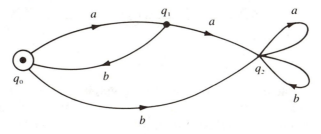

Then $L(M)$ consists of the empty string together with all strings of even length whose symbols alternate between a and b, starting with a and ending with b. Thus,

$$ababab. \; . \; .ab \in L(M).$$

(e) Now let's construct an automaton M that excludes the empty string but otherwise accepts the same language as the one in (d). We can insist on an initial "syllable" ab by making the following modification of the preceding automaton:

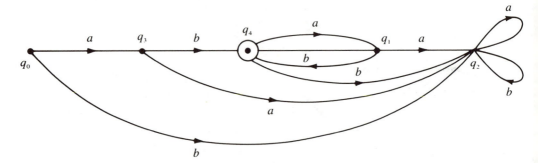

(f) We seek an automaton M such that

$$L(M) = \{w \in \{a, b\}^* \mid \text{no two consecutive letters in } w \text{ are the same}\}.$$

The following diagram gives an appropriate M:

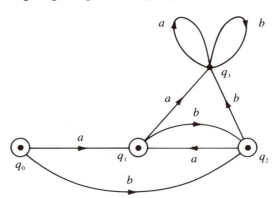

Notice that in every completed state graph, each vertex has exactly one edge emanating from it corresponding to each alphabet symbol. For if *no* edge from q were marked with σ, then $\delta(q, \sigma)$ would not be defined by the graph; and if *two* edges from q were marked with σ, then δ would not be well-defined. (That is, $\delta(q, \sigma)$ would have two different values, which is taboo because δ is a function.)

We are likely to find that networks of long and multi-claused sentences result from an attempt to describe a complex language verbally. And if such a description is all we have for a language, it can be a real challenge to check a particular string to see whether it satisfies the criteria for membership. But once we know a transition diagram for the language, it is a triviality to test a particular string's membership credentials.

   4.59 DEFINITION.   A language L is **regular** if $L = L(M)$ for some finite automaton M.

The preceding examples display several regular languages, without formally proving regularity. *Proving* that a particular language is or is not regular can be a difficult matter, and we will only touch briefly on the subject here. It can be shown that every finite language is regular. (We omit the details, but to get the flavor of the argument try this exercise: draw the graph of a finite automaton M with alphabet $\Sigma = \{a, b, c\}$, having the property that $L(M) = \{cab\}$. Then do Exercise 5 at the end of this section.)

The situation for infinite languages is considerably more complicated. The *Pumping Lemma,* which we are about to examine, is an important tool for proving that certain languages are not regular. First we need some notation and terminology. If Σ is an alphabet and x and y are words in Σ^*, then the **concatenation** of x and y, denoted xy, is the word obtained upon following x by y. Thus if $x =$ cat and $y =$ nips, then $xy =$ catnips. Concatenation of a word with itself is usually abbreviated with exponential notation; so with x and y as just given, the concatenation denoted x^2y^3x is shorthand for

   catcatnipsnipsnipscat.

   4.60 PUMPING LEMMA.   Let Σ be an alphabet, and let $L \subseteq \Sigma^*$ be an infinite regular language. Then there are strings x, y, z in Σ^*, with $y \neq \epsilon$, such that for every $n \in \mathbb{N}$ the string xy^nz belongs to L.

   PROOF OUTLINE.   Because L is regular, there is a finite automaton M such that $L = L(M)$, and let's suppose that M has exactly k states. Since L is infinite, there is a word $w \in L$ of length greater than k. Now, when M reads the symbols of w, it shifts from state to state; and, since w has more letters than M has states, there must be some state q that M enters twice during the reading (by the pigeonhole principle).

Thus there are strings x, y, z, with $y \neq \epsilon$, such that $w = xyz$ (so the strings x, y, z can be viewed as "syllables" of w) and such that M is in state q before and after reading y. Then M also accepts the longer word $xyyz$; for, after reading the middle section yy, the machine is in state q, and then reading z takes M into an accepting state. (We know this from the fact that M accepts w.) More generally, an induction argument shows that M accepts every word of the form xy^nz for $n \in \mathbb{N}$. Therefore every such word belongs to L (since L consists of precisely those words accepted by M), and the argument is complete.   □

4.61 EXAMPLES.

(a) Consider the language $L \subseteq \{a, b\}^*$ consisting of all strings formed by following a string of a's with a string of b's of the same length:

$$L = \{a^n b^n \mid n > 0\}.$$

We assert that L is not regular. If we suppose otherwise, then by the Pumping Lemma there exist strings x, y, $z \in \{a, b\}^*$, with $y \neq \epsilon$, such that $xy^nz \in L$ for all $n > 0$. But y cannot be a string of a's, since otherwise when $n > 1$ the string xy^nz would have more a's than b's and hence not belong to L. Similarly, y cannot be a string of b's. Finally, if y were to have both a's and b's, then the string xy^2z would not have the form $a^n b^n$, and therefore it could not belong to L. So the assumption that L is regular has led to a contradiction, and the assertion is verified.

In this example our intuition is that no finite automaton has enough memory to keep count of an arbitrarily large number of a's in order to check that each string of a's is followed by a string of b's of the same length. To accept L, a more sophisticated piece of "hardware" called a *pushdown automaton* is used. Roughly speaking, this is an automaton with an expandable memory modelled after the plate-stacking devices used by cafeterias.

(b) Consider the patterns of parentheses used in arithmetic expressions. For instance, the expression

$$((a + b)c + d(ab))f + (g)(ac + b)$$

has the parentheses pattern

$$(()())()()$$

and can be understood; whereas

$$a + (c + d)) + f(g + c(d((\ $$

has the parentheses pattern

$$())(((($$

and leaves us confused. Define L to be the set of acceptable parentheses patterns. (We won't write a formal definition.) So

$$L \subseteq \{), (\}^*.$$

It can be shown that L is not regular; therefore no finite automaton has the capacity to scan *every* arithmetic expression and check whether its parenthesization is legitimate. As in part (a), a pushdown automaton can do the job. Alternatively, we can use an automaton with a countably infinite number of states, diagrammed as follows:

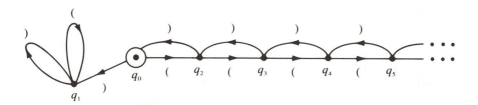

If an upper bound on the expression lengths is given in advance, then a *finite* automaton can be constructed that will accept the legitimate parenthesizations.

In addition to their work as language acceptors, automata also have important applications in other aspects of computer science, particularly in the analysis of algorithms and in the study of the field called *computational complexity*.

Exercises

1. Let $x = a_1 \ldots a_m$ and $y = b_1 \ldots b_n$ be strings over the alphabet Σ. Then the concatenation xy is the string $a_1 \ldots a_m b_1 \ldots b_n$. Give a more formal definition of the concatenation xy as a kind of function, in the manner of Definition 4.47.

2. Describe the language accepted by the automaton with this state diagram:

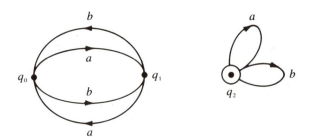

   Explain briefly. Then go on to assert (without proof) some general principle that this example supports.

3. Imagine that the rectangle in this figure

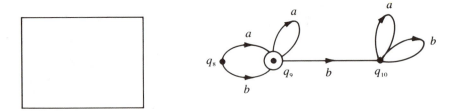

   covers up the transition graph for a finite-state machine M, and assume that q_8, q_9, q_{10} are not states of M. Now consider a new machine N whose transition graph consists of the hidden graph together with the visible graph in the figure. How are the languages $L(M)$ and $L(N)$ related? Explain.

4. Under what condition(s) on a finite automaton M is the empty string accepted by M?

5. Suppose L_1 and L_2 are regular languages over the alphabet $\Sigma = \{a, b\}$; say, $L_1 = L(M_1)$ and $L_2 = L(M_2)$, where M_1 and M_2 are finite automata. Assume that every string in L_1 starts with a, and every string in L_2 starts with b. Without proof, but with some explanation, give a way to use the transition diagrams of M_1 and M_2 to construct a transition diagram for an automaton M such that

$$L(M) = L_1 \cup L_2.$$

6. In each case, describe the language accepted by the automaton with the given transition diagram.

(a)

(b)

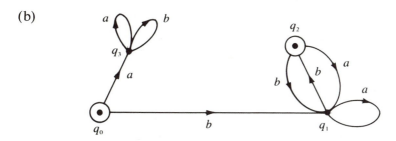

7. In each of the following parts, give a state graph for a finite automaton that accepts the described language in $\{a, b\}^*$.

   (a) The set of all words that start with a (as we read from left to right).
   (b) The set of all words that end with b.
   (c) The set of all words that have exactly one b.
   (d) The set of all words that have exactly two b's.
   (e) All the words with an even number of a's.
   (f) All the words in which three consecutive b's don't occur.

8. In the proof of Theorem 4.49, we gave a procedure for listing the elements of Σ^* based on a given ordering for Σ. The resulting ordering of Σ^* is called the *lexicographic ordering* of Σ^*. For the purpose of this exercise, think of Σ as our standard alphabet:

$$\Sigma = \{a, b, c, \ldots, z\}.$$

   (a) The words in a dictionary are listed in what we will call *alphabetic order*. Explain the differences between alphabetic order and lexicographic order as the terms are being used here.
   (b) Could the proof of Theorem 4.49 have been carried out by listing the words in alphabetic order?

CHAPTER **5**

Permutations
and Combinations

Let me count the wheys.
Little Miss Muffet

5.1 COMBINATORIAL PROBLEMS

Imagine that you are gathering data for a major report on the government's economic policies. An influential friend arranges for your access to the most up-to-date files in the Bureau of the Census. Because the report is due in a few days, you are hoping that these files will give you the information you need to complete your report. Unfortunately, the files contain raw data that has not been organized or cross-referenced in any way; there is no way you can extract enough coherent information in the available time. You lose your job. Moral: *having* a set is not enough for most purposes; we must understand its *structure*.

In this chapter our focus will be on topics from the branch of mathematics known as *combinatorics,* a subject concerned with counting, selection, and arrangement. These are processes by which we organize sets so that we can interpret and apply the data they contain. Generally speaking, combinatorial questions ask whether a subset of a given set can be chosen and arranged in a way that conforms with certain constraints and, if so, in how many ways it can be done. As we proceed we will extend and refine the counting procedures introduced in Chapter 4, but our concern here will be almost entirely with finite sets. As we have seen, counting the elements of a set can be a difficult task. While we can all count a tubful of marbles, given the time and the will, it is another matter to count a set that has not presented itself to our senses.

We have already discussed some combinatorial topics: we counted the subsets of finite sets in Chapter 2; we chose systems of distinct representatives in Chapter 3; and we considered the product rule and countability questions in Chapter 4. (Showing that a set is countable can amount to giving a procedure for arranging the elements of that set in a line.) Combinatorics also includes the theory of permutations and combinations and probability theory. These topics have an enormous range of applications in pure and applied mathematics and computer science. For example, combinatorial methods can be used to estimate the time and storage requirements of complex computer programs. Combinatorics also deals with questions about *configurations* of sets: families of finite sets that overlap according to some prescribed numerical or geometrical conditions.

Here are some representative combinatorial problems.

5.1 EXAMPLES.

(a) Ten people are seated in a ten-seat row in a movie theater, one person per seat: say p_1 is in seat 1, p_2 is in seat 2, and so on. Trouble begins when p_1 is offended by p_2 and insists that p_2 move to seat 10. Meanwhile p_3 falls in love with p_8 and demands to sit on p_8's right. A few other such needs are expressed, and after some heated discussion a configuration is agreed upon that will make everyone happy. Someone offers a valid plan whereby the rearrangement can be carried out in 14 steps, where each step consists of two adjacent people getting up and switching seats. Then someone offers a second plan in which only 11 such switches are needed to achieve the same result, but there isn't time to check the details before the movie begins. Should they try the second plan?

(b) A standard checkerboard has eight rows and eight columns of squares, and the squares are alternately black and white. Suppose we remove the two diagonally opposite white corners from the board. Can we cover the remaining squares with dominoes? (Assume that each domino is a rectangle that covers exactly two

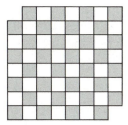

squares, and also assume that dominoes are not allowed to over-lap.) This one is easy, so we give the answer immediately: Each domino will cover a white square and a black square, therefore any distribution of dominoes will cover the same number of squares of both colors. But because we have removed the two white corners, there are more black squares than white ones to be covered, hence we cannot achieve the desired coverage.

(c) Now picture a complete checkerboard, but ignore the colors and view the lines (including those along the outer border) as streets and the little squares as city blocks. How many routes are there from the southwest corner to the northeast corner, assuming that the only allowable directions of travel are east and north?

Once we view the situation from the appropriate perspective, we will use an elementary computation to show that there are 12,870 acceptable routes.

(d) A whiz at Rubik's Cube wants to obtain a certain configuration. Rubik's Cube appears to be constructed by slicing a given cube into 27 little cubes; these are called the **cubies**.† There are three kinds of cubies with at least one visible face: corner cubies, center cubies, and edge cubies. (We can also imagine an invisible cubie in the middle of the Cube, but that won't enter our discussion.)

† We are using the terminology in A. H. Frey, Jr., and D. Singmaster, *Handbook of Cubik Math* (Hillside, N.J.: Enslow, 1982).

The visible faces of the cubies are **facelets**. Question: can the whiz manipulate a Cube, without disassembling it, into a configuration that differs from the original one only in that a single edge cubie has been flipped around?

? ?

(e) Ten joyful music lovers meet on December 16 to celebrate Beethoven's birthday. They raise their glasses in a toast, and each insists upon clinking his or her glass with everyone else's. How many clinks occur? Later in the evening they will select from among them the best possible team of three people to represent them in the National Three-Person Minuet Championships. How many teams must they check?

(f) Here is a famous problem that was posed by W. S. B. Woolhouse in *The Ladies' and Gentlemen's Diary* in 1844 and solved (affirmatively) by Rev. P. T. Kirkman in 1850. It is referred to as "Kirkman's schoolgirls problem," and it is a nice representative of the subject of combinatorial configurations (or *designs*). Suppose a teacher has a class of 15 girls that she wants to take out walking on seven consecutive days. She wants the girls to walk in five rows, with three girls per row. Unfortunately, by the end of a day's walk, the girls in each row no longer get along. Here's the problem: Is it possible to reorganize the girls every day so that no two girls share the same row more than once?

Among the preceding examples, (a) has a quick solution when formulated in the language of *permutations*. The solution to example (d) also follows from basic properties of permutations, although the formulation of it has some subtleties. Examples (c) and (e) also have quick solutions, but they use *combinations,* not permutations. We'll see solutions to all of these; but Kirkman's schoolgirl problem, (f), is best handled as part of a general study of combinatorial designs, which we leave for more advanced courses.†

† For a solution to the problem, see Richard A. Brualdi, *Introductory Combinatorics* (New York: Elsevier, 1977), especially p. 219.

5.2 THE ADDITION AND PRODUCT RULES (review)

We will be making a great many choices in this chapter, so we will restate the two fundamental rules of selection that were developed in Chapter 4. The first is a rewording of 4.15; the second is a verbatim copy of 4.22.

ADDITION RULE. If an element is to be chosen from one of n pairwise disjoint finite sets that contain $m_1, \ldots, m_n$ elements, respectively, then there are altogether $m_1 + \cdots + m_n$ possibilities for that choice.

PRODUCT RULE. Suppose that n consecutive choices are to be made. Suppose further that there are m_1 possibilities for the first choice and, for each k satisfying $2 \leqslant k \leqslant n$, that there are m_k possibilities for the kth choice, no matter how the first $k - 1$ choices have been made. Then there are exactly $m_1 m_2 \cdots m_n$ possibilities for the sequence of n choices.

Before moving into new material, it may be helpful to see these rules in action again, so here is an example for that purpose.

5.2 EXAMPLE. Determine the number of odd three-digit integers that have no repeated digits.

SOLUTION 1. The idea is to represent the selection of such an integer as a succession of choices, each from a set whose size is independent of the preceding choices. Then we can invoke the product rule. We do this in three steps. First we choose the units digit: there are five possibilities, since it must be odd. Next we choose the hundreds digit: there are eight possibilities, because it must be nonzero and different from the units digit. Last we choose the tens digit: there are eight possibilities, because it must not be one of the two digits already chosen. Therefore the answer, by the product rule, is $5 \cdot 8 \cdot 8 = 320$.

SOLUTION 2. We proceed in a way similar to Solution 1, but we first choose the units digit, then the tens digit, then the hundreds digit. As before, there are five possibilities for the units digit. After that there are nine possibilities for the tens digit. Finally, how many possibilities are left for the hundreds digit? That depends on whether the tens digit was zero, so the conditions are not appropriate for application of the product rule. We resolve this difficulty by separating the selections into two families: those numbers with zero for the tens digit, and those with a nonzero tens digit. In the first family there are five possible units digits, then one choice for tens digit (zero), and eight possible hundreds digits; in all, $5 \cdot 1 \cdot 8 = 40$ possibilities, by the product rule. In the second family there are five possible units digits, then eight choices for tens digits (zero and the units digit are unavailable), and seven possible hundreds digits; in all, $5 \cdot 8 \cdot 7 = 280$ possibilities. Since the two families are disjoint, the addition rule yields a grand total of $40 + 280 = 320$ possibilities.

OTHER PROCEDURES. If we choose the digits from left to right, the number of choices available for the units digit will depend on the *parities* (the oddness or evenness) of the first two choices. Once again it is necessary to separate the entire collection into pairwise disjoint batches. The details are left to the reader.

The crudest way to proceed is to list all the possibilities,

$$103, 105, 107, 109, 123, 125, 127, 129, 135, 137, 139, \ldots, 987,$$

and count them.

Exercises

1. An **n-bit string** is an n-tuple of 0s and 1s. Prove that for each $n \geqslant 1$, exactly half the n-bit strings have an even number of 1s. (Use induction on n.)

2. In each of the following cases, determine the number of four-digit integers that satisfy the given condition.
   (a) Unrestricted.
   (b) The digits are all even.
   (c) There are no repeated digits.
   (d) No two consecutive digits are equal.
   (e) The integer is even and no two consecutive digits are equal.

3. The number $n = 15,246,000$ has the prime factorization

$$2^4 \cdot 3^2 \cdot 5^3 \cdot 7 \cdot 11^2.$$

   (a) How many positive divisors does n have? (Use the Fundamental Theorem of Arithmetic.)
   (b) How many positive divisors does n have that are not divisible by 10?

4. A host offers his party guests five varieties of crackers, six cheeses, and three jams. Regrettably one of the cheeses (a great delicacy) has the

property that when combined with two or more jams, it produces a lethal result. How many nonlethal snacks are possible? [Assume that a snack consists of a cracker with at most one cheese and any combination of jams (possibly none).]

5. Twelve horses are in a race. The only results that matter are the first three finishers. How many possibilities are there?

5.3 INTRODUCTION TO PERMUTATIONS

5.3 DEFINITION. Let A be a nonempty set. A **permutation** of A is a bijection from A to A.

5.4 EXAMPLE. Let $A = \{1, 2, 3, 4\}$. Then the mapping $\sigma: A \to A$ given by the table

i	1	2	3	4
$\sigma(i)$	4	1	3	2

is a permutation of A.

If $A = \{a_1, \ldots, a_n\}$, then we usually write

$$(5.5) \qquad \sigma = \begin{pmatrix} a_1 & a_2 & \cdots & a_n \\ \sigma(a_1) & \sigma(a_2) & \cdots & \sigma(a_n) \end{pmatrix}$$

for the permutation $\sigma: A \to A$ that takes a_i to $\sigma(a_i)$, for $1 \leq i \leq n$. For instance, in this notation the permutation σ in Example 5.4 is written

$$\begin{pmatrix} 1 & 2 & 3 & 4 \\ 4 & 1 & 3 & 2 \end{pmatrix}.$$

Here we have followed the standard custom of listing the domain elements in ascending order in the top row. This is not essential; for instance, according to 5.5, the symbol

$$\begin{pmatrix} 3 & 1 & 4 & 2 \\ 3 & 4 & 2 & 1 \end{pmatrix}$$

also represents σ.

The set of all permutations of a set A is called the **symmetric group** on A, denoted S_A. If $A = \mathbb{N}_n = \{1, 2, \ldots, n\}$ we write S_n instead of $S_{\mathbb{N}_n}$. Finally, a set with n elements is called an **n-set**.

5.6 THEOREM.   If A and B are n-sets, then there are exactly $n!$ bijections from A to B. In particular, $\#S_n = n!$.

PROOF 1.   Let $A = \{a_1, \ldots, a_n\}$ and $B = \{b_1, \ldots, b_n\}$. A bijection pairs a_1 with one of the n elements of B. Once we choose that element of B, there remain $n - 1$ elements of B from which we can choose an element to pair with a_2, and then $n - 2$ candidates for pairing with a_3, and so on. Altogether there are $n(n - 1)(n - 2) \cdots 2 \cdot 1 = n!$ possibilities, by the product rule.

The second statement follows from the first by taking $A = B = \mathbb{N}_n$.

PROOF 2.   We use induction on n. (The product rule will not be needed.) If $n = 1$, there is only one bijection (in fact, only one function) from $\{a_1\}$ to $\{b_1\}$, and $1! = 1$. Now assume there are $k!$ bijections from one k-set to another, and suppose $\#A = \#B = k + 1$; say $A = \{a_1, \ldots, a_{k+1}\}$ and $B = \{b_1, \ldots, b_{k+1}\}$. Let J_1 be the set of all bijections from A to B that take a_1 to b_1. Then each $\sigma \in J_1$ is completely determined by the images $\sigma(a_2), \ldots, \sigma(a_{k+1})$, that is, by σ's restriction to $\{a_2, \ldots, a_{k+1}\}$. Because that restriction is a bijection from the k-set $\{a_2, \ldots, a_{k+1}\}$ to the k-set $\{b_2, \ldots, b_{k+1}\}$, the induction hypothesis tells us that there are $k!$ possibilities for it. Thus $\#J_1 = k!$. Similarly, if J_i denotes the set of all bijections from A to B that take a_1 to b_i, with $2 \leqslant i \leqslant k + 1$, then $\#J_i = k!$. Notice that the set of bijections from A to B is equal to $\cup_{i=1}^{k+1} J_i$, a union of pairwise disjoint sets; therefore by Theorem 4.15 the union has $(k + 1) \cdot k! = (k + 1)!$ elements, completing the induction argument.   □

5.7 EXAMPLE.   For every $n \geqslant 1$, the identity mapping on $\mathbb{N}_n$ is an element of S_n. When $n = 1$ there are no other mappings, so

$$S_1 = \left\{ \begin{pmatrix} 1 \\ 1 \end{pmatrix} \right\}.$$

We have

$$S_2 = \left\{ \begin{pmatrix} 1 & 2 \\ 1 & 2 \end{pmatrix}, \begin{pmatrix} 1 & 2 \\ 2 & 1 \end{pmatrix} \right\},$$

and the symmetric group S_3 consists of the following six permutations:

$$\begin{pmatrix} 1 & 2 & 3 \\ 1 & 2 & 3 \end{pmatrix}, \quad \begin{pmatrix} 1 & 2 & 3 \\ 2 & 1 & 3 \end{pmatrix}, \quad \begin{pmatrix} 1 & 2 & 3 \\ 3 & 2 & 1 \end{pmatrix},$$

$$\begin{pmatrix} 1 & 2 & 3 \\ 1 & 3 & 2 \end{pmatrix}, \quad \begin{pmatrix} 1 & 2 & 3 \\ 2 & 3 & 1 \end{pmatrix}, \quad \begin{pmatrix} 1 & 2 & 3 \\ 3 & 1 & 2 \end{pmatrix}.$$

From Theorem 5.6 we know that $\#S_3 = 3! = 6$, so this list is complete. (Without 5.6 it would take careful bookkeeping to assure that all the permutations of $\{1, 2, 3\}$ have been included.)

If σ and λ are in S_A, then so is their composition, $\sigma \circ \lambda$, since the composition of two bijections from A to A is again a bijection from A to A. The permutation $\sigma \circ \lambda$ is usually written $\sigma\lambda$; we call it a **product**, and we say we have **multiplied** σ and λ. (As you might expect, σ and λ are called the **factors** of the product $\sigma\lambda$.) Thus $\sigma\lambda(x) = \sigma(\lambda(x))$ for all $x \in A$. To illustrate, consider these two permutations in S_4:

$$\sigma = \begin{pmatrix} 1 & 2 & 3 & 4 \\ 3 & 1 & 4 & 2 \end{pmatrix} \quad \text{and} \quad \lambda = \begin{pmatrix} 1 & 2 & 3 & 4 \\ 4 & 1 & 3 & 2 \end{pmatrix}.$$

Then

$$\sigma\lambda = \begin{pmatrix} 1 & 2 & 3 & 4 \\ 2 & 3 & 4 & 1 \end{pmatrix}.$$

In detail:

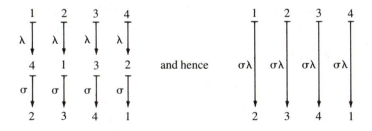

In practice we don't write all these arrows to compute $\sigma\lambda$. We write

$$\sigma\lambda = \underbrace{\begin{pmatrix} 1 & 2 & 3 & 4 \\ 3 & 1 & 4 & 2 \end{pmatrix}}_{\sigma} \underbrace{\begin{pmatrix} 1 & 2 & 3 & 4 \\ 4 & 1 & 3 & 2 \end{pmatrix}}_{\lambda} = \begin{pmatrix} 1 & 2 & 3 & 4 \\ & & & \end{pmatrix}.$$

Then we think, "1 goes to 4 (by means of λ), and then 4 goes to 2 (by means of σ)," and we write 2 in the blank space under the 1. Then we proceed similarly with the other elements. Since a permutation is a bijection, every entry of the top row will appear exactly once in the bottom row after a computation of this kind. For example,

$$\begin{pmatrix} 1 & 2 & 3 & 4 \\ 2 & 1 & 4 & 2 \end{pmatrix}$$

does *not* denote a permutation, since the simultaneous actions $1 \mapsto 2$ and $4 \mapsto 2$ defy injectivity. Also remember: *to compute $\sigma\lambda$, first apply λ, then σ.*

5.8 EXAMPLE. With

$$\sigma = \begin{pmatrix} 1 & 2 & 3 & 4 \\ 3 & 1 & 4 & 2 \end{pmatrix} \quad \text{and} \quad \lambda = \begin{pmatrix} 1 & 2 & 3 & 4 \\ 4 & 1 & 3 & 2 \end{pmatrix}$$

as before, we have

$$\lambda\sigma = \begin{pmatrix} 1 & 2 & 3 & 4 \\ 4 & 1 & 3 & 2 \end{pmatrix}\begin{pmatrix} 1 & 2 & 3 & 4 \\ 3 & 1 & 4 & 2 \end{pmatrix} = \begin{pmatrix} 1 & 2 & 3 & 4 \\ 3 & 4 & 2 & 1 \end{pmatrix} \neq \sigma\lambda.$$

Therefore, *the result of multiplying permutations can be expected to depend on the order in which the permutations are listed.* (We say that permutation multiplication is not a **commutative** operation.) Notice, however, that sometimes the order doesn't matter. For instance,

$$\begin{pmatrix} 1 & 2 & 3 & 4 \\ 4 & 2 & 3 & 1 \end{pmatrix}\begin{pmatrix} 1 & 2 & 3 & 4 \\ 1 & 3 & 2 & 4 \end{pmatrix} = \begin{pmatrix} 1 & 2 & 3 & 4 \\ 4 & 3 & 2 & 1 \end{pmatrix}$$

$$= \begin{pmatrix} 1 & 2 & 3 & 4 \\ 1 & 3 & 2 & 4 \end{pmatrix}\begin{pmatrix} 1 & 2 & 3 & 4 \\ 4 & 2 & 3 & 1 \end{pmatrix}.$$

5.9 REMARK. Some authors write functions to the *right* of the domain elements on which they act: $(x)\sigma$ instead of $\sigma(x)$. With that notation the product $\sigma\lambda$ has the action $(x)\sigma\lambda = ((x)\sigma)\lambda$; that is, *first σ then λ*. If you encounter permutation computations in other books or papers, check to see whether products should be read from right to left, as we do here, or from left to right. (*It matters,* as Example 5.8 shows.) Suggestion: start by locating a couple of worked examples and multiplying them for yourself, and then compare your results with the author's.

An element of S_n is represented by a symbol of the form

$$\begin{pmatrix} 1 & 2 & 3 & \cdots & n \\ \sigma(1) & \sigma(2) & \sigma(3) & \cdots & \sigma(n) \end{pmatrix},$$

and the bottom row of this symbol is just a rearrangement of the top row. Conversely, every arrangement of the integers from 1 to n in a row can be viewed as the bottom row of a permutation symbol. Hence there is a one-to-one correspondence between S_n and the collection of all linear arrangements of the numbers from 1 to n. For this reason, permutations are often described as "ordered arrangements" in the mathematical literature.

Permutations appear in a variety of mathematical contexts and in the modelling of physical phenomena. In general, permutations arise when the essential data in the description of an event can be summarized by observing how n objects shift among n locations, one object per location. If the locations are numbered from 1 to n, a process that moves the object in location i to location j can be associated with a permutation σ that takes i to j.

5.10 EXAMPLE. A major athletic competition has contests in five water sports, six running events, four bicycling events, and seven varieties of self-defense.

    (a) Each participant chooses one event from each of the four categories. In how many ways can the selections be made?

SOLUTION. This is a straightforward application of the product rule: altogether there are $5 \cdot 6 \cdot 4 \cdot 7 = 840$ possibilities.

    (b) Now suppose that besides selecting events, each participant also must specify the order in which he will engage in the events (perhaps bicycling first, then a water sport, and so on). How many possibilities are there for an individual's event program?

SOLUTION. After a participant has selected his four events, he must then specify an ordering of the events. But each such ordering corresponds to an element of S_4, so there are $4! = 24$ possible orderings. Therefore, by the product rule, there are $840 \cdot 24 = 20{,}160$ program possibilities.

5.11 EXAMPLE.

(a) In how many ways can eight indistinguishable rooks be placed on a chess board so that none of them can attack another? To answer this question, first recall that a rook can attack any other piece located in its row or column. (Rows are horizontal, and row 1 is on top; columns are vertical, and column 1 is at the left.) To say that two rooks are indistinguishable means that given any config-uration of rooks on the board, if any two rooks are interchanged, then the new configuration is viewed as the same as the original.

Note that there is at least one solution: position all the rooks on the main diagonal, as pictured.

Also note that if there were nine or more rooks there could be no solution, because some row would have at least two rooks, by the pigeonhole principle.

In an arrangement of eight rooks satisfying the given condi-tions, each row contains a rook, and so does each column. (If all eight rooks were situated in seven or fewer rows, then two would be in the same row, again by the pigeonhole principle; similarly for columns.) With each such arrangement we associate a permu-tation as follows: for each i satisfying $1 \le i \le 8$, let $\sigma(i)$ be the number of the column occupied by the rook in row i. So, for in-stance, the arrangement corresponds to the permutation

$$\sigma = \begin{pmatrix} 1 & 2 & 3 & 4 & 5 & 6 & 7 & 8 \\ 3 & 1 & 4 & 8 & 7 & 6 & 2 & 5 \end{pmatrix}.$$

This procedure gives a one-to-one correspondence between the set of acceptable rook configurations and the symmetric group S_8. (Check this.) Therefore there are $8! = 40{,}320$ such configurations.

(b) Now let's revise the conditions of part (a) and suppose that the rooks *are* distinguishable; for example, suppose they are all differently colored. Again we ask for the number of ways in which the rooks can be arranged on the board in a nonattacking mode. Each such arrangement can be viewed as the result of two successive choices: first we decide which squares will hold the rooks, then we decide the pattern of colors for the rooks on those squares. By the first part of this example, there are $8!$ ways to choose an acceptable configuration of squares. Then, having fixed such a configuration, there are $8!$ ways of distributing the eight colors among those squares. Therefore, by the product rule, there are altogether $(8!)^2 = 1{,}625{,}702{,}400$ possibilities.

It is sometimes difficult to apply the product rule directly to a given combinatorial construction. In that event, it may be helpful to regard the construction as a part of a larger process to which the product rule *does* apply. The following example illustrates this approach.

5.12 EXAMPLE. A necklace is made by stringing seven differently colored beads on a cord and then tying the ends of the cord together. The knot will be small enough to slide through the holes in the beads. Determine the number of different necklaces that can be made in this way from a given set of beads, assuming that necklaces are called "different" if their color patterns are different.

SOLUTION 1. The temptation is to blurt out a fast answer: "The ordered strings of seven beads are in one-to-one correspondence with S_7, hence there are $7! = 5040$ possibilities." But, for example, the untied beaded strings in the figure (see page 220) will all produce the same necklace once the cord is tied, so we need to count more carefully.

The statement of the example is concerned with the construction of necklaces from beaded strings. On the other hand, given a collection of necklaces, we can produce and display a beaded string like the ones in the figure by first choosing a necklace, then cutting it, and then displaying the result. For the latter two parts of the construction, we first decide which of the necklace's seven beads is going to be on the left in the final display, and then we decide (before we cut) which of that bead's two neighbors is going to be on the right in the display. As we have already discussed, there are 7! possible results. Thus by the product rule we have

$$7! = (\text{number of necklaces}) \cdot 7 \cdot 2,$$

and therefore the number of necklaces is $7!/14 = 360$.

SOLUTION 2.   There are 7! ways of ordering seven beads from left to right. Call two such strings of beads *equivalent* if they yield the same necklace. This is an equivalence relation, and we assert that each equivalence class has 14 strings in it. For, given any string, removing the bead from the right end and putting it on the left produces an equivalent string. Iterating (repeating again and again) this maneuver produces seven equivalent strings. Moreover, reversing the bead order of these seven strings produces seven more equivalent strings. Thus we have 14 left-to-right strings in the equivalence class of the given string. We leave it to the reader to determine that these constitute the entire equivalence class (that is, no other strings produce the same necklace). We have 7! strings altogether and 14 strings in each equivalence class, where each equivalence class corresponds to a necklace. Therefore

$$7! = 14 \cdot (\text{number of necklaces}),$$

which gives the same result as Solution 1.

Exercises

1. Which of the following functions are permutations? Explain briefly.
    (a) The identity function on the set $\mathbb{Z}$ of integers.
    (b) The function on the unit circle that shifts each point on the circle clockwise by an angle of $\pi/16$ radians.

(c) The function from the closed interval [0, 8] to itself given by the formula $x \mapsto x/3$.

(d) The function from the open interval (0, 8) to itself given by $x \mapsto x/3$.

(e) The function from $\mathbb{R}$ to $\mathbb{R}$ given by $x \mapsto x/3$.

2. Use the language of functions to explain why

$$\begin{pmatrix} 1 & 2 & 3 & 4 \\ 3 & 4 & 1 & 3 \end{pmatrix} \quad \text{and} \quad \begin{pmatrix} 1 & 2 & 3 & 4 & 5 \\ 3 & 1 & 4 & 6 & 2 \end{pmatrix}$$

do not represent permutations.

3. List the members of S_4.

4. Compute the indicated products:

(a) $\begin{pmatrix} 1 & 2 & 3 & 4 & 5 \\ 3 & 1 & 5 & 2 & 4 \end{pmatrix}\begin{pmatrix} 1 & 2 & 3 & 4 & 5 \\ 4 & 1 & 5 & 2 & 3 \end{pmatrix}$

(b) $\begin{pmatrix} 1 & 2 & 3 & 4 \\ 2 & 4 & 1 & 3 \end{pmatrix}\begin{pmatrix} 1 & 2 & 3 & 4 \\ 3 & 4 & 1 & 2 \end{pmatrix}$

5. Suppose m and n are positive integers, with $m < n$. Strictly speaking, if $\lambda \in S_m$ and $\sigma \in S_n$, then the product $\sigma\lambda$ is not defined.

(a) Suggest a reasonable way to view S_m as a subset of S_n in such a way that the product $\sigma\lambda$ makes sense.

(b) Use the idea in part (a) to obtain the product $\sigma\lambda$, where

$$\sigma = \begin{pmatrix} 1 & 2 & 3 & 4 & 5 \\ 5 & 3 & 1 & 2 & 4 \end{pmatrix} \quad \text{and} \quad \lambda = \begin{pmatrix} 1 & 2 & 3 \\ 3 & 1 & 2 \end{pmatrix}.$$

6. Suppose $\sigma, \lambda, \mu \in S_n$. Cite a theorem in Chapter 3 that justifies the assertion

$$(\sigma\lambda)\mu = \sigma(\lambda\mu)$$

and explain briefly.

7. It can be shown that the assertion in Exercise 6 can be extended to lengthier products of permutations. Thus, given $\sigma_1, \ldots, \sigma_t \in S_n$, no matter how parentheses are inserted in the expression

$$\sigma_1\sigma_2 \cdots \sigma_t$$

to create a meaningful product, the results will be the same. For example,

$$(\sigma_1(\sigma_2\sigma_3))\sigma_4 = (\sigma_1\sigma_2)(\sigma_3\sigma_4) = \sigma_1((\sigma_2\sigma_3)\sigma_4).$$

Therefore it is customary to omit parentheses from products of this kind, unless there is a special reason to prescribe the order in which the products are to be performed. Now define

$$\sigma_1 = \begin{pmatrix} 1 & 2 & 3 & 4 & 5 \\ 3 & 5 & 4 & 1 & 2 \end{pmatrix}, \qquad \sigma_2 = \begin{pmatrix} 1 & 2 & 3 & 4 & 5 \\ 2 & 4 & 1 & 5 & 3 \end{pmatrix},$$

$$\sigma_3 = \begin{pmatrix} 1 & 2 & 3 & 4 & 5 \\ 5 & 4 & 3 & 2 & 1 \end{pmatrix}.$$

Compute each of the following products.
(a) $\sigma_1\sigma_2\sigma_3$ (b) $\sigma_1\sigma_3\sigma_2$ (c) $\sigma_2\sigma_2\sigma_3$ (d) $\sigma_3\sigma_3\sigma_3$

8. Let $\sigma \in S_n$. Define σ^0 to be the identity permutation:

$$\sigma^0 = \begin{pmatrix} 1 & 2 & \cdots & n \\ 1 & 2 & \cdots & n \end{pmatrix}.$$

For $k > 0$ define

$$\sigma^k = \underbrace{\sigma\sigma \cdots \sigma}_{k \text{ factors}} \qquad \text{(the } k\text{th \textbf{power} of } \sigma\text{).}$$

Now let

$$\sigma = \begin{pmatrix} 1 & 2 & 3 & 4 & 5 \\ 3 & 5 & 4 & 1 & 2 \end{pmatrix}.$$

(a) Determine all the members of S_5 that are powers of σ, and explain your reasoning.
(b) Compute σ^{268}, and explain.

9. Consider the permutations

$$\sigma = \begin{pmatrix} 1 & 2 & 3 & 4 & 5 & 6 & 7 & 8 \\ 7 & 2 & 4 & 1 & 5 & 6 & 3 & 8 \end{pmatrix}$$

$$\text{and} \qquad \lambda = \begin{pmatrix} 1 & 2 & 3 & 4 & 5 & 6 & 7 & 8 \\ 1 & 6 & 3 & 4 & 5 & 8 & 7 & 2 \end{pmatrix}.$$

By discussing how each of these permutations acts on the set $\mathbb{N}_8$, give an intuitive explanation for why, without actually multiplying, you would expect that

$$\sigma\lambda = \lambda\sigma.$$

(While you're at it, verify that this equation is true.)

10. Imagine that you have made it your mission to reduce eyestrain among students of mathematics and computer science. Use this as a base for arguing that functions should be written to the *right* of their domain elements. (See Remark 5.9.) Include the function diagram

$$A \xrightarrow{f} B \xrightarrow{g} C \xrightarrow{h} D$$

as part of your discussion.

11. How many permutations are there in S_{10} that hold the numbers 2 and 8 fixed?

12. How many permutations σ are there in S_{10} having the property that the restriction $\sigma|_{\mathbb{N}_4}$ belongs to S_4?

13. How many seven-letter "words" can be obtained by arranging the letters of the word "anagram"? (You may use the fact that a set of seven elements has 35 three-element subsets.)

14. A riverboat gambler has five inverted cups, with a bean hidden under one of them. A *move* is an interchange of two cups, and a *repositioning* is the result of a finite sequence of moves.
    (a) Associate the gambler's actions with elements of S_5.
    (b) How many moves are there?
    (c) How many repositionings are there?
    (d) How many repositionings are there that take the bean from the position on the gambler's far left to the position on his far right?

15. Eight married couples are to form one large circle for a folk dance. How many circles can be formed under each of the following conditions?
    (a) Each woman's husband is on her immediate right.
    (b) The people in the circle alternate by sex.
    (c) There are no restrictions.
    (d) There are two men who loathe each other and refuse to hold each other's hand.

16. A necklace is to be made with seven differently colored beads, as in Example 5.12. Unfortunately the string is quite thick, and the knot will be too large to pass through the holes in the beads and too ugly to be worn where it can be seen. (It will be worn behind the neck, under the collar.) How many different necklaces are there?

5.4 PERMUTATIONS AND GEOMETRIC SYMMETRY

Symmetry . . . is one idea by which man through the ages has tried to comprehend and create order, beauty, and perfection.

HERMANN WEYL
Symmetry

What makes a daisy beautiful? We cannot answer this here beyond saying that there is something about the balance, harmony, and regularity of the daisy's structure that we find at once startling, mysterious, and glorious. What we *can* do is describe certain aspects of the "regularity" of a daisy's structure. The idealized daisy has 16 identically shaped petals. Here we

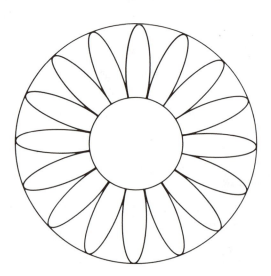

have inscribed the daisy in a circle, so its petal tips mark the vertices of a regular 16-gon, that is, a polygon with 16 sides of equal length. Now if the whole plane is rotated by an angle of $2\pi/16 = \pi/8$ radians around the center of the floral design, the daisy is carried onto itself. Such a rotation is called a **symmetry** of the daisy. Repeated applications of this rotation continue to carry the daisy onto itself. Thus if σ denotes the rotation by $\pi/8$, then σ^n, meaning the composition

$$\underbrace{\sigma\sigma \ \cdot \ \cdot \ \cdot \ \sigma,}_{n}$$

is also a symmetry of the daisy.

Now consider the following two five-petalled flowers:

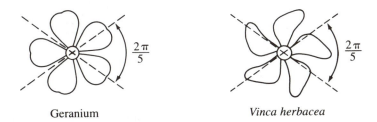

Geranium   *Vinca herbacea*

Here a rotation by $2\pi/5$ radians about either flower's center is a symmetry of the flower. But the geranium has more symmetry than the *Vinca herbacea*. For, if L is a line that passes through the geranium's center and splits one of the petals in half, then the flower is carried onto itself by a **reflection** through L. (This reflection is the function from the plane to itself that carries each point to its "mirror image" on the other side of L.) But *Vinca herbacea* has no such reflective property, because of the irregularity of its petal shape.

A symmetry of a daisy acts as a permutation on the vertex set of the corresponding regular polygon. Moreover, if we know the symmetry's action on the vertices, then we know what this symmetry does to the flower, and indeed to the whole plane. This observation allows us to study the geometric concept of symmetry by studying certain families of permutations.†

5.13 DEFINITION.   A **geometric figure** is a subset of the plane or of ordinary three-dimensional space. A **symmetry** of a geometric figure F is a bijection s of the plane or space to itself such that $s(F) = F$ and such that *distances are preserved*. In other words, for any two points x and y, the

---

† For an extensive treatment of this topic, see Hermann Weyl, *Symmetry* (Princeton: Princeton University Press, 1952).

distance between $s(x)$ and $s(y)$ is equal to the distance between x and y. More briefly, we say that a symmetry of F is a **rigid motion** that carries F onto F.

A **polygon** is a geometric figure in the plane consisting of a finite sequence of points (the **vertices**) $v_1, v_2, \ldots, v_n$, together with the line segments (the **edges** or **sides**) connecting v_i and v_{i+1}, for $1 \leqslant i \leqslant n$. (Here it is understood that $v_{n+1} = v_1$.)

Now consider the symmetries of a polygon. We leave to your geometric intuition these facts: a symmetry of a polygon takes vertices to vertices, and we know what it does to every point if we know its action on the vertices.

If a polygon P has n vertices in locations numbered 1 through n, then a symmetry of P determines a permutation $\sigma \in S_n$ as follows: for each $i \in \mathbb{N}_n$, the number $\sigma(i)$ is the location to which the symmetry carries the vertex that started in location i.

5.14 EXAMPLE. Let P be the equilateral triangle with vertices at locations labelled 1, 2, 3 in the accompanying figure.

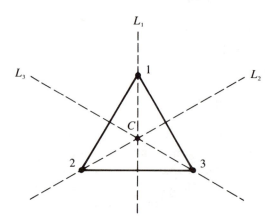

In this figure, each line L_i is the perpendicular bisector of the edge opposite the vertex at location i. A counterclockwise rotation of $2\pi/3$ radians about the center C is a symmetry of the triangle, and it sends the vertex initially at location 1 to location 2, the one at location 2 to location 3, and so on. Thus the corresponding permutation in S_3 is

$$\begin{pmatrix} 1 & 2 & 3 \\ 2 & 3 & 1 \end{pmatrix}.$$

Similarly, reflecting the triangle through the line L_2 corresponds to the permutation

$$\begin{pmatrix} 1 & 2 & 3 \\ 3 & 2 & 1 \end{pmatrix}.$$

Altogether, we have the following correspondence between symmetries of the triangle and permutations in S_3.

(5.15)	Symmetry		Corresponding permutation
	s_1:	identity function	$\begin{pmatrix} 1 & 2 & 3 \\ 1 & 2 & 3 \end{pmatrix} = \sigma_1$
	s_2:	reflection through L_1	$\begin{pmatrix} 1 & 2 & 3 \\ 1 & 3 & 2 \end{pmatrix} = \sigma_2$
	s_3:	reflection through L_2	$\begin{pmatrix} 1 & 2 & 3 \\ 3 & 2 & 1 \end{pmatrix} = \sigma_3$
	s_4:	reflection through L_3	$\begin{pmatrix} 1 & 2 & 3 \\ 2 & 1 & 3 \end{pmatrix} = \sigma_4$
	s_5:	counterclockwise rotation by $2\pi/3$	$\begin{pmatrix} 1 & 2 & 3 \\ 2 & 3 & 1 \end{pmatrix} = \sigma_5$
	s_6:	counterclockwise rotation by $4\pi/3$	$\begin{pmatrix} 1 & 2 & 3 \\ 3 & 1 & 2 \end{pmatrix} = \sigma_6$

We have listed six distinct permutations in Table 5.15. Since $\#S_3 = 3! = 6$ and every symmetry of the triangle corresponds to an element of S_3, we know that we have found *all* the symmetries of the triangle.

From our association of elements in S_n with symmetries of an n-gon (a polygon with n vertices), we conclude that an n-gon can have at most $n!$ symmetries. Moreover, if s_1 and s_2 are symmetries corresponding to permutations σ_1 and σ_2, respectively, then the composition $s_2 \circ s_1$ corresponds to the permutation $\sigma_2\sigma_1$. (To check this, use the fact that a symmetry and its associated permutation both have essentially the same action. More precisely, if a symmetry takes the vertex at location i to location j, then the associated permutation takes i to j.) It follows that we can determine the effect of applying a succession of symmetries by computing the product of the corresponding permutations. So once we establish the correspondence between symmetries and permutations, we can answer geometric questions of the form, "What is the effect of successively applying the symmetries $s_1, s_2, \ldots, s_t$?" by computing permutation products ("Compute $\sigma_t \cdots \sigma_2 \sigma_1$").

5.16 EXAMPLE. With the symmetries $s_1, \ldots, s_6$ as in Table 5.15, compute the composition

$$s_4 \circ s_6 \circ s_2 \circ s_5 \circ s_3$$

while keeping your mind free of geometric thoughts.

SOLUTION. We take the corresponding product of permutations:

$$\sigma_4 \sigma_6 \sigma_2 \sigma_5 \sigma_3 = \begin{pmatrix} 1 & 2 & 3 \\ 2 & 1 & 3 \end{pmatrix} \begin{pmatrix} 1 & 2 & 3 \\ 3 & 1 & 2 \end{pmatrix} \begin{pmatrix} 1 & 2 & 3 \\ 1 & 3 & 2 \end{pmatrix} \begin{pmatrix} 1 & 2 & 3 \\ 2 & 3 & 1 \end{pmatrix} \begin{pmatrix} 1 & 2 & 3 \\ 3 & 2 & 1 \end{pmatrix}$$

$$= \begin{pmatrix} 1 & 2 & 3 \\ 3 & 2 & 1 \end{pmatrix} = \sigma_3.$$

Therefore the given composition of symmetries is equal to s_3.

Example 5.16 illustrates an important reason to represent symmetries by permutations: it allows us to compute symmetry compositions in a purely mechanical fashion, without having to struggle with the complexities of geometric imagery. This is very much in the spirit of the quotation from Whitehead on page 56.

The symmetries of a square correspond to permutations in S_4. But

notice that no symmetry of the square corresponds to the permutation $\begin{pmatrix} 1 & 2 & 3 & 4 \\ 1 & 3 & 2 & 4 \end{pmatrix}$, since a symmetry that holds the vertex at location 1 fixed can't take the vertex at location 2 to location 3. More generally, each symmetry of the **regular n-gon** (an equilatral and equiangular polygon) takes the vertex initially at location 1 to one of the n vertex locations, call it $\sigma(1)$. Then, because a symmetry is distance-preserving, it must take the vertex initially at location 2 to one of the two neighboring locations of

$\sigma(1)$. It follows from the product rule that there are precisely $2n$ symmetries of the regular n-gon, and these correspond to a set of $2n$ permutations in S_n.

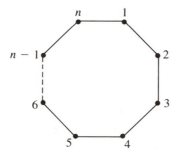

5.17 EXAMPLE. A regular 9-gon has 18 symmetries. If the vertices are at locations numbered counterclockwise from 1 to 9, then there are exactly two symmetries that take the vertex at location 1 to the vertex at location 5, namely, those associated with the following permutations in S_9:

$$\begin{pmatrix} 1 & 2 & 3 & 4 & 5 & 6 & 7 & 8 & 9 \\ 5 & 6 & 7 & 8 & 9 & 1 & 2 & 3 & 4 \end{pmatrix} \text{ and } \begin{pmatrix} 1 & 2 & 3 & 4 & 5 & 6 & 7 & 8 & 9 \\ 5 & 4 & 3 & 2 & 1 & 9 & 8 & 7 & 6 \end{pmatrix}.$$

The first of these corresponds to a counterclockwise rotation of the 9-gon by $8\pi/9$ radians. The second corresponds to a reflection through the line that passes through the center of the 9-gon and the vertex at location 3. (Suggestion: Draw and label the regular 9-gon to check these two assertions.)

The set of permutations associated with the symmetries of the regular n-gon is usually denoted D_n; it is called the **dihedral group of degree n**. Our discussion here shows that working with symmetries of the regular n-gon is essentially the same as computing with permutations in D_n. The collection of symmetries of the n-gon and the dihedral group D_n are said to be *isomorphic* structures. (From the Greek: *iso* means equal and *morphē* means form.)

We have seen that, with the help of permutations, the geometry of polygonal symmetries can be analyzed by tracking the paths of vertices as they move from one location to another. It turns out that the movie theater problem and the Rubik's Cube problem from Example 5.1 can also be formulated (and then solved) as permutation problems. In the movie theater problem, the locations are the seats. (They hold people instead of vertices, but otherwise we proceed as in the geometric setting.) A permutation

$$\sigma = \begin{pmatrix} 1 & 2 & \cdots & 10 \\ \sigma(1) & \sigma(2) & \cdots & \sigma(10) \end{pmatrix}$$

corresponds to the rearrangement in which the person in seat i ends up in seat $\sigma(i)$, for $1 \leqslant i \leqslant 10$. As for Rubik's Cube, each of the Cube's six faces is partitioned into nine facelets. An operation on the Cube consists of a succession of face rotations, and we can describe the effect of an operation by judiciously recording the position of each facelet before and after the operation. When we looked at symmetries of the n-gon, we saw that not every permutation of vertex positions corresponds to a symmetry. Similarly, not every permutation of the set of facelet positions corresponds to a Cube operation. For instance, a facelet on a corner cubie cannot be moved to the center of a face. We'll return to both of these problems later in this chapter.

Exercises

1. A **scalene** triangle is a triangle that is not isosceles. Show that the identity function is the only symmetry of a scalene triangle.

2. Let X be the set of all symmetries of the unit circle. Without proof, give a geometric description of the members of X. Also, is X finite, countably infinite, or uncountable?

3. (a) List the members of the dihedral group D_4 (which corresponds to the set of symmetries of the square).
   (b) List the members of D_6 (corresponding to the symmetries of the regular hexagon), and give a geometric explanation of a way in which D_3 can be viewed as a subset of D_6.
   (c) Show (without all the details) how for each $n \geqslant 2$, the dihedral group D_n can be viewed as a subset of the set of symmetries of the unit circle.

4. Let F be a geometric figure.
   (a) Prove that if s_1 and s_2 are symmetries of F, then so is the composition $s_2 \circ s_1$.
   (b) Prove that if s is a symmetry of F, then so is s^{-1}.

5. In each case describe the set of all symmetries of the given geometric figure in the coordinatized plane. Include the cardinality of the symmetry set as part of your description.

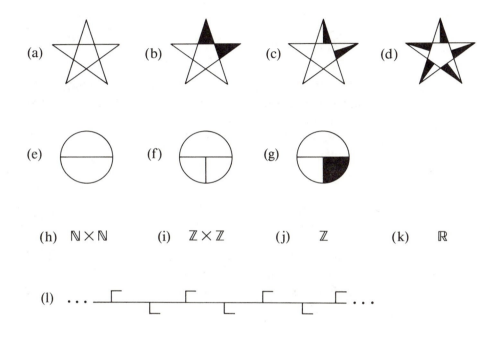

(a)   (b)   (c)   (d)

(e)   (f)   (g)

(h) $\mathbb{N} \times \mathbb{N}$ (i) $\mathbb{Z} \times \mathbb{Z}$ (j) $\mathbb{Z}$ (k) $\mathbb{R}$

(l) ...

6. Describe the symmetries of each of the following figures in three-dimensional space, and determine the number of symmetries.

   (a) A regular tetrahedron
   (b) A pyramid (with a square base and four sides, each an equilateral triangle)
   (c) A regular octahedron (formed by gluing together the bases of two pyramids of the same size)
   (d) A cube
   (e) A rectangular noncubic solid
   (f) A sphere

7. Let $Q = (a, b)$ be a point in the coordinatized xy-plane $\mathbb{R}^2 = \mathbb{R} \times \mathbb{R}$. For each point $P \in \mathbb{R}^2$, let $R_Q(P)$ denote the mirror image of P on the other side of Q. The function $R_Q: \mathbb{R}^2 \to \mathbb{R}^2$ is called **reflection in Q**.

   (a) Give a precise formula for R_Q (that is, give the coordinates of $R_Q(P)$ in terms of those of Q and P).
   (b) Prove that R_Q preserves distances. (See Definition 5.13.)
   (c) Prove that R_Q is a symmetry of any circle centered at Q.

8. The notation in this exercise is that of Table 5.15. Use permutations (and not geometry) to determine the result of applying the symmetry composition $s_6 \circ s_3 \circ s_5$ a total of 787 times.

5.5 DECOMPOSITION INTO CYCLES

In this section and the next, we will see how to decompose elements of S_n into products of especially simple kinds of permutations called *cycles*, and we will consider some applications. But first consider this: in focusing on S_n, the set of permutations of $\{1, \ldots, n\}$, aren't we neglecting the study of permutations of *other* finite sets? The response to this is that a typical n-set $A = \{a_1, \ldots, a_n\}$ has permutations of the form

$$(*) \qquad \begin{pmatrix} a_1 & a_2 & \cdots & a_n \\ a_{i_1} & a_{i_2} & \cdots & a_{i_n} \end{pmatrix},$$

where the subscripts in the bottom row are some arrangement of $1, 2, \ldots, n$. Properties of such a permutation are completely determined by the corresponding permutation of the subscripts:

$$(**) \qquad \begin{pmatrix} 1 & 2 & \cdots & n \\ i_1 & i_2 & \cdots & i_n \end{pmatrix}.$$

(If you like, view (**) as an abbreviation for (*) obtained by erasing all the a's.) Thus we can focus our attention on S_n with the knowledge that our results will carry over to S_A for any n-set A. In the terminology introduced near the end of the preceding section, if A has n elements, then the symmetric groups S_A and S_n are *isomorphic*. (We will have a more detailed look at the concept of isomorphic structures in Section 6.3.)

From now on we let e denote the identity permutation:

$$e = \begin{pmatrix} 1 & 2 & \cdots & n \\ 1 & 2 & \cdots & n \end{pmatrix}.$$

Strictly speaking, there is a different identity permutation for each value of n, but it will always be clear which identity is relevant.

A permutation is a bijection and therefore has an inverse that is also a bijection. That is,

$$\sigma \in S_n \implies \sigma^{-1} \in S_n,$$

and σ^{-1} takes each element $\sigma(i)$ back to i. The standard permutation notation displays $\sigma(i)$ directly below i, and therefore σ^{-1} can be obtained by
interchanging the two rows of σ. For example, if

$$\sigma = \begin{pmatrix} 1 & 2 & 3 & 4 \\ 3 & 1 & 4 & 2 \end{pmatrix} \quad \text{then} \quad \sigma^{-1} = \begin{pmatrix} 3 & 1 & 4 & 2 \\ 1 & 2 & 3 & 4 \end{pmatrix} = \begin{pmatrix} 1 & 2 & 3 & 4 \\ 2 & 4 & 1 & 3 \end{pmatrix}.$$

Here the second equality reflects the standard practice of listing the top
row in ascending order, a visual aid in computations.

5.18 EXAMPLE.   Let

$$\sigma = \begin{pmatrix} 1 & 2 & 3 & 4 & 5 & 6 & 7 & 8 \\ 3 & 7 & 1 & 6 & 4 & 8 & 2 & 5 \end{pmatrix}.$$

Check the equation

$$\sigma^{-1} = \begin{pmatrix} 1 & 2 & 3 & 4 & 5 & 6 & 7 & 8 \\ 3 & 7 & 1 & 6 & 8 & 4 & 5 & 6 \end{pmatrix}.$$

SOLUTION 1.   Compute σ^{-1} directly by the row-interchange procedure and compare it with the candidate:

$$\sigma^{-1} = \begin{pmatrix} 3 & 7 & 1 & 6 & 4 & 8 & 2 & 5 \\ 1 & 2 & 3 & 4 & 5 & 6 & 7 & 8 \end{pmatrix} = \begin{pmatrix} 1 & 2 & 3 & 4 & 5 & 6 & 7 & 8 \\ 3 & 7 & 1 & 5 & 8 & 4 & 2 & 6 \end{pmatrix}$$
$$\neq \begin{pmatrix} 1 & 2 & 3 & 4 & 5 & 6 & 7 & 8 \\ 3 & 7 & 1 & 6 & 8 & 4 & 5 & 6 \end{pmatrix}.$$

Therefore the equation is false.

SOLUTION 2.   The permutation σ^{-1} must satisfy the equation
$\sigma\sigma^{-1} = e$; so the strategy is to multiply the two given permutations and
observe whether the answer is e:

$$\begin{pmatrix} 1 & 2 & 3 & 4 & 5 & 6 & 7 & 8 \\ 3 & 7 & 1 & 6 & 4 & 8 & 2 & 5 \end{pmatrix} \begin{pmatrix} 1 & 2 & 3 & 4 & 5 & 6 & 7 & 8 \\ 3 & 7 & 1 & 6 & 8 & 4 & 5 & 6 \end{pmatrix}$$
$$= \begin{pmatrix} 1 & 2 & 3 & 4 & 5 & 6 & 7 & 8 \\ 1 & 2 & 3 & 8 & 5 & 6 & 4 & 8 \end{pmatrix} \neq e.$$

False again.

SOLUTION 3.   Notice that our given candidate for σ^{-1} has two appearances of 6 in the bottom row, and it is therefore not a permutation. In particular, it is not σ^{-1}.

If $\sigma \in S_n$ and $t \geqslant 0$, we define σ^t recursively by

(5.19)
$$\begin{cases} \sigma^0 = e, \\ \sigma^{k+1} = \sigma^k \sigma. \end{cases}$$

Thus $\sigma^1 = \sigma$ and, more generally, if $t > 0$ then

$$\sigma^t = \underbrace{\sigma\sigma \cdots \sigma}_{t}.$$

We also can define negative powers of σ: if $t > 0$ define

(5.20)
$$\sigma^{-t} = (\sigma^{-1})^t.$$

5.21 EXAMPLE.   If

$$\sigma = \begin{pmatrix} 1 & 2 & 3 & 4 & 5 \\ 3 & 5 & 1 & 2 & 4 \end{pmatrix},$$

then

$$\sigma^2 = \begin{pmatrix} 1 & 2 & 3 & 4 & 5 \\ 1 & 4 & 3 & 5 & 2 \end{pmatrix},$$

$$\sigma^{-1} = \begin{pmatrix} 1 & 2 & 3 & 4 & 5 \\ 3 & 4 & 1 & 5 & 2 \end{pmatrix},$$

$$\sigma^{-2} = \begin{pmatrix} 1 & 2 & 3 & 4 & 5 \\ 1 & 5 & 3 & 2 & 4 \end{pmatrix}.$$

A quick computation shows that in Example 5.21 we have $\sigma^{-2} = (\sigma^2)^{-1}$. Perhaps this is not a surprise; it is a special case of the next theorem, whose statement will seem plausible because of our previous experience in multiplying real numbers. The complete proofs are unexpectedly technical (for instance, there are special cases according to the signs of the exponents), and we'll give only a partial proof. Notice that part (c) is a special case of part (b).

5.22 THEOREM (LAWS OF EXPONENTS). If $\sigma \in S_n$, then for all $s, t \in \mathbb{Z}$,

(a) $\sigma^s \sigma^t = \sigma^{s+t}$,
(b) $(\sigma^s)^t = \sigma^{st}$,
(c) $(\sigma^s)^{-1} = \sigma^{-s}$.

PARTIAL PROOF OF (a). We will prove that $\sigma^s \sigma^t = \sigma^{s+t}$ when s and t are natural numbers. Fix $s \in \mathbb{N}$; the proof will be by induction on t. For $t = 1$ we have, by 5.19,

$$\sigma^s \sigma^1 = \sigma^s \sigma = \sigma^{s+1}.$$

Now assume that $\sigma^s \sigma^k = \sigma^{s+k}$ for some $k \in \mathbb{N}$, and remember that permutation product is actually composition of functions. We have

$$
\begin{aligned}
\sigma^s \sigma^{k+1} &= \sigma^s (\sigma^k \sigma) &&\text{(by 5.19)} \\
&= (\sigma^s \sigma^k) \sigma &&\text{(by the associative law for composition of functions)} \\
&= (\sigma^{s+k}) \sigma &&\text{(by the induction hypothesis)} \\
&= \sigma^{(s+k)+1} &&\text{(by 5.19)} \\
&= \sigma^{s+(k+1)}.
\end{aligned}
$$

This completes the proof of the assertion.   □

We can associate a graph with each permutation $\sigma \in S_n$ as follows. Represent each element $i \in \{1, 2, \ldots, n\}$ by a point; these are the vertices of the graph. Then, for each i, draw a directed edge from vertex i to vertex $\sigma(i)$. So, for instance, the permutation

$$\sigma = \begin{pmatrix} 1 & 2 & 3 & 4 & 5 & 6 & 7 & 8 & 9 & 10 \\ 5 & 8 & 7 & 10 & 3 & 4 & 1 & 2 & 9 & 6 \end{pmatrix}$$

has this graph:

(5.23)

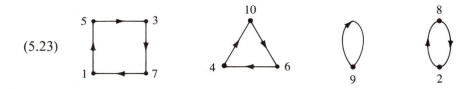

An application of σ then corresponds to moving each vertex in the direction of the arrow to the location of the next vertex. Each vertex has only one edge leading *away* from it, because a permutation is a function, and so takes each point in its domain to a unique image. Also, each vertex has exactly one edge leading *into* it, because a permutation is bijective. Thus, graphs like the following three do not represent permutations:

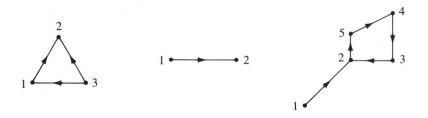

5.24 DEFINITION.   Suppose $1 \leqslant r \leqslant n$. A permutation $\sigma \in S_n$ is called a **cycle of length r** or an **r-cycle** if there are r different numbers $a_1, \ldots, a_r$ in $\mathbb{N}_n$ such that

$$\sigma(a_i) = \begin{cases} a_{i+1} & \text{if } 1 \leqslant i \leqslant r-1 \\ a_1 & \text{if } i = r \end{cases}$$

and $\sigma(x) = x$ for all $x \notin \{a_1, \ldots, a_r\}$. The set $\{a_1, \ldots, a_r\}$ is called the **orbit** of σ (and of each a_i).

A cycle that fits the description in 5.24 is usually denoted by

$$\sigma = (a_1 \quad a_2 \quad \cdots \quad a_r)$$

and its graph is an r-gon with directed edges:

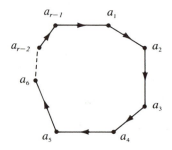

Here we have not bothered to write the little loops like

3

that correspond to fixed points. We will continue to omit them in future permutation graphs in order to avoid clutter.

5.25 EXAMPLE. The permutation

$$\sigma = \begin{pmatrix} 1 & 2 & 3 & 4 & 5 & 6 & 7 \\ 7 & 3 & 5 & 4 & 1 & 6 & 2 \end{pmatrix}$$

is a 5-cycle, namely, $\sigma = (1\ \ 7\ \ 2\ \ 3\ \ 5)$. Its graph looks like this:

5.26 REMARKS.

(a) In writing the symbol for a cycle of length $r \geqslant 2$, the list inside the parentheses can start with any element in the orbit. For instance, in Example 5.25 we can also write $\sigma = (7\ \ 2\ \ 3\ \ 5\ \ 1) = (2\ \ 3\ \ 5\ \ 1\ \ 7)$. (The underlying intuition: the description of a wheel can begin at any of its spokes.)

(b) Consider a 1-cycle, say, $\sigma = (a_1)$. Careful reading of Definition 5.24 shows that σ is actually the identity permutation.

(c) If σ is an r-cycle, so is σ^{-1}. In fact, it is easily checked that if $\sigma = (a_1\ \ a_2\ \cdots\ a_r)$, then $\sigma^{-1} = (a_r\ \ a_{r-1}\ \cdots\ a_1)$. If σ corresponds to a clockwise movement around a polygon, then σ^{-1} corresponds to a counterclockwise movement of the same magnitude.

5.27 DEFINITION. A family of cycles is **disjoint** if their orbits are pairwise disjoint; that is, no cycle in the family moves any element moved by another cycle in the family.

5.28 EXAMPLE.

$$\alpha = \begin{pmatrix} 1 & 2 & 3 & 4 & 5 & 6 & 7 \\ 1 & 6 & 3 & 5 & 2 & 4 & 7 \end{pmatrix} \quad \text{and} \quad \beta = \begin{pmatrix} 1 & 2 & 3 & 4 & 5 & 6 & 7 \\ 1 & 2 & 7 & 4 & 5 & 6 & 3 \end{pmatrix}$$

are disjoint cycles, since $\alpha = (2\ 6\ 4\ 5)$, $\beta = (3\ 7)$, and $\{2, 6, 4, 5\} \cap \{3, 7\} = \emptyset$.

The formal statement and proof of the following theorem is technical, but the underlying idea is easily stated and easily believed: If Frank moves objects from one position to another in his backyard A, and George moves objects around in his backyard B, and neither moves anything in the other person's yard, then the result is independent of the order in which they act.

5.29 THEOREM. Let A and B be nonempty sets. Suppose f and g are functions from $A \cup B$ to $A \cup B$ satisfying

$$f(A) \subseteq A \quad \text{and} \quad f(x) = x, \quad \forall x \in B,$$
$$g(B) \subseteq B \quad \text{and} \quad g(x) = x, \quad \forall x \in A.$$

Then $g \circ f = f \circ g$.

PROOF. We must show that $(g \circ f)(x) = (f \circ g)(x)$ for all $x \in A \cup B$. If $x \in A$ then $(g \circ f)(x) = g(f(x)) = f(x)$, since $f(x) \in A$; also, $(f \circ g)(x) = f(g(x)) = f(x)$, since $g(x) = x$ for all $x \in A$. This shows that $(g \circ f)(x) = (f \circ g)(x)$ for all $x \in A$. The same equation holds for all $x \in B$ by a similar argument (check this), and hence for all $x \in A \cup B$.   □

5.30 COROLLARY. If α and β are disjoint cycles, then $\alpha\beta = \beta\alpha$. (More briefly: disjoint cycles commute.)

PROOF. Let A and B be the orbits of α and β, respectively. Since α and β are disjoint, we have $A \cap B = \emptyset$; so $\alpha(x) = x$ for all $x \in B$ and $\beta(x) = x$ for all $x \in A$. Now apply Theorem 5.29.   □

5.31 EXAMPLE. In S_{10} we have

$$(3\ 1\ 5\ 7\ 9)(2\ 8\ 6) = (2\ 8\ 6)(3\ 1\ 5\ 7\ 9)$$
$$= \begin{pmatrix} 1 & 2 & 3 & 4 & 5 & 6 & 7 & 8 & 9 & 10 \\ 5 & 8 & 1 & 4 & 7 & 2 & 9 & 6 & 3 & 10 \end{pmatrix}.$$

This permutation can be graphed as follows:

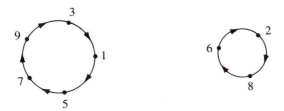

(Following our convention, vertices corresponding to the fixed values 4 and 10 are omitted from our graph.)

The action of a product of disjoint cycles is easy to picture. Is *every* permutation in S_n a product of disjoint cycles? (A single cycle is viewed as such a product.) We claim that the answer is yes. Although our examples may make the claim seem evident, we need a proof to be sure that our examples were not atypical. In fact, at first glance the action of an arbitrary permutation $\sigma \in S_n$ on the set $\mathbb{N}_n = \{1, 2, \ldots, n\}$ might appear to be too chaotic for such a controlled description. Our goal is to extract order from this apparent chaos.

5.32 CYCLE DECOMPOSITION THEOREM.   Every permutation in S_n is a product of disjoint cycles.

PROOF.   We begin by following the trail of an element in $\mathbb{N}_n$ under repeated action by σ. Consider the sequence

$$1 = \sigma^0(1), \quad \sigma(1), \quad \sigma^2(1), \quad \sigma^3(1), \ldots .$$

Since $\mathbb{N}_n$ is finite this list must contain repetitions. Suppose the first repetition is

(*) $\sigma^r(1) = \sigma^{t_1}(1),$

where $0 \leqslant r < t_1$ and t_1 is minimal. If r were positive, then applying σ^{-r} to both sides of (*) would produce an earlier repetition, which would contradict the minimality of t_1. Therefore $r = 0$, so we have shown that σ acts as a t_1-cycle on the elements $1, \sigma(1), \ldots, \sigma^{t_1-1}(1)$.

If σ moves no other elements of $\mathbb{N}_n$, then σ is equal to the t_1-cycle $\gamma_1 = (1 \quad \sigma(1) \quad \cdots \quad \sigma^{t_1-1}(1))$, and we are finished. But if σ *does* move other elements, choose some element b moved by σ but not by γ_1 and consider the sequence $b, \sigma(b), \sigma^2(b), \ldots$. By the same argument as before, this yields a t_2-cycle $\gamma_2 = (b \quad \sigma(b) \quad \cdots \quad \sigma^{t_2-1}(b))$.

We claim that γ_1 and γ_2 are disjoint cycles. To see this, suppose $\sigma^i(1) = \sigma^j(b)$ for some nonnegative i and j, with j as small as possible. We must have $j > 0$, since b is outside the orbit of γ_1. Also, since $\sigma^0(1) = \sigma^{t_1}(1) = 1$, we can suppose that $i > 0$. But then application of σ^{-1} to the equation $\sigma^i(1) = \sigma^j(b)$ gives us $\sigma^{i-1}(1) = \sigma^{j-1}(b)$, contradicting the minimality of j. So γ_1 and γ_2 must be disjoint.

Now, if no other elements of $\mathbb{N}_n$ are moved by σ, then $\sigma = \gamma_2\gamma_1$. Otherwise we repeat the preceding argument on the remaining elements moved by σ. Eventually we get a factorization $\sigma = \gamma_m \cdots \gamma_2\gamma_1$, where γ_i is a t_i-cycle and the cycles $\gamma_1, \ldots, \gamma_m$ are disjoint.   □

5.33 REMARKS.

(a) A completely formal version of the proof of Theorem 5.32 would use induction on n. The statements involving the words "repeat" and "eventually" in the last paragraph of the proof would be replaced by an application of the induction hypothesis to the complement of the first cycle's orbit.

(b) The proof of Theorem 5.32 actually gives an algorithm for factoring a permutation σ into disjoint cycles: Start with any element $x \in \mathbb{N}_n$ and keep applying σ until x reappears; this gives the first cycle. Repeat this process on any element that is not in the first cycle, and continue until all elements of $\mathbb{N}_n$ have appeared in cycles.

(c) As noted in 5.26, a 1-cycle is just the identity permutation, e. In expressing a nontrival permutation as a product of cycles, it is customary to list only cycles of length 2 or more, since factors equal to e have no effect on the product. For instance, instead of

$$(1)(2 \quad 3 \quad 4)(5)(7)(6 \quad 8)$$

we write

$$(2 \quad 3 \quad 4)(6 \quad 8).$$

(d) The factorization of a permutation into disjoint cycles is unique, except that the order in which the cycles are listed as factors is unimportant, in view of the fact that disjoint cycles commute (see Corollary 5.30). Also, a cycle of length 2 or more can be written in more than one way, according to our use of notation. For instance,

$$(1 \quad 2 \quad 3) = (3 \quad 1 \quad 2) = (2 \quad 3 \quad 1) = \begin{pmatrix} 1 & 2 & 3 \\ 2 & 3 & 1 \end{pmatrix}.$$

Thus we have

$$(1 \quad 2 \quad 3)(4 \quad 5)(6 \quad 7 \quad 8 \quad 9) = (7 \quad 8 \quad 9 \quad 6)(5 \quad 4)(3 \quad 1 \quad 2)$$
$$= (5 \quad 4)(2 \quad 3 \quad 1)(9 \quad 6 \quad 7 \quad 8).$$

(e) When we mention the *orbits* of a permutation σ, we mean the orbits of the disjoint cycles in its decomposition. The orbits of σ are the paths followed by the elements of $\mathbb{N}_n$ under repeated application of σ. Each orbit corresponds to a closed polygon in the graph of σ.

5.34 EXAMPLE.

(a) $\begin{pmatrix} 1 & 2 & 3 & 4 & 5 & 6 & 7 & 8 & 9 \\ 4 & 9 & 2 & 5 & 1 & 7 & 6 & 8 & 3 \end{pmatrix} = (1 \quad 4 \quad 5)(2 \quad 9 \quad 3)(6 \quad 7)$

(b) $\begin{pmatrix} 1 & 2 & 3 & 4 & 5 & 6 & 7 & 8 & 9 & 10 & 11 & 12 \\ 9 & 6 & 8 & 7 & 2 & 5 & 4 & 1 & 3 & 10 & 12 & 11 \end{pmatrix}$
$$= (1 \quad 9 \quad 3 \quad 8)(2 \quad 6 \quad 5)(11 \quad 12)(4 \quad 7)$$

(c) Let $\sigma = \alpha\beta$, with $\alpha = (1 \quad 2 \quad 3)$ and $\beta = (2 \quad 3 \quad 4)$. Express σ as a product of disjoint cycles.

SOLUTION.   As in parts (a) and (b), we follow the algorithm of the cycle decomposition theorem. The only variation here is that σ is now given as a product of two cycles. We must remember to apply the right-hand cycle first.

$$\left. \begin{array}{l} 1 \overset{\beta}{\mapsto} 1 \overset{\alpha}{\mapsto} 2 \\ \\ 2 \overset{\beta}{\mapsto} 3 \overset{\alpha}{\mapsto} 1 \end{array} \right\} \quad \text{This yields the 2-cycle } (1 \quad 2).$$

$$\left. \begin{array}{l} 3 \overset{\beta}{\mapsto} 4 \overset{\alpha}{\mapsto} 4 \\ \\ 4 \overset{\beta}{\mapsto} 2 \overset{\alpha}{\mapsto} 3 \end{array} \right\} \quad \text{This yields the 2-cycle } (3 \quad 4).$$

Therefore $\sigma = (1 \quad 2)(3 \quad 4)$. Once this process is mastered, the answer can be written directly (without all the arrows).

Exercises

1. Compute the inverse of each of the following permutations.

(a) $\begin{pmatrix} 1 & 2 & 3 & 4 & 5 & 6 & 7 \\ 3 & 1 & 7 & 4 & 6 & 5 & 2 \end{pmatrix}$ (b) e (c) $\begin{pmatrix} 1 & 2 & 3 & 4 \\ 3 & 1 & 4 & 2 \end{pmatrix}^{-1}$

2. Suppose

$$\sigma^5 = \begin{pmatrix} 1 & 2 & 3 & 4 & 5 \\ 5 & 1 & 4 & 3 & 2 \end{pmatrix} \text{ and } \sigma^3 = \begin{pmatrix} 1 & 2 & 3 & 4 & 5 \\ 1 & 2 & 4 & 3 & 5 \end{pmatrix}.$$

Compute each of the following permutations, and try to keep the amount of computation to a minimum. Show your method.

(a) σ^{-5} (b) σ^8 (c) σ^{-15} (d) σ^{23} (e) σ

3. Let $\sigma = \begin{pmatrix} 1 & 2 & 3 & 4 & 5 \\ 3 & 1 & 5 & 2 & 4 \end{pmatrix}$.

(a) Compute σ^{34} by performing at most six permutation products. (Use nothing more than the standard procedure for multiplying permutations.)
(b) Compute σ^{-32}. [After part (a) has been completed, this should require only the technique for finding inverses.]

4. Draw the graph of the permutation

$$\begin{pmatrix} 1 & 2 & 3 & 4 & 5 & 6 & 7 & 8 & 9 & 10 & 11 \\ 5 & 3 & 7 & 11 & 1 & 10 & 4 & 6 & 9 & 8 & 2 \end{pmatrix}.$$

5. How does the graph of a permutation compare with the graph of its inverse?

6. Prove that if $\sigma \in S_n$ then

$$(\sigma^s)^t = \sigma^{st}, \qquad \forall s, t \in \mathbb{N}.$$

(You may use Theorem 5.22(a) in your proof. Proceed by induction on t. Because this is a special case of 5.22(b), do not *assume* that result in your proof.).

7. (a) Does the cycle notation $\sigma = (1 \quad 3 \quad 7 \quad 5 \quad 2)$ tell us the domain of σ? Explain.

(b) Compute the indicated products.

$$(1 \quad 3 \quad 5)\begin{pmatrix} 1 & 2 & 3 & 4 & 5 & 6 & 7 \\ 3 & 1 & 4 & 2 & 7 & 5 & 6 \end{pmatrix} = \begin{pmatrix} 1 & 2 & 3 & 4 & 5 & 6 & 7 \\ & & & & & & \end{pmatrix}$$

$$(3 \quad 2 \quad 1 \quad 5)^{-1}(1 \quad 7 \quad 3 \quad 4)^2(1 \quad 2) = \begin{pmatrix} 1 & 2 & 3 & 4 & 5 & 6 & 7 & 8 \\ & & & & & & & \end{pmatrix}$$

8. Let σ be an r-cycle. Without proof, obtain formulas showing the following.

(a) If r is even then σ^2 is a product of two disjoint $r/2$-cycles.

(b) If r is odd then σ^2 is an r-cycle.

[Suggestion: First experiment with the cycles $(1 \quad 2 \quad 3 \quad 4)$ and $(1 \quad 2 \quad 3 \quad 4 \quad 5)$.]

9. Express each of the following permutations as a product of disjoint cycles.

(a) $\begin{pmatrix} 1 & 2 & 3 & 4 & 5 & 6 & 7 \\ 5 & 6 & 1 & 4 & 3 & 7 & 2 \end{pmatrix}$

(b) $(1 \quad 2 \quad 3 \quad 4)(4 \quad 5 \quad 6 \quad 1)(1 \quad 2)$

(c) $(1 \quad 2)^{50}(2 \quad 3 \quad 4)^{101}$

(d) $\begin{pmatrix} 1 & 2 & 3 & 4 & 5 & 6 & 7 & 8 & 9 \\ 5 & 8 & 9 & 2 & 3 & 6 & 7 & 4 & 1 \end{pmatrix}^{-1}$

10. If $\sigma \in S_n$ then $\sigma\sigma^{-1} = e$, from the definition of inverse function. Use this fact to prove the following:

(a) Each $\sigma \in S_n$ has only one inverse.

(b) If $\sigma, \lambda \in S_n$ and $\lambda^{-1} = \sigma^{-1}$, then $\lambda = \sigma$.

11. Show that if $\sigma, \tau \in S_n$ then $(\sigma\tau)^{-1} = \tau^{-1}\sigma^{-1}$.

12. Let $\sigma \in S_n$. Define a relation $\sim$ on $\mathbb{N}_n$ as follows:

$$x \sim y \quad \Leftrightarrow \quad \sigma^k(x) = y \quad \text{for some } k \in \mathbb{Z}.$$

(a) Prove that $\sim$ is an equivalence relation on $\mathbb{N}_n$.

(b) Use the terminology of this section to describe the associated equivalence classes.

13. Let

$$\sigma = \begin{pmatrix} 1 & 2 & 3 & 4 \\ 3 & 1 & 4 & 2 \end{pmatrix} \quad \text{and} \quad \lambda = (1 \quad 2 \quad 3 \quad 4).$$

(a) Find a permutation $\alpha \in S_4$ such that $\sigma^{-1}\alpha\lambda = (2 \quad 1 \quad 4)$.

(b) Show that there is only one such permutation α.

14. Suppose $2 \leqslant k \leqslant n$. How many k-cycles are there in S_n?

15. Find all the values of n for which S_n consists of cycles. (Recall that the identity $e \in S_n$ can be viewed as a 1-cycle.)

5.6 THE ORDER OF A PERMUTATION; A CARD-SHUFFLING EXAMPLE

Suppose we shuffle a deck of cards, perhaps in a very intricate way, and then we perform *exactly* the same maneuver again. And again. If we continue indefinitely must the cards eventually return to their original configuration? If so, when?

Our goal in this section is to achieve a more complete understanding of permutation behavior, with particular attention to features that can simplify large-scale computations. Even with computers at our disposal, knowing how to find an answer *somehow* is often not enough. We also need to consider the *efficiency* of our procedures; therefore, any insights that reduce many steps to a few are welcome.† The cycle decomposition theorem from Section 5.5 will be an important tool in our work here. At the end of the section we'll settle the card-shuffling problem posed in the preceding paragraph.

Suppose we want to compute a high power of a given permutation. For example, suppose

$$\sigma = \begin{pmatrix} 1 & 2 & 3 & 4 & 5 & 6 & 7 & 8 & 9 \\ 4 & 9 & 2 & 5 & 1 & 7 & 6 & 8 & 3 \end{pmatrix}$$

and we want to compute $\sigma^{100,000,000}$ by hand. Is this hopeless? No, but we need a few results before we can do the computation.

5.35 THEOREM. Let $\alpha, \beta \in S_n$, and suppose $\alpha\beta = \beta\alpha$. Then $(\alpha\beta)^t = \alpha^t\beta^t$ for all $t \in \mathbb{Z}$.

---

† The reader interested in computer programs that multiply permutations should see D. E. Knuth, *The Art of Computer Programming*, vol. 1, 2d ed. (Reading, Mass: Addison-Wesley, 1973), especially pages 162–172.

PROOF OUTLINE.   First prove by induction on k that $\alpha\beta^k = \beta^k\alpha$ for all $k \in \mathbb{N}$. Next prove the theorem for $t \geq 0$ by induction on t. Then use Theorem 5.22(c) to complete the proof for $t < 0$.   □

5.36 COROLLARY.   If $\sigma = \gamma_1 \cdots \gamma_m$ is a factorization of σ into disjoint cycles, then $\sigma^t = \gamma_1^t \cdots \gamma_m^t$ for all $t \in \mathbb{Z}$.

PROOF OUTLINE.   Use induction on m together with 5.30 and 5.35.   □

5.37 DEFINITION.   Let $\sigma \in S_n$. The **order** of σ, denoted ord σ, is the smallest positive integer k such that $\sigma^k = e$.

5.38 EXAMPLE.   Let

$$\sigma = \begin{pmatrix} 1 & 2 & 3 & 4 & 5 \\ 5 & 3 & 4 & 2 & 1 \end{pmatrix}.$$

Then

$$\sigma^2 = \begin{pmatrix} 1 & 2 & 3 & 4 & 5 \\ 1 & 4 & 2 & 3 & 5 \end{pmatrix}, \qquad \sigma^3 = \begin{pmatrix} 1 & 2 & 3 & 4 & 5 \\ 5 & 2 & 3 & 4 & 1 \end{pmatrix},$$

$$\sigma^4 = \begin{pmatrix} 1 & 2 & 3 & 4 & 5 \\ 1 & 3 & 4 & 2 & 5 \end{pmatrix}, \qquad \sigma^5 = \begin{pmatrix} 1 & 2 & 3 & 4 & 5 \\ 5 & 4 & 2 & 3 & 1 \end{pmatrix},$$

and $\sigma^6 = e$. Therefore ord $\sigma = 6$.

We claim that every $\sigma \in S_n$ has an order. To see this, first note that the powers $e, \sigma, \sigma^2, \sigma^3, \ldots$ are not all different, since S_n is finite. So we have $\sigma^i = \sigma^j$ for some i and j satisfying $0 \leq i < j$. Multiplication by σ^{-i} then gives $e = \sigma^{j-i}$. Thus $e = \sigma^k$ for some $k > 0$, and the smallest such k is ord σ.

5.39 THEOREM. If ord $\sigma = k$ then the permutations $e, \sigma, \ldots, \sigma^{k-1}$ are all different, and every power of σ is equal to one of these k permutations.

PROOF.   First a reminder of the *division algorithm* from grade school arithmetic: Given any two positive integers, one will divide into the other, yielding a quotient and a remainder that is strictly less than the divisor. More generally, in Section 6.5 we will show that if k and s are integers and $k > 0$, then we can write $s = kq + r$ for some integers q and r, with $0 \leqslant r < k$. With this notation, if σ^s is an arbitrary power of σ and $k = \text{ord } \sigma$, then

$$\sigma^s = \sigma^{kq+r} = (\sigma^k)^q \sigma^r \quad \text{(by the laws of exponents)}$$
$$= \sigma^r, \quad \text{(since } \sigma^k = e\text{)}$$

where $0 \leqslant r \leqslant k - 1$. Therefore every power of σ is equal to one of $e, \sigma, \ldots, \sigma^{k-1}$.

If the permutations $e, \sigma, \ldots, \sigma^{k-1}$ were not all different, then there would be integers i, j satisfying $0 \leqslant i < j \leqslant k - 1$ such that $\sigma^i = \sigma^j$. Multiplication of this equation by σ^{-i} would then yield a contradiction of the minimality of k.   □

There is a more efficient way to compute ord σ than by squandering our days computing $\sigma^2, \sigma^3, \ldots$ and waiting until $\sigma^k = e$, and we'll now give a slick method that uses the cycle decomposition of σ.

First consider the order of a k-cycle $\gamma = (a_1 \ \cdots \ a_k)$. The cycle's graph makes it clear that ord $\gamma = k$, since a forward movement of k positions returns each element to its initial position, while a move of less than k positions fails to do so. More formally, it follows from Definition 5.24 that if $1 \leqslant t \leqslant k$, then

$$(5.40) \qquad \gamma^t(a_i) = \begin{cases} a_{i+t} & \text{if } i + t \leqslant k, \\ a_{i+t-k} & \text{if } i + t > k. \end{cases}$$

(Use an induction argument to see this, or at least convince yourself by trying a few values of t.) So $\gamma^k(a_i) = a_i$ for $1 \leqslant i \leqslant k$, therefore $\gamma^k = e$; moreover $\gamma^t \neq e$ when $0 < t < k$. Therefore ord $\gamma = k$.

5.41 EXAMPLE.   Consider the 5-cycle $\gamma = (3 \ 1 \ 7 \ 2 \ 4)$, and let's compute γ^{72}. We know that ord $\gamma = 5$, by what we have just shown. Employing the division algorithm, we have $72 = 5(14) + 2$, and therefore

$$\gamma^{72} = \gamma^{5(14)+2} = (\gamma^5)^{14}\gamma^2 = \gamma^2 = (3 \ 7 \ 4 \ 1 \ 2).$$

Now we can settle a problem posed earlier.

5.42 EXAMPLE.   Compute $\sigma^{100,000,000}$ if

$$\sigma = \begin{pmatrix} 1 & 2 & 3 & 4 & 5 & 6 & 7 & 8 & 9 \\ 4 & 9 & 2 & 5 & 1 & 7 & 6 & 8 & 3 \end{pmatrix}.$$

SOLUTION.   From 5.34(a) we have $\sigma = (1 \quad 4 \quad 5)(2 \quad 9 \quad 3)(6 \quad 7)$, a product of disjoint cycles. Therefore

$$\sigma^{100,000,000} = (1 \quad 4 \quad 5)^{100,000,000}(2 \quad 9 \quad 3)^{100,000,000}(6 \quad 7)^{100,000,000},$$

by Corollary 5.36. But ord$(6 \quad 7) = 2$ and $100,000,000$ is divisible by 2, so $(6 \quad 7)^{100,000,000} = e$. Also, $100,000,000 = 3(33,333,333) + 1$, so $(1 \quad 4 \quad 5)^{100,000,000} = (1 \quad 4 \quad 5)$ and $(2 \quad 9 \quad 3)^{100,000,000} = (2 \quad 9 \quad 3)$. Therefore

$$\sigma^{100,000,000} = (1 \quad 4 \quad 5)(2 \quad 9 \quad 3) = \begin{pmatrix} 1 & 2 & 3 & 4 & 5 & 6 & 7 & 8 & 9 \\ 4 & 9 & 2 & 5 & 1 & 6 & 7 & 8 & 3 \end{pmatrix}.$$

5.43 DEFINITION.   Let S be a set of integers that contains at least one nonzero integer. The **least common multiple** of S is the smallest positive integer that is a multiple of every member of S. The least common multiple of $\{r_1, r_2, \ldots, r_m\}$ is usually denoted $\mathrm{lcm}(r_1, \ldots, r_m)$ or $[r_1, \ldots, r_m]$.

5.44 EXAMPLES.

$$\mathrm{lcm}(-6, 8, 5) = 120$$
$$\mathrm{lcm}(1, 2, 3, 4, 5, 6) = 60$$
$$\mathrm{lcm}(32, -36) = 288$$

We will look more closely at least common multiples when we discuss *number theory* in Chapter 6. Now we can present the algorithm for determining the order of an arbitrary permutation in S_n.

5.45 THEOREM.   Suppose $\sigma \in S_n$ has the disjoint cycle decomposition $\sigma = \gamma_1 \cdots \gamma_m$, where γ_i is a k_i-cycle, for $1 \leqslant i \leqslant m$. Then

$$\mathrm{ord}\ \sigma = \mathrm{lcm}(k_1, \ldots, k_m).$$

(Therefore, to compute the order of a permutation in S_n, express the permutation as a product of disjoint cycles and then compute the least common multiple of the orders of those cycles.)

PROOF SKETCH. Consider the graph of σ. An application of σ corresponds to each vertex moving to the position of the next vertex in its directed polygon. Since the ith polygon has k_i vertices, after k_i applications of σ, but no fewer, the objects in the ith polygon are back in their initial position. So if $\sigma^s = e$ then s must be a multiple of k_i, and this is true for each value of i satisfying $1 \leqslant i \leqslant m$. Therefore the smallest nonnegative s for which $\sigma^s = e$ is the least common multiple of $k_1, \ldots, k_m$.

ACTUAL PROOF. We assert:

(*) $\quad \sigma^s = e \quad \Leftrightarrow \quad s$ is a multiple of k_i, for each i satisfying $1 \leqslant i \leqslant m$.

To prove "$\Leftarrow$", assume that s is a multiple of each k_i; say, $s = q_i k_i$ for $1 \leqslant i \leqslant m$. Then

$$\sigma^s = \gamma_1^s \cdots \gamma_m^s = (\gamma_1^{k_1})^{q_1} \cdots (\gamma_m^{k_m})^{q_m} = e,$$

as desired.

Now for "$\Rightarrow$". Assume that $\sigma^s = e$. Then we can write $s = k_1 q + t$, with $0 \leqslant t < k_1$; thus $\gamma_1^s = (\gamma_1^{k_1})^q \gamma_1^t = \gamma_1^t$. Let a_1 denote an element in the orbit of γ_1. If t were positive we would have

$$\sigma^s(a_1) = \gamma_1^s(a_1) = \gamma_1^t(a_1) \neq a_1.$$

(The first equality follows from the fact that none of $\gamma_2, \ldots, \gamma_m$ moves a_1, since the cycles are disjoint, and the nonequality follows from formula 5.40.) But this contradicts our assumption that $\sigma^s = e$. Therefore $t = 0$, and so $s = k_1 q$; that is, s is a multiple of k_1. A similar argument shows that s is also a multiple of $k_2, \ldots, k_m$, so we have proved assertion (*).

The theorem now follows from the "smallest" condition in the definition of ord σ, and from the definition of least common multiple. $\quad\square$

5.46 EXAMPLE.

(a) Compute the order of

$$\sigma = \begin{pmatrix} 1 & 2 & 3 & 4 & 5 & 6 & 7 & 8 & 9 & 10 & 11 & 12 & 13 & 14 & 15 \\ 12 & 5 & 8 & 3 & 13 & 9 & 6 & 4 & 10 & 7 & 14 & 1 & 11 & 2 & 15 \end{pmatrix}.$$

SOLUTION. We first write σ as a product of disjoint cycles:

$$\sigma = (1 \quad 12)(2 \quad 5 \quad 13 \quad 11 \quad 14)(3 \quad 8 \quad 4)(6 \quad 9 \quad 10 \quad 7).$$

Therefore ord $\sigma = \text{lcm}(2, 5, 3, 4) = 60$.

(b) Compute the order of

$$\sigma = (1 \quad 8 \quad 4 \quad 3)(1 \quad 4 \quad 7)(1 \quad 4 \quad 6 \quad 2)(1 \quad 4 \quad 6)(1 \quad 4 \quad 5).$$

SOLUTION. To apply Theorem 5.45, we must first express σ as a product of disjoint cycles. (We are given σ as a product of cycles, but they are not disjoint.) If we do this in the manner of Example 5.34(c), we obtain $\sigma = (1 \quad 2 \quad 3)(4 \quad 5 \quad 6 \quad 7 \quad 8)$. Therefore ord $\sigma = \text{lcm}(3, 5) = 15$.

Now we can apply all of this to the analysis of card shuffling, the topic that opened this section.

5.47 EXAMPLE.

(a) Perform a "perfect shuffle" on a standard deck of 52 playing cards as follows. Cut the deck exactly in half, place one half in each hand, and then *riffle* the cards; that is, release the cards (using your thumbs) from the two half-decks in an alternating way so that they overlap, then push them together to constitute one deck. (See the picture.)

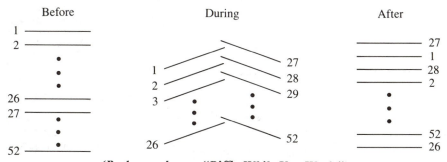

(Background song: "Riffle While You Work.")

Now, with respect to this shuffle, we repeat the question from the start of this section: If a perfect shuffle is performed repeatedly, will the cards eventually return to their original order and, if so, how soon?

SOLUTION. In the shuffle, the top 26 cards go to the respective positions 2, 4, 6, 8, . . . , and the next 26 cards (in initial positions $26 + 1$, $26 + 2$, $26 + 3$, . . .) go to the respective positions 1, 3, 5, 7, ... This position change defines a permutation $\sigma \in S_{52}$ by

$$\left. \begin{array}{r} i \mapsto 2i \\ 26 + i \mapsto 2i - 1 \end{array} \right\} \text{ for } 1 \leqslant i \leqslant 26,$$

that is,

$$\sigma = \begin{pmatrix} 1 & 2 & 3 & \cdots & 26 & 27 & 28 & 29 & \cdots & 52 \\ 2 & 4 & 6 & \cdots & 52 & 1 & 3 & 5 & \cdots & 51 \end{pmatrix}.$$

This has the disjoint cycle decomposition

$$\sigma = (1 \quad 2 \quad 4 \quad 8 \quad 16 \quad 32 \quad 11 \quad 22 \quad 44 \quad 35 \quad \cdots \quad 40 \quad 27),$$

(Exercise: fill in the blank.)

a 52-cycle!! Therefore ord $\sigma = 52$, so the cards will return to their original order after precisely 52 perfect shuffles.

(b) *Experiment.* Based on the preceding discussion you might guess that if a succession of perfect shuffles is performed on a deck of n cards, where n is any even natural number, then exactly n shuffles are required to return the cards to their original order. Test this conjecture by determining the order of the perfect shuffle for decks of 2, 4, 6, and 8 cards.

(c) Now let's return to a 52-card deck. Among all possible shuffles, does the perfect shuffle have largest order? In other words, is there a shuffle that will require *more* than 52 executions to return the deck to its original configuration?

SOLUTION. Every shuffle corresponds to a permutation in S_{52}. Each $\sigma \in S_{52}$ has a factorization into disjoint cycles, and the lengths $k_1, \ldots, k_t$ of these cycles determine ord σ. Therefore, searching for a permutation of large order in S_{52} amounts to searching for positive integers $k_1, \ldots, k_t$ with these properties:

$$k_1 + k_2 + \cdots + k_t \leqslant 52,$$
$$\text{lcm}(k_1, k_2, \ldots, k_t) \text{ is large.}$$

With this in mind, consider the permutation $\sigma = \gamma_1 \gamma_2 \gamma_3 \gamma_4 \gamma_5 \gamma_6$, where

$$\gamma_1 = (1 \quad 2 \quad 3 \quad 4) \qquad\qquad \text{(4-cycle)}$$
$$\gamma_2 = (5 \quad 6 \quad \cdots \quad 9) \qquad\qquad \text{(5-cycle)}$$
$$\gamma_3 = (10 \quad 11 \quad \cdots \quad 16) \qquad\qquad \text{(7-cycle)}$$
$$\gamma_4 = (17 \quad 18 \quad \cdots \quad 25) \qquad\qquad \text{(9-cycle)}$$
$$\gamma_5 = (26 \quad 27 \quad \cdots \quad 36) \qquad\qquad \text{(11-cycle)}$$
$$\gamma_6 = (37 \quad 38 \quad \cdots \quad 49) \qquad\qquad \text{(13-cycle)}$$

Then

$$\text{ord } \sigma = \text{lcm}(4, 5, 7, 9, 11, 13) = 180,180.$$

(Here we are computing the least common multiple of six numbers that have no common factors except 1, so their least common multiple is their product. We'll spend more time on this kind of thing in Section 6.5.) Thus the shuffle corresponding to the permutation σ constructed here must be performed 180,180 times before the deck returns to its original order!

Exercises

1. Give an informal definition of the notion of *order* of a permutation. (That is, say Definition 5.37 in words.)

2. Compute the order of each of the following permutations.

   (a) $\begin{pmatrix} 1 & 2 & 3 & 4 & 5 & 6 \\ 3 & 5 & 1 & 6 & 4 & 2 \end{pmatrix}$ (b) $(1 \quad 3 \quad 5 \quad 8 \quad 2)(2 \quad 8 \quad 3 \quad 1 \quad 6)$

   (c) $\begin{pmatrix} 1 & 2 & 3 & 4 & 5 & 6 & 7 & 8 & 9 & 10 & 11 & 12 \\ 7 & 4 & 12 & 8 & 9 & 10 & 11 & 3 & 6 & 5 & 1 & 2 \end{pmatrix}$

3. Let

$$\sigma = \begin{pmatrix} 1 & 2 & 3 & 4 & 5 & 6 & 7 \\ 2 & 1 & 4 & 5 & 3 & 7 & 6 \end{pmatrix}.$$

   Compute σ^{1057}.

4. Prove that a permutation and its inverse have the same order.

5. Exhibit a permutation in S_{30} that has order 2,520. Justify your answer.

6. Exhibit a permutation in S_{10} that has the biggest possible order. Justify your answer.

7. (a) Give an example of integers m and n with $1 \leqslant m < n$ such that the largest order of an element in S_m is equal to the largest order of an element in S_n.

   (b) If $1 \leqslant m < n$, is it possible for S_m to have an element of larger order than that of any element of S_n? Explain.

8. The Schlimazel Truck Corporation manufactures a three-axle truck, with two wheels per axle, that also carries two spares. The company recommends that owners periodically perform an elaborate tire interchange in order to keep the tire wear even. The pattern for the recommended maneuver is shown in the accompanying diagram. (Tires 7 and 8 are the spares.)

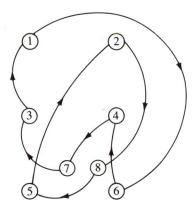

   (a) Do you agree that the company's plan will keep the wear on the tires approximately uniform over many miles of driving?

   (b) After how many of these tire interchanges will the tires first return to their initial configuration?

9. Suppose the "perfect shuffle" of Example 5.47 is performed with a deck of 26 cards. After how many repetitions will the deck first return to its original order?

10. Outline an algorithm (a computer program, if you like) for finding the least common multiple of two integers a and b (that is, the lcm of the set $\{a, b\}$) based on Definition 5.43.

11. Write a detailed proof of Theorem 5.35.

5.7 ODD AND EVEN PERMUTATIONS; APPLICATIONS TO CONFIGURATIONS

In this section we will classify permutations into two types, called *even* and *odd,* and we will use this classification to show the impossibility of certain rearrangement tasks. Most of our applications will involve some familiar plastic puzzles, but the underlying ideas also have a wide range of applications in mathematics and physics, and they should not be sneeringly dismissed as child's play.

Everyone instinctively knows that any rearrangement of people in a row of theater seats (one person per seat) can be carried out by a succession of two-person switches. This observation suggests another way to decompose permutations in S_n, and our next theorem will make this explicit.

First some terminology: a 2-cycle is also called a **transposition**. The transposition $\tau = (i \quad j)$ interchanges i and j [that is, $\tau(i) = j$ and $\tau(j) = i$] and holds everything else fixed. We observe that a transposition τ satisfies $\tau^2 = e$; so $\tau^{-1} = \tau$. A transposition is the simplest possible nontrivial permutation, for if a permutation σ is not the identity, it moves some element x, say $\sigma(x) = y \neq x$. Then $\sigma(y) \neq y$, since σ is injective; so σ moves at least two elements. A transposition moves *exactly* two elements. The mathematical formulation of our theater observation in the preceding paragraph is as follows.

5.48 THEOREM. If $n \geq 2$, then every permutation in S_n is a product of transpositions.

PROOF. Let $\sigma \in S_n$. If $\sigma = e$, we have $\sigma = (1 \quad 2)(1 \quad 2)$. If $\sigma \neq e$, there is a disjoint cycle decomposition $\sigma = \gamma_1 \cdots \gamma_m$, so it suffices to show that a typical r-cycle $\gamma = (a_1 \quad a_2 \quad \cdots \quad a_r)$ with $r \geq 2$ is a product of transpositions. The following formula does the job:

$$(5.49) \qquad (a_1 \quad a_2 \quad \cdots \quad a_r) = (a_1 \quad a_r)(a_1 \quad a_{r-1}) \cdots (a_1 \quad a_3)(a_1 \quad a_2).$$

Check this formula by applying the permutations on both sides of the alleged equation to each of $a_1, \ldots, a_r$. (A completely formal proof would use Definition 5.24 and induction on r.) ☐

5.50 EXAMPLE. Express

$$\sigma = \begin{pmatrix} 1 & 2 & 3 & 4 & 5 & 6 & 7 & 8 & 9 & 10 \\ 6 & 7 & 10 & 5 & 8 & 2 & 1 & 4 & 3 & 9 \end{pmatrix}$$

as a product of transpositions.

SOLUTION. We first write σ as a product of disjoint cycles, then apply 5.49 to each of those cycles:

$$\sigma = (1 \quad 6 \quad 2 \quad 7)(3 \quad 10 \quad 9)(4 \quad 5 \quad 8)$$
$$= (1 \quad 7)(1 \quad 2)(1 \quad 6)(3 \quad 9)(3 \quad 10)(4 \quad 8)(4 \quad 5).$$

5.51 REMARK. Here are some observations about factorization into products of transpositions.

(a) We cannot expect the transpositions in such a product to be disjoint. For instance, a 3-cycle moves exactly three elements, whereas a product of disjoint transpositions moves an even number of elements.

(b) The inequality

$$(1 \quad 3)(1 \quad 2) \neq (1 \quad 2)(1 \quad 3)$$

shows that the order in which transpositions are listed as factors is important. [It is only when permutations are *disjoint* that we can be sure (without further checking) that their relative position doesn't matter.]

Suppose a nontrivial permutation is applied to a set of people who are waiting in line. This changes the ordering, so that now there is at least one person who is ahead of someone he used to be behind. We say that the **orientation** of these two people has been **reversed**, or that a **reversal** has occurred. Similarly, if $\sigma \in S_n$ and $\{i, j\} \subseteq \mathbb{N}_n$, with $i \neq j$, we say that σ reverses the orientation of i and j if either

$$i < j \quad \text{and} \quad \sigma(i) > \sigma(j)$$

or

$$i > j \quad \text{and} \quad \sigma(i) < \sigma(j).$$

More compactly: σ reverses i and j if

$$\frac{\sigma(i) - \sigma(j)}{i - j} < 0.$$

Incidentally, notice that

$$\frac{\sigma(i) - \sigma(j)}{i - j} = \frac{\sigma(j) - \sigma(i)}{j - i},$$

so that this quotient depends only on the pair of integers under consideration, not on the way they are labelled.

For some applications we may need to know whether the total number of reversals caused by a given permutation is even or odd. (See Example 5.60.) For this purpose it is useful to define the **sign** of σ, denoted sgn σ, by

(5.52) $\operatorname{sgn} \sigma = \begin{cases} 1 & \text{if the number of reversals is even,} \\ -1 & \text{if the number of reversals is odd.} \end{cases}$

Otherwise stated:

(5.53) $\operatorname{sgn} \sigma = (-1)^r,$ where r is the number of reversals caused by σ.

The following theorem gives the basic properties of the sign function.

5.54 THEOREM.

(a) $\operatorname{sgn} \sigma = \prod_{i<j} \dfrac{\sigma(i) - \sigma(j)}{i - j}$

(b) $\operatorname{sgn} \sigma\lambda = (\operatorname{sgn} \sigma)(\operatorname{sgn} \lambda)$

(c) If τ is a transposition, then $\operatorname{sgn} \tau = -1$.

(d) If $\sigma = \tau_1 \cdots \tau_s$, where the τ_i are transpositions, then $\operatorname{sgn} \sigma = (-1)^s$.

PROOF.

(a) Here the indicated product is understood to represent the product of all the quotients of the form

$$\frac{\sigma(i) - \sigma(j)}{i - j}$$

for which the inequality $1 \le i < j \le n$ holds. From our earlier discussion we know that such a quotient is negative if and only if σ reverses i and j. Therefore, the product under scrutiny has the same sign (as a real number) as sgn σ. We will be done if we can show that the product has absolute value 1. We have

$$\left| \prod_{i<j} \frac{\sigma(i) - \sigma(j)}{i - j} \right| = \prod_{i<j} \left| \frac{\sigma(i) - \sigma(j)}{i - j} \right| = \frac{\prod |\sigma(i) - \sigma(j)|}{\prod |i - j|}.$$

But the factors in the denominator on the right are all the absolute values of differences of two integers between 1 and n, one factor for each integer pair; and the same is true of the numerator, since σ is a bijection. Therefore the quotient is equal to 1, which is what we wanted to show.

(b) $\displaystyle \text{sgn } \sigma\lambda = \prod_{i<j} \frac{\sigma\lambda(i) - \sigma\lambda(j)}{i - j}$

$\displaystyle \qquad = \prod_{i<j} \frac{\sigma\lambda(i) - \sigma\lambda(j)}{\lambda(i) - \lambda(j)} \cdot \frac{\lambda(i) - \lambda(j)}{i - j}$

$\displaystyle \qquad = \prod_{i<j} \frac{\sigma\lambda(i) - \sigma\lambda(j)}{\lambda(i) - \lambda(j)} \cdot \prod_{i<j} \frac{\lambda(i) - \lambda(j)}{i - j}$

$\qquad = (\text{sgn } \sigma)(\text{sgn } \lambda)$

(The last equality follows from the fact that as $\{i, j\}$ varies through the two-element subsets of $\mathbb{N}_n$, so does $\{\lambda(i), \lambda(j)\}$.)

(c) Let $\tau = (r \ \ s)$, with $r < s$; we will count the reversals. Under τ's action, r reverses with every element strictly between r and s, and so does s. That's two reversals for every number strictly between r and s, making an even number of reversals so far. Finally, τ reverses the orientation of r with respect to s. That makes the total number of reversals odd, so sgn $\tau = -1$.

(d) This follows immediately from (b) and (c), by induction on s. □

5.55 COROLLARY. If $\sigma = \tau_1 \cdots \tau_s$ and also $\sigma = \tau_1' \cdots \tau_t'$, where each τ_i and τ_i' is a transposition, then s and t are both even or both odd.

PROOF. From the theorem we have sgn $\sigma = (-1)^s = (-1)^t$. The conclusion follows. □

5.56 DEFINITION. A permutation $\sigma \in S_n$ is **odd** if sgn $\sigma = -1$ and **even** if sgn $\sigma = 1$. The evenness or oddness of σ is called its **parity**.

A permutation $\sigma \in S_n$ ($n \geqslant 2$) can be written as a product of transpositions in infinitely many ways. For example, we have $e = (1\ \ 2)(1\ \ 2) = (1\ \ 2)(1\ \ 2)(1\ \ 2)(1\ \ 2)$ and so on, and such expressions can be appended to any factorization of σ to get a new factorization, or we can insert them between any two factors of a given factorization. We can also manipulate expressions in other ways; for example,

$$(2\ \ 3)(3\ \ 4) = (2\ \ 3\ \ 4) = (4\ \ 2\ \ 3) = (4\ \ 3)(4\ \ 2).$$

But however the symbols are tormented to produce a new factorization into transpositions, parity must be preserved.

The sign of a permutation can be computed directly from Equation 5.52, but it is useful to have a more explicit algorithm for the job. We'll give two.

First notice that if γ is an r-cycle then sgn $\gamma = (-1)^{r+1}$ by formula 5.49, Theorem 5.54(d), and the fact that $(-1)^{r-1} = (-1)^{r+1}$. Therefore:

(5.57)   If $\sigma = \gamma_1 \cdots \gamma_t$, where γ_i is an r_i-cycle, then
$$\text{sgn } \sigma = (-1)^{r_1 + \cdots + r_t + t}.$$

For example, if

$$\sigma = \begin{pmatrix} 1 & 2 & 3 & 4 & 5 \\ 3 & 1 & 5 & 2 & 4 \end{pmatrix},$$

we have $\sigma = (1\ \ 3\ \ 5\ \ 4\ \ 2)$, and therefore sgn $\sigma = (-1)^6 = 1$. And if

$$\lambda = \begin{pmatrix} 1 & 2 & 3 & 4 & 5 & 6 & 7 & 8 & 9 & 10 & 11 & 12 \\ 3 & 1 & 9 & 7 & 8 & 4 & 6 & 5 & 2 & 12 & 11 & 10 \end{pmatrix},$$

then $\lambda = (1\ \ 3\ \ 9\ \ 2)(4\ \ 7\ \ 6)(5\ \ 8)(10\ \ 12)$, and therefore

$$\text{sgn } \lambda = (-1)^{4+3+2+2+4} = -1.$$

The next method can be easily converted to a computer program if so desired.

(5.58)   Given a permutation in its standard form

$$\begin{pmatrix} 1 & 2 & \cdots & n \\ \sigma(1) & \sigma(2) & \cdots & \sigma(n) \end{pmatrix},$$

scan the bottom row from left to right, and determine
how many times there is a number in that row with a
smaller number to its right (possibly with other numbers
in between). If there are t such instances, then

$$\text{sgn } \sigma = (-1)^t.$$

(Each occurrence of a number to the left of a smaller number corresponds
to a reversal, so statement 5.53 gives the result.) Let's use this method on
the permutations σ and λ to which we applied 5.57. In the bottom row of
σ we find two numbers to the left of smaller numbers:

3 is to the left of 1 and 2,
5 is to the left of 2 and 4.

Therefore 5.58 gives sgn $\sigma = (-1)^4 = 1$. The bottom row of λ has more
numbers to examine, so we'll make a table.

Number	Smaller number to the right
3	1, 2
9	7, 8, 4, 6, 5, 2
7	4, 6, 5, 2
8	4, 6, 5, 2
4	2
6	5, 2
5	2
12	11, 10
11	10

Therefore sgn $\lambda = (-1)^{23} = -1$.

We conclude this section by using the results about signs to demonstrate that certain configurations cannot be achieved.

5.59 EXAMPLE. In the theater problem (Example 5.1(a)), each
neighborly seat switch corresponds to a transposition. Since 14 switches
will achieve the desired rearrangement, the corresponding permutation in
S_{10} is even. Therefore it *cannot* be expressed as a product of 11 transpositions, as suggested, so the second plan should be rejected.

5.60 EXAMPLE. A familiar puzzle consists of a square grid of 16
squares, with movable tiles numbered 1 through 15 occupying fifteen of
those squares and the remaining square blank. A legal move involves slid-

ing a tile onto the blank location from an adjacent location. Given the puzzle in this configuration,

1	2	3	4
5	6	7	8
9	10	11	12
13	14	15	■

is it possible to rearrange the tiles to the opposite order by a sequence of legal moves?

15	14	13	12
11	10	9	8
7	6	5	4
3	2	1	■

SOLUTION. Think of the blank location as being occupied by a black tile bearing the number 16. With this intuition, our puzzle consists of sixteen tiles in sixteen locations, and a legal move interchanges the black tile with one of its neighbors, keeping all the other tiles fixed. Thus a legal move is a transposition. The attempted rearrangement corresponds to this permutation:

$$\sigma = \begin{pmatrix} 1 & 2 & 3 & 4 & 5 & 6 & 7 & 8 & 9 & 10 & 11 & 12 & 13 & 14 & 15 & 16 \\ 15 & 14 & 13 & 12 & 11 & 10 & 9 & 8 & 7 & 6 & 5 & 4 & 3 & 2 & 1 & 16 \end{pmatrix}.$$

Then σ is odd, since $\sigma = (1 \ 15)(2 \ 14)(3 \ 13)(4 \ 12)(5 \ 11)(6 \ 10)(7 \ 9)$. Therefore any effort that produces the desired rearrangement must involve an odd number of legal moves. On the other hand, each legal move requires a horizontal or vertical move of the black tile. For that tile to return to its initial position, it must be subjected to an even number of vertical moves and an even number of horizontal moves, hence to an even number of moves altogether. Thus the desired rearrangement is impossible.

5.61 EXAMPLE. A detailed analysis of the Rubik's Cube problem (Example 5.1(d)) would take us too long here, so we will make do with a brief sketch of the argument. (For more details, including the development of more insightful notation, see the book by Frey and Singmaster cited on p. 209.) We view a Cube operation as a permutation of facelet positions. The flip operation in question constitutes an interchange of two facelets; that is, a transposition, which is an *odd* permutation. Now consider a 90° clockwise twist (looking down) of the upper face. To describe the corresponding permutation, we must account for the movements of all twenty-one facelets on the upper layer of the Cube.

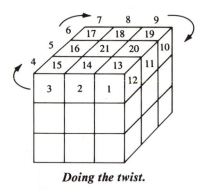

Doing the twist.

The movements of the sides of the layer are summarized by three 4-cycles:

$$(1 \quad 4 \quad 7 \quad 10)(2 \quad 5 \quad 8 \quad 11)(3 \quad 6 \quad 9 \quad 12).$$

Movements of the outer facelets of the top face are given by the product

$$(13 \quad 15 \quad 17 \quad 19)(14 \quad 16 \quad 18 \quad 20).$$

Finally, while the center facelet of the top face remains in the center, its position changes by rotation, and we record its movement by a specially denoted 4-cycle:

$$(\uparrow \quad \rightarrow \quad \downarrow \quad \leftarrow)$$

Altogether we have six disjoint 4-cycles, hence an *even* permutation. But every cube operation is a succession of face rotations. Therefore every cube operation corresponds to an even permutation, so the desired flip operation is *impossible.*

Exercises

1. Express each of the following permutations as a product of transpositions.

(a) $\begin{pmatrix} 1 & 2 & 3 & 4 & 5 & 6 & 7 & 8 \\ 3 & 1 & 7 & 5 & 2 & 4 & 8 & 6 \end{pmatrix}$

(b) $\begin{pmatrix} 1 & 2 & 3 & 4 & 5 & 6 & 7 & 8 \\ 4 & 2 & 7 & 1 & 8 & 6 & 3 & 5 \end{pmatrix}$

(c) $(1 \quad 2 \quad 3)(2 \quad 3 \quad 4)$

(d) $(3 \quad 1 \quad 4)\begin{pmatrix} 1 & 2 & 3 & 4 & 5 & 6 & 7 \\ 7 & 1 & 4 & 3 & 5 & 2 & 6 \end{pmatrix}$

(e) $(3 \quad 5 \quad 7 \quad 1 \quad 4 \quad 8)^{-1}$

2. (a) Describe the orbits of a permutation that is a product of disjoint transpositions. Use a graph to illustrate your remarks.
   (b) Let $\sigma \in S_n$, with $n \geqslant 2$. State a necessary and sufficient condition on the factorization of σ into disjoint cycles in order for σ to be expressible as a product of disjoint transpositions, and justify your claim.

3. In each case determine the number of reversals and then use 5.53 to compute the sign of the given permutation.

(a) $\begin{pmatrix} 1 & 2 & 3 \\ 3 & 1 & 2 \end{pmatrix}$ (b) $\begin{pmatrix} 1 & 2 & 3 & 4 & 5 \\ 2 & 3 & 4 & 5 & 1 \end{pmatrix}$ (c) $\begin{pmatrix} 1 & 2 & 3 & 4 & 5 & 6 \\ 2 & 3 & 1 & 4 & 5 & 6 \end{pmatrix}$

4. The notation $(1 \quad 3 \quad 5)$ can be used to represent infinitely many cycles, one in each S_n for which $n \geqslant 5$. Show that all these cycles cause exactly the same reversals, so that there is no ambiguity in the phrase, "the reversals caused by $(1 \quad 3 \quad 5)$."

5. Show that for each $n \geqslant 2$, there is a permutation for which the number of reversals is $n(n-1)/2$. (Suggestion: First verify the assertion for $n = 2, 3,$ and 4.)

6. (a) Use induction on n to prove that if $n \geqslant 2$, every element of S_n can be expressed as a product of transpositions from the set

$$\{(1 \quad 2), (1 \quad 3), \ldots, (1 \quad n)\}.$$

(As part of the induction step, view S_k as the subset of S_{k+1} consisting of all the permutations that fix $k + 1$. Then show that if

$\sigma \in S_{k+1}$ moves $k+1$, then there is a suitable transposition τ such that $\tau\sigma \in S_k$.)

(b) Interpret the result in part (a) in the context of theater-seat arrangements (Example 5.1(a)).

7. The set of even permutations in S_n is called the **alternating group**, denoted A_n.

   (a) Show that if σ, $\lambda \in A_n$ then $\sigma\lambda \in A_n$. (We therefore say that A_n is *closed under the product operation*.)
   (b) Show that if $\sigma \in A_n$ then $\sigma^{-1} \in A_n$.
   (c) Show that if $\sigma \in A_n$ and $\lambda \in S_n$ then $\lambda^{-1}\sigma\lambda \in A_n$.

8. Every rearrangement of people in a row of theater seats can be achieved by successive switches of *neighbors*. Prove this assertion's mathematical formulation: If $n \geqslant 2$, every permutation in S_n is a product of transpositions of the form $(i \quad i+1)$. (Hint: By Theorem 5.48, it suffices to show that every *transposition* can be written as such a product. Think of how it would be done in a theater.)

9. Show that if $\sigma = \tau_1\tau_2$, where τ_1 and τ_2 are disjoint transpositions, then also $\sigma = \tau_2\tau_1$, but there is no *other* way to express σ as a product of two transpositions.

10. Suppose $n \geqslant 2$, and let τ be a fixed transposition in S_n. Let A_n denote the alternating group. (See Exercise 7.) Define a function $f: A_n \to S_n$ by

$$f(\sigma) = \tau\sigma, \qquad \forall \sigma \in A_n.$$

   (a) Show that f is a bijection from A_n to the set of odd permutations in S_n.
   (b) Use part (a) to conclude that $\#A_n = n!/2$.

11. In each of the following cases, compute the sign of the given permutation.

   (a) $\begin{pmatrix} 1 & 2 & 3 & 4 & 5 & 6 & 7 & 8 & 9 \\ 3 & 6 & 9 & 8 & 2 & 4 & 5 & 7 & 1 \end{pmatrix}$

   (b) $(2 \quad 8 \quad 4)^{755}$ (c) $(1 \quad 3 \quad 8 \quad 5)^{-621}$

   (d) $(3 \quad 8 \quad 7 \quad 2 \quad 1)(4 \quad 3 \quad 6 \quad 2)(5 \quad 7)(1 \quad 2 \quad 3 \quad 4 \quad 5)$

12. For which n is the cycle $(1 \quad 2 \quad 3 \quad \cdots \quad n)$ even?

13. (a) Is there a procedure that will take the puzzle in Example 5.60 from its initial configuration (the first picture in 5.60) to the following configuration in exactly 23 moves? Explain.

	1	3	4
5	2	6	8
9	10	7	11
13	14	15	12

(b) Is there a way to manipulate the puzzle into the following configuration from the initial one in 5.60? Explain.

1	2	3	4
5	14	7	
9	10	11	8
13	6	15	12

14. Exercise 7(a) shows that the set of even permutations in S_n is closed
under product. Is the set of *odd* permutations in S_n also closed under
product?

15. For which values of $n \geq 3$ does the regular n-gon have a symmetry
whose associated permutation is a transposition? Explain.

5.8   BINOMIAL AND
MULTINOMIAL COEFFICIENTS

In Chapter 2 we learned that an n-element set has 2^n subsets. But in most
contexts in which we must choose subsets, there are restrictions on the
sizes of subsets to be chosen. We'll start this section by considering a question that is important in combinatorics and probability theory, as well as
in algebraic computations: If $0 \leq k \leq n$, how many k-element subsets are
contained in a set of n elements? Then we will obtain intriguing connections between some of the numbers (the so-called binomial coefficients)
that arise in the discussion, and we will give some applications.

In what follows, a set with k elements will be called a **k-set**, as before,
or a **combination** of k elements; a subset that is a k-set is a **k-subset**.

5.62 DEFINITION. Let k and n be integers, with $0 \leqslant k \leqslant n$. The number of k-subsets of an n-set is written

$$\binom{n}{k}$$

and called the **binomial coefficient** n **over** k. (Other notations are $C(n, k)$ and $_nC_k$; both are pronounced "n choose k.")

5.63 THEOREM. Let n be a nonnegative integer. Then

(a) $\binom{n}{0} = 1$,

(b) $\binom{n}{n} = 1$,

(c) $\binom{n}{k} = 0$ if $k > n$,

(d) $\binom{n}{k} = \binom{n}{n-k}$ if $0 \leqslant k \leqslant n$,

(e) if $n \geqslant 1$ then $\binom{n}{1} = \binom{n}{n-1} = n$,

(f) $2^n = \binom{n}{0} + \binom{n}{1} + \cdots + \binom{n}{n}$.

PROOF OUTLINE.

(a) An n-set contains only one 0-set, namely, $\emptyset$.

(b) The only n-subset of an n-set S is S itself. (Franklin Roosevelt was widely acclaimed for a statement something like this.)

(c) If $k > n$ then an n-set contains no k-subsets.

(d) Given an n-set S, complementation gives a bijection between the collection of k-subsets of S and the collection of $(n - k)$-subsets of S. Alternatively stated, the number of ways of choosing a k-subset of S is equal to the number of ways of choosing the complement of a k-subset of S.

(e) The fact that $\binom{n}{1} = n$ is clear from the definition, and the equality $\binom{n}{1} = \binom{n}{n-1}$ follows from (d).

(f) We know (from Theorem 2.71) that if S is an n-set then $\#P(S) = 2^n$. On the other hand, $P(S)$ has as its members $\emptyset$, the 1-subsets of S, the 2-subsets of S, and so on. Counting all these subsets gives

$$\#P(S) = \binom{n}{0} + \binom{n}{1} + \binom{n}{2} + \cdots + \binom{n}{n},$$

and this yields the desired equation. $\square$

The next theorem provides an explicit formula for the binomial coefficients, and we will give two very different proofs. The strategy of the first proof is to count something two different ways, then to equate the results and deduce the conclusion. The second proof is by induction.

5.64 THEOREM.   If $0 \leqslant k \leqslant n$, then

$$\binom{n}{k} = \frac{n!}{k!(n-k)!}.$$

PROOF 1.   First consider how many repetition-free sequences of length k can be chosen from a given set of n elements. There are n possibilities for the first element, then $n - 1$ possibilities for the second element, and so on; eventually there are $n - (k - 1)$ possibilities for the kth element. Altogether there are

(*) $n(n - 1) \cdots (n - k + 1)$

sequences, by the product rule.

Now let's make the selection in another way. First choose k elements from the given n-set: there are $\binom{n}{k}$ ways to do this, by definition of the symbol $\binom{n}{k}$. Then, having chosen these k elements, there are $k!$ ways to arrange them in a sequence of length k. Thus the product rule yields a total of

(**) $\binom{n}{k} \cdot k!$

repetition-free sequences of length k.

Since (*) and (**) both count the same set, they are equal; therefore

(***)
$$\binom{n}{k} = \frac{n(n-1)\cdots(n-k+1)}{k!}$$

$$= \frac{n(n-1)\cdots(n-k+1)}{k!} \cdot \frac{(n-k)!}{(n-k)!}$$

$$= \frac{n!}{k!(n-k)!}.$$

This completes the first proof.

PROOF 2.   We will use induction on n. If $n = 0$ then also $k = 0$; and $\binom{0}{0} = 1$ because the empty set is the only 0-set. On the other hand,

$$\frac{0!}{0!(0-0)!} = 1$$

because $0! = 1$ by definition. This establishes the basis step of the induction. Now suppose the theorem is true for $n = t$. That is, assume

$$\binom{t}{k} = \frac{t!}{k!(t-k)!}, \qquad \text{for } 0 \leq k \leq t.$$

To complete the induction we must verify that the theorem holds for $t + 1$; that is, that

$$\binom{t+1}{k} = \frac{(t+1)!}{k!(t+1-k)!}, \qquad \text{for } 1 \leq k \leq t+1.$$

Let S be a $(t+1)$-set, and fix an element $x \in S$. Then we can write $S = \{x\} \cup S_0$, where S_0 is a t-set. Now, a k-subset of S is either a k-subset of S_0 or has the form $A \cup \{x\}$ for some $(k-1)$-subset $A \subseteq S_0$. But from the definition of binomial coefficient, there are $\binom{t}{k}$ of the former and $\binom{t}{k-1}$ of the latter. Thus

(5.65) $$\binom{t+1}{k} = \binom{t}{k} + \binom{t}{k-1}$$ (by the addition rule)

$$= \frac{t!}{k!(t-k)!} + \frac{t!}{(k-1)!\,(t-k+1)!}$$ (by the induction hypothesis)

$$= \frac{t!(t - k + 1) + t!k}{k!(t - k + 1)!}$$ (we have found a common denominator)

$$= \frac{t!(t + 1)}{k!(t - k + 1)!}$$ (factor $t!$ out of both numerator terms)

$$= \frac{(t + 1)!}{k!(t + 1 - k)!}. \quad \square$$

Formula (***) in Proof 1 of Theorem 5.64 is useful for computational purposes. It has fewer multiplications than the formula in the statement of the theorem.

5.66 COROLLARY (PASCAL'S FORMULA). If n and k are integers satisfying $1 \leqslant k \leqslant n - 1$, then

$$\binom{n}{k} = \binom{n-1}{k} + \binom{n-1}{k-1}.$$

PROOF.   Substitute $t = n - 1$ in Equation 5.65. $\square$

Pascal's formula gives a recurrence relation for the computation of binomial coefficients, given the initial data $\binom{n}{0} = \binom{n}{n} = 1$ for all n. Notice that no multiplication is needed for this computation (unlike the formula in Theorem 5.64). The recursion yields the following table, known as **Pascal's triangle**. To reproduce it, write the left column of 1s and the diagonal of 1s. Then, from top to bottom, plug each hole in the triangle with the sum of the number directly above the hole and the number to the upper left of the hole.

BINOMIAL COEFFICIENTS (PASCAL'S TRIANGLE)

n \ k	0	1	2	3	4	5	6	7	8	$\cdots$
0	1									
1	1	1								
2	1	2	1							
3	1	3	3	1						
4	1	4	6	4	1					
5	1	5	10	10	5	1				
6	1	6	15	20	15	6	1			
7	1	7	21	35	35	21	7	1		
8	1	8	28	56	70	56	28	8	1	
$\vdots$	$\vdots$	$\vdots$	$\vdots$	$\vdots$	$\vdots$	$\vdots$	$\vdots$	$\vdots$	$\vdots$	$\cdots$

5.67 EXAMPLE.   Now we can quickly settle a couple of the problems posed at the start of the chapter.

SOLUTION TO 5.1(e).   The number of clinks is equal to the number of 2-sets in a 10-set, which is

$$\binom{10}{2} = \frac{10!}{2!8!} = 45.$$

Similarly, the number of minuet teams that must be checked is

$$\binom{10}{3} = \frac{10!}{3!7!} = 120.$$

SOLUTION TO 5.1(c).   Each route from southwest to northeast consists of 16 one-block legs; exactly eight of these legs are easterly and eight are northerly. Each route can be represented by a string of sixteen E's and N's, using eight copies of each letter. For instance, the route shown on page 209 is represented by

NEEENNENNNNEENEE.

Such a string is completely specified by indicating which eight of the sixteen positions will contain N's. Otherwise put: each route corresponds to an eight-element subset of $\mathbb{N}_{16}$. Therefore, altogether there are

$$\binom{16}{8} = \frac{16!}{8!8!} = \frac{16 \cdot 15 \cdot 14 \cdot 13 \cdot 12 \cdot 11 \cdot 10 \cdot 9 \cdot 8!}{8!8!} = 12{,}870$$

allowable routes. More generally, the same argument can be used to show that for an $n \times n$ array of city blocks, the number of allowable southwest-to-northeast routes is $\binom{2n}{n}$. [It can also be shown (by methods beyond our treatment here) that when n is large, $\binom{2n}{n}$ is approximately $4^n/\sqrt{\pi n}$. For instance, when $n = 100$, the number of routes is approximately 9×10^{58}, which is nothing to sniff at.]

If the problem setting is generalized still further, to an $m \times n$ grid, then the same reasoning as before yields $\binom{m+n}{n}$ routes. (See figure.)

5.68 EXAMPLE.   Electronically transmitted information is often coded into strings of 0s and 1s. (Each digit is called a **bit**.) The **weight** of a string is its number of 1s. Interference or mechanical problems can introduce errors into transmitted messages, and the detection and correction of errors is a major concern of the mathematical subject known as *coding*

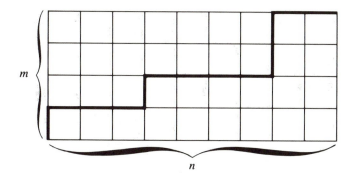

theory. Consider a code whose words are all the 16-bit strings whose weight is divisible by 4. Let's compute the number of codewords. If k is a multiple of 4 then there are $\binom{16}{k}$ codewords of weight k, since each word is determined by the set of k locations that hold 1s. So altogether the number of codewords is

$$\binom{16}{0} + \binom{16}{4} + \binom{16}{8} + \binom{16}{12} + \binom{16}{16}.$$

With the help of Theorem 5.63, this simplifies to

$$\binom{16}{8} + 2\left[1 + \binom{16}{4}\right].$$

Then computing $\binom{16}{8}$ and $\binom{16}{4}$ by means of Theorem 5.64 gives 16,512 for the total number of codewords.

Incidentally, while the code described in Example 5.68 allows certain errors to be detected, it doesn't allow error *correction.* For instance, an error (a bit received as 1 when 0 was intended, or vice versa) in an odd number of bits in a transmitted word will have odd weight and hence will not be a codeword. But even though we may know a word to be wrong, we have no way to know what was intended. For this reason it is desirable for codewords to be sufficiently different from each other so that if a small number of errors are made in transmitting a word, it will be clear which codeword was intended. This is a big subject and we won't go into it (but see Exercise 11 on page 142 for the notion of *distance* between codewords). Our message here is only that a thorough description of a code involves a lot of counting, and that binomial coefficients play a very active role.

5.69 DEFINITION.   Let S be a finite nonempty set. If $A \subseteq S$, define the **probability** of A (with respect to S), denoted $\Pr(A)$, to be the quotient $\#A/\#S$. (A more complete notation would be $\Pr_S(A)$, but if S has been fixed throughout a given discussion, we usually agree to remember it without writing it.)

5.70 EXAMPLES.   With the notation of Definition 5.69, for any finite nonempty set S we have

$$\Pr(\emptyset) = 0, \qquad \Pr(S) = 1,$$

and, in general,

$$0 \leqslant \Pr(A) \leqslant 1.$$

The probability of a set A can be interpreted as follows. View S as the set of possible outcomes of some experiment or the set of possible choices in some decision process, with each outcome or choice equally likely. (Imagine yourself blindfolded and choosing a card from a well-shuffled deck.) Then $\Pr(A)$ represents the fraction of the time that we can expect an experimental outcome or choice to be in A.

5.71 EXAMPLES.

(a) What is the probability of being dealt a flush in poker?

SOLUTION.   A poker deck has 52 cards, and it is partitioned into four 13-card suits. A poker hand has five cards, so the number of possible poker hands is $\binom{52}{5}$. A flush hand consists of cards from only one suit. There are $\binom{13}{5}$ flush hands of each suit, hence $4 \cdot \binom{13}{5}$ flush hands altogether. The desired probability is therefore the fraction

$$\frac{\#(\text{flush hands})}{\#(\text{poker hands})} = \frac{4 \cdot \binom{13}{5}}{\binom{52}{5}} = \frac{5{,}148}{2{,}598{,}960} \approx \frac{1}{500}.$$

(In the formal language of Definition 5.69, this is the probability of the set of flushes with respect to the set of all poker hands.)

(b) What is the probability of being dealt a full house in poker?

SOLUTION.   A full house consists of a triplet of cards of one numerical value together with a pair of cards of another value. There are 13 possibilities for the value of the triplet; having fixed that, there are $\binom{4}{3}$ ways to select three cards with that value. Then there remain 12 possibilities for the value of the pair; having chosen that value there are $\binom{4}{2}$ ways to choose the two cards. Therefore, by the product rule, we deduce that the probability of being dealt a full house is

$$\frac{\#(\text{full houses})}{\#(\text{poker hands})} = \frac{13 \cdot \binom{4}{3} \cdot 12 \cdot \binom{4}{2}}{\binom{52}{5}}$$

$$= \frac{13 \cdot 4 \cdot 12 \cdot 6}{2,598,960} \approx \frac{1}{694}.$$

By comparing this with the probability of a flush, we see that the full house is less likely and hence more highly prized.

We have seen equations (in 5.63 and 5.66) that give connections between different binomial coefficients. This is to be expected. For instance, in choosing a k-subset of an n-set, along the way we will have chosen a k'-subset, for some $k' < k$, so it is natural to expect an association between $\binom{n}{k}$ and $\binom{n}{k'}$. Here are some further samples of such phenomena.

5.72 THEOREM.

(a) If $1 \le k \le n$, then $\binom{n}{k} = \frac{n}{k}\binom{n-1}{k-1}$.

(b) If m and n are nonnegative integers, then

$$\sum_{k=0}^{n}\binom{m+k}{k} = \binom{m+n+1}{n}.$$

We'll start with two different proofs of (a). The first proof will come from playing with the formula for binomial coefficients given by Theorem 5.64. It is arithmetically easy and requires no set-theoretical baggage. On the other hand, it gives no insight into the meaning of the equation in terms of decision procedures. The second proof gives that insight, but some perspiration is lost along the way. (We must not be too critical of the first proof. Once a result is discovered *somehow*, perhaps then it is possible to interpret it and mine its riches.) We will also give two proofs of (b). The first has a geometric flavor, while the second relies on arithmetic. Both proofs use our earlier results on binomial coefficients.

PROOF 1 OF (a).

$$\binom{n}{k} = \frac{n!}{k!(n-k)!} = \frac{n}{k} \cdot \frac{(n-1)!}{(k-1)!(n-k)!}$$

$$= \frac{n}{k} \cdot \frac{(n-1)!}{(k-1)!((n-1)-(k-1))!} = \frac{n}{k} \binom{n-1}{k-1}$$

PROOF 2 OF (a). Choosing a k-subset of an n-set S can be done by first choosing an element $x \in S$ and then choosing $k-1$ elements from the remaining $n-1$ elements. Altogether there are $n\binom{n-1}{k-1}$ possibilities for this, by the product rule. But let's be careful. We have done *more* than pick a k-subset: we have also singled out one element x for special attention (as the first element chosen). More precisely, we have chosen an ordered pair (x, B), where B is a k-subset of S and $x \in B$. There is another way to do this: first choose B and *then* choose x. There are $\binom{n}{k}$ ways to choose B and, having done that, there are then k ways to choose x in B. This gives $k\binom{n}{k}$ possibilities for the choice of (x, B). We have counted something in two ways (a strategy familiar from the first proof of 5.64), and therefore the results are equal:

$$n\binom{n-1}{k-1} = k\binom{n}{k}.$$

The conclusion follows.

PROOF 1 OF (b). In the labelled graph pictured here, consider all the

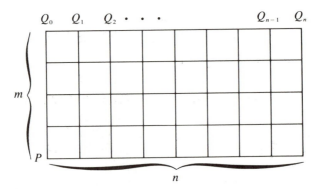

paths that start at P and end at a vertex on the upper boundary, moving up or to the right at each step. By the discussion at the end of Example 5.67, for each k satisfying $0 \leq k \leq n$, the number of paths from P to Q_k is

$\binom{m+k}{k}$. Hence the total number of such paths is

(*)
$$\sum_{k=0}^{n} \binom{m+k}{k}.$$

Now extend the original diagram by adding a new row at the top. Each

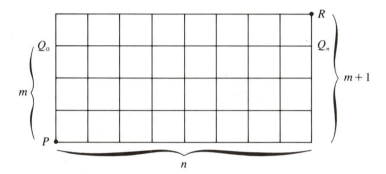

path already counted has exactly one extension to a path from P to R that touches no additional vertices on the original graph. (Move up from the original endpoint, then move right to R.) Moreover, *every* path from P to R is such an extension. Therefore the quantity (*) is equal to the number of paths from P to R. But the number of paths from P to R is also equal to

$$\binom{m+n+1}{n},$$

again by the argument of Example 5.67; therefore the stated equation is true.

PROOF 2 OF (b).   Let m be any nonnegative integer. We must prove that the asserted equation holds for every nonnegative integer n, and we will do this by induction on n.

If $n = 0$ then the equation to be verified is

$$\binom{m+0}{0} = \binom{m+1}{0};$$

and this is true by Theorem 5.63(a).

Now suppose the result is true for $n = t$; we must show that

$$\sum_{k=0}^{t+1} \binom{m+k}{k} = \binom{m+t+2}{t+1}.$$

But

$$\sum_{k=0}^{t+1} \binom{m+k}{k} = \sum_{k=0}^{t} \binom{m+k}{k} + \binom{m+t+1}{t+1}$$

$$= \binom{m+t+1}{t} + \binom{m+t+1}{t+1} \qquad \text{(by the induction hypothesis)}$$

$$= \binom{m+t+2}{t+1}. \qquad \text{(by Pascal's formula)} \qquad \square$$

A **binomial** is an algebraic expression that is a sum of two terms. The following theorem gives us a formula for the coefficients of the terms in the expansion of a power of a binomial; hence the name *binomial coefficient*.

5.73 BINOMIAL THEOREM. Let $a, b \in \mathbb{R}$, and let $n \in \mathbb{N}$. Then

$$(a + b)^n = \sum_{k=0}^{n} \binom{n}{k} a^k b^{n-k}.$$

Preamble to proof. When we multiply together n factors, each equal to $a + b$, we get a sum of terms of the form $a^k b^{n-k}$. The number of times a particular $a^k b^{n-k}$ appears as a term in the product is equal to the number of times exactly k of the a's have been chosen from the n factors $a + b$ to use as a multiplier. There are $\binom{n}{k}$ ways to make this choice. Therefore, it is reasonable to believe that the coefficient of $a^k b^{n-k}$ in the product is $\binom{n}{k}$, and now we will prove it.

PROOF. We will use the following piece of notational magic: given numbers $c_0, c_1, \ldots, c_t$, then

(∗) $$\sum_{k=0}^{t} c_k = \sum_{k=1}^{t+1} c_{k-1}.$$

(The magic becomes believable by means of an induction argument on t.)

Now we will prove the theorem by induction on n. If $n = 1$ the result is trivially true, so suppose the result is true for $n = t$; that is,

$$(a + b)^t = \sum_{k=0}^{t} \binom{t}{k} a^k b^{t-k}.$$

We must prove the result for $n = t + 1$:

$$(a + b)^{t+1} = \sum_{k=0}^{t+1} \binom{t+1}{k} a^k b^{(t+1)-k}.$$

We compute:

$(a + b)^{t+1} = (a + b)(a + b)^t$

$$= (a + b) \sum_{k=0}^{t} \binom{t}{k} a^k b^{t-k} \qquad \text{(by the induction hypothesis)}$$

$$= \sum_{k=0}^{t} \binom{t}{k} a^{k+1} b^{t-k} + \sum_{k=0}^{t} \binom{t}{k} a^k b^{t-k+1}$$

$$= \sum_{k=1}^{t+1} \binom{t}{k-1} a^k b^{t-k+1} + \sum_{k=0}^{t} \binom{t}{k} a^k b^{t-k+1} \qquad \text{(by (*))}$$

$$= a^{t+1} + \sum_{k=1}^{t} \binom{t}{k-1} a^k b^{t-k+1} + \sum_{k=1}^{t} \binom{t}{k} a^k b^{t-k+1} + b^{t+1}$$

$$= a^{t+1} + \sum_{k=1}^{t} \left[\binom{t}{k-1} + \binom{t}{k} \right] a^k b^{t-k+1} + b^{t+1}$$

$$= a^{t+1} + \sum_{k=1}^{t} \binom{t+1}{k} a^k b^{t-k+1} + b^{t+1} \qquad \text{(by Pascal's formula)}$$

$$= \sum_{k=0}^{t+1} \binom{t+1}{k} a^k b^{(t+1)-k}. \qquad \square$$

5.74 COROLLARY. $2^n = \binom{n}{0} + \binom{n}{1} + \cdots + \binom{n}{n-1} + \binom{n}{n}$

PROOF. Let $a = b = 1$ in the binomial theorem. (This is our second proof of this fact. The first appeared in Theorem 5.63). $\square$

5.75 EXAMPLE. Suppose x is a real number. Compute the coefficient of x^6 in the expansion of $(x - 3)^{11}$.

SOLUTION. By the binomial theorem,

$$(x - 3)^{11} = (x + (-3))^{11} = \sum_{k=0}^{11} \binom{11}{k} x^k (-3)^{11-k}.$$

So the term containing x^6 is $\binom{11}{6}(-3)^5 x^6$. But $\binom{11}{6} = 462$ and $(-3)^5 = -243$, therefore the answer is $-112,266$.

5.76 EXAMPLE. Imagine that you are at the market and that you have purchased the following items:

4 bananas (not attached)
5 cans of tuna
2 boxes of cereal
4 lemons
3 bottles of a particular classic cola
6 light bulbs.

Because some of the items are easily damaged, you care about the order in which they are put into the grocery bag, but you don't distinguish one banana from another, one cereal box from another, etc., for this purpose. In how many essentially different ways can the clerk fill the bag?

Notice that each pattern of bag loading corresponds to a sequence of length 24 of the form

(banana, cola, cereal, bulb, bulb, tuna, . . .).

So we are really counting sequences of length 24, subject to certain conditions on the terms, with the understanding that certain sequences are equivalent. (For example, we will not interpret the interchange of two bananas, while fixing the positions of the other objects, as a different packing arrangement.)

Let's state this problem a bit more formally: We are given n objects to be listed in sequence. The objects are partitioned into t types, with k_i objects of the ith type, for $1 \leq i \leq t$. How many essentially different sequences are there, if we do not distinguish between sequences whose respective terms are of the same type?

We can settle this with the help of binomial coefficients and the product rule. In a sequence of length n, there are $\binom{n}{k_1}$ ways to select the positions that will be filled by objects of the first type (say bananas). Having done this, there are $\binom{n-k_1}{k_2}$ ways to select the positions for the objects of the second type, and so on. Altogether, the product rule gives

$$\binom{n}{k_1}\binom{n-k_1}{k_2}\binom{n-k_1-k_2}{k_3} \cdots \binom{n-k_1-\cdots-k_{t-1}}{k_t}$$

different results. By the formula for binomial coefficients, this is equal to

$$\frac{n!}{k_1!(n-k_1)!} \cdot \frac{(n-k_1)!}{k_2!(n-k_1-k_2)!} \cdot \frac{(n-k_1-k_2)!}{k_3!(n-k_1-k_2-k_3)!} \cdot$$
$$\cdots \frac{(n-k_1-k_2-\cdots-k_{t-1})!}{k_t!0!}$$

Observe that the right-hand factor in each denominator is equal to the numerator of the next factor. So, after cancellation, the whole product dwindles to

$$\frac{n!}{k_1!k_2! \cdots k_t!}$$

Now we can calculate the answer to the grocery-store problem:

$$\frac{24!}{4!5!2!4!3!6!}$$

It is now natural to make the following definition.

5.77 DEFINITION. Let n and $k_1, \ldots, k_t$ be nonnegative integers, with $k_1 + \cdots + k_t = n$. Define the **multinomial coefficient**

$$\binom{n}{k_1 \ k_2 \ \ldots \ k_t} = \frac{n!}{k_1!k_2! \cdots k_t!}$$

A formal interpretation of multinomial coefficients proceeds as follows. Let A be an n-set, and suppose A is partitioned into blocks A_1, $A_2, \ldots, A_t$, where each A_i is a k_i-set, for $1 \leq i \leq t$. Consider the collection of all sequences of length n whose terms are all the elements of A. (That is, consider the set of surjections $\mathbb{N}_n \rightarrow A$.) Call two such sequences f and g *equivalent* if, for each i satisfying $1 \leq i \leq n$, the elements $f(i)$ and $g(i)$ belong to the same block. This is an equivalence relation on this collection of sequences, and the multinomial coefficient

$$\binom{n}{k_1 \ k_2 \ \ldots \ k_t}$$

is the number of equivalence classes.

5.78 EXAMPLE. Consider the number of different words that can be made using all the letters of "Mississippi." (By a *word* we just mean a string of letters.) There are four i's, four s's, two p's, and one M; eleven letters in all. The solution is the multinomial coefficient

$$\binom{11}{4 \ 4 \ 2 \ 1} = \frac{11!}{4!4!2!1!} = 34{,}650.$$

Like the term "binomial coefficient," the term "multinomial coefficient" comes from considering algebraic expressions. Given real numbers $x_1, x_2, \ldots, x_t$, consider the power

$$(x_1 + \cdots + x_t)^n = (x_1 + \cdots + x_t)(x_1 + \cdots + x_t) \cdots (x_1 + \cdots + x_t).$$

After performing this product but before collecting like terms, a typical term in this product has the form

$$x_1^{k_1} x_2^{k_2} \cdots x_t^{k_t}.$$

The coefficient of $x_1^{k_1} \cdots x_t^{k_t}$ *after* collecting like terms is equal to the number of ways of picking k_1 factors equal to x_1, and k_2 factors equal to x_2, and so on, as we multiply the n copies of $x_1 + x_2 + \cdots + x_t$. This is precisely the multinomial coefficient

$$\binom{n}{k_1 \ k_2 \ \ldots \ k_t}.$$

5.79 EXAMPLE. In the expansion of $(x_1 + x_2 + x_3 + x_4 + x_5)^7$, the coefficient of $x_1^2 x_3 x_4^3 x_5$ is

$$\binom{7}{2 \ 0 \ 1 \ 3 \ 1} = \frac{7!}{2!0!1!3!1!} = 420.$$

Exercises

1. Compute the following binomial coefficients.

   (a) $\binom{9863}{1}$ (e) $\binom{63}{60}$

   (b) $\binom{805}{2}$ (f) $\binom{940}{939}$

   (c) $\binom{14}{4}$ (g) $\binom{0}{0}$

   (d) $\binom{12^{800}}{0}$ (h) $\binom{5}{8}$

2. A team of nine starting baseball players will be chosen from a group of thirteen hopefuls.

   (a) How many possibilities are there for the team?
   (b) Suppose the coach's child is guaranteed a position on the team. Now how many possibilities are there for the team?

3. Without referring to the text, write the first ten lines of Pascal's triangle.

4. Determine *all* the nonnegative integers n that satisfy the equation

$$\binom{n}{4} = \binom{n}{6}.$$

5. A local restaurant offers a special salad lunch subject to these conditions: Each diner chooses exactly five ingredients from a menu of eight possibilities and then tops it with a blend of exactly two of the restaurant's four dressings.

   (a) How many dressed salad combinations are possible?
   (b) Jones loves the lunch, but he cannot tolerate the simultaneous inclusion of radishes (one of the eight items on the menu) and peanut butter dressing (one of the four choices), though all other possibilities are acceptable to him. How many acceptable choices does he have?
   (c) In the statement of the lunch conditions, change each "exactly" to "at most." *Now* answer (a) again. (Assume that the diner chooses at least one salad ingredient, but allow for the possibility of no dressing.)

6. The definition of $\binom{n}{k}$ mentions "an n-set." Show that this definition is independent of the n-set used. That is, show that if A_1 and A_2 are both n-sets, then the number of k-subsets of A_1 is equal to the number of k-subsets of A_2.

7. Write out the expression $(x + y)^7$ as a sum of integer multiples of terms of the form $x^i y^j$.

8. What is the coefficient of y^3 in the expansion of $(2x - 5y)^6$?

9. Prove that for nonnegative integers m and n the following formula holds:

$$\sum_{k=0}^{n} \binom{k}{m} = \binom{n+1}{m+1}$$

Use induction on n, starting with $n = 0$. (Refer to Definition 5.62 in the case $n = 0$.) Pascal's formula may be helpful.

10. Suppose $0 \leqslant k_1 + k_2 \leqslant n$. Verify the equation

$$\binom{n}{k_1}\binom{n - k_1}{k_2} = \binom{n}{k_1 + k_2}\binom{k_1 + k_2}{k_1}$$

in two ways:

(a) by direct application of the formula for binomial coefficients (Theorem 5.64);
(b) by an intuitive argument similar to that used in the text to verify 5.72(a).

11. Prove the equation $\binom{n+1}{3} - \binom{n-1}{3} = (n - 1)^2$ for each $n \geqslant 1$.

12. Eight varieties of plant seeds are to be selected for a garden. Altogether 15 varieties of vegetables and 10 flower varieties are available. Determine the number of ways the selection can be made under each of the following conditions.

(a) Four vegetables and four flowers must be chosen.
(b) There must be more flower varieties then vegetable varieties.
(c) There must be at least two flower varieties.

13. Determine the number of paths from P to Q in the accompanying graph under each of the following conditions. (Assume each leg of the path moves either up or to the right.)
(a) The path must go through vertex R.

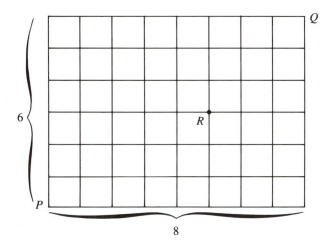

(b) The path must not go through vertex R.

(c) Every one-directional section of each path has even length. (An edge of a small square is one unit long.)

14. In how many ways can a collection of 25 objects be partitioned into two blocks of ten elements and one block of five elements?

15. What is the probability of being dealt four-of-a-kind in poker (that is, a five-card hand containing four cards of the same value, such as four aces or four 5s).

16. Show that if A is a nonempty finite set with subsets A_1 and A_2, then

$$\Pr(A_1 \cup A_2) = \Pr(A_1) + \Pr(A_2) - \Pr(A_1 \cap A_2).$$

17. (a) What is the probability that a fair coin will land on heads exactly four times if it is flipped seven times?

(b) What is the probability that a fair coin will land on heads either three times or four times if it is flipped seven times?

18. In baseball's World Series, two teams play a sequence of games, and the first team to win four games is the champion. Suppose evenly matched teams A and B are to play against each other in the Series. Explain why the probability that A will win the series is $\frac{1}{2}$; and go on to explain why this is larger than the probability that a fair coin will land on heads exactly four times in seven flips. [Refer to Exercise 17(a).]

19. With words and without formulas, explain why it is that if $n = k_1 + k_2$, then

$$\binom{n}{k_1 \; k_2} = \binom{n}{k_1} = \binom{n}{k_2}.$$

20. With words and without formulas, explain the equation

$$\binom{n}{1 \; 1 \; \ldots \; 1} = n!$$

21. In the setting of Example 5.76, suppose you insist that light bulbs be packed on top, just above the bananas, but the order in which the products in the lower part of the bag are put in doesn't matter. Under those conditions, determine the number of essentially different acceptable orderings in which the products can be put into the bag.

22. (a) How many different words can be constructed using (all) the letters of the word "magnesium"?
    (b) How many different words can be constructed from "magnesium" if the vowels are to remain in their original order?

23. (a) How many words can be constructed using all the letters of the word "zebratious"?
    (b) Answer the same question, except require the vowels to appear in alphabetical order.
    (c) Answer the same question, but require the vowels to appear in alphabetical order and the consonants to do likewise.

24. When the expression $(x_1 + x_2 + x_3 + x_4 + x_5)^9$ is multiplied out, what is the coefficient of $x_1^4 x_4^2 x_5^3$?

25. Determine the coefficient of $x^2 yz$ in the expansion of

$$(2x - y + z + 1)^7.$$

CHAPTER 6

Algebraic Structures

6.1 OPERATIONS

We have seen a variety of ways in which two members of a given set can be used as raw materials to produce a single element of the set. For instance, given two permutations σ_1 and σ_2, we obtain their *product* $\sigma_1\sigma_2$; given two subsets of a set A, their *union* is also a subset of A; given two real numbers, we can take their *sum*; and so on.

There are dramatic differences among the operations that we have just listed, because the objects they involve are quite different. But there are also similarities. For instance, each of these operations satisfies an associative law, and there are other common features as well. One of our goals in this chapter is to study operations in enough generality that we can know immediately, upon proving certain properties of an operation, that certain others must follow automatically.

The study of individual operations will lead us into the consideration of **algebraic structures**, that is, sets with one or more operations that may be linked in certain ways. In subsequent sections we will focus our attention on certain classes of structures whose operations satisfy particularly useful axiom systems (for example, groups, rings, and Boolean algebras), and especially to the following important and remarkable examples: the integers, polynomials, and complex numbers. We can give only the rudiments, because a thorough treatment would require a massive volume for each. In fact *no* treatment can be complete, because each topic is an active area of current research.

Consider the job done by subtraction of real numbers: to each ordered pair $(a, b) \in \mathbb{R} \times \mathbb{R}$, subtraction assigns a number denoted $a - b$. That is, subtraction is a function from $\mathbb{R} \times \mathbb{R}$ to $\mathbb{R}$. This motivates the following definition.

6.1 DEFINITION. Let S be a nonempty set. A **binary operation** $*$ on S is a function

$$*: S \times S \rightarrow S.$$

(The adjective *binary*, which we will usually omit, informs us that $*$ acts on ordered *pairs*. This distinguishes it from other kinds of operations that we won't focus on here; for instance, a function from S to S is sometimes called a *unary operation* on S.) If $(a, b) \in S \times S$, we write $a * b$ instead of the more cumbersome function notation $*((a, b))$ for the result of applying $*$ to (a, b).

6.2 EXAMPLES.

(a) Ordinary addition, subtraction, and multiplication are operations on $\mathbb{R}$, with the respective actions

$$(a, b) \overset{+}{\mapsto} a + b$$

$$(a, b) \overset{-}{\mapsto} a - b$$

$$(a, b) \overset{\cdot}{\mapsto} a \cdot b.$$

Division, given by $(a, b) \mapsto a \div b$, is strictly speaking *not* an operation on $\mathbb{R}$, since $a \div 0$ has no meaning. (That is, ordered pairs of the form $(a, 0)$ are not in the operation's domain.) But $\div$ *is* an operation on the set $\mathbb{R}$ of nonzero real numbers.

(b) If A is any set, then union and intersection are operations on the power set $P(A)$.

(c) Consider the set Σ^* of all strings over a finite alphabet Σ, as defined in the last section of Chapter 4. The **concatenation** of two strings is the string obtained by following one by the other. More precisely: if $w_1 = a_1 \ldots a_m$ and $w_2 = b_1 \ldots b_n$, with the a's and b's in Σ, define $w_1 w_2 = a_1 \ldots a_m b_1 \ldots b_n$; if $w \in \Sigma^*$ and ϵ is the empty word, define $\epsilon w = w \epsilon = w$. So if Σ is our usual alphabet, $w_1 = $ ybar, and $w_2 = $ cand, then $w_1 w_2 = $ ybarcand and $w_2 w_1 = $ candybar.

(d) Let S be any nonempty set. Define an operation $*$ on S by declaring $a * b = a$ for all a, $b \in S$. (That is, $*$ takes each ordered pair to its first coordinate.)

(e) Define an operation $*$ on $\mathbb{R}$ by $a * b = \max\{a, b\}$, the larger of a and b. This operation is useful in "sorting" problems. For instance, if $\{a_1, \ldots, a_n\}$ is a set of real numbers then

$$((((a_1 * a_2) * a_3) * \cdots) * a_{n-1}) * a_n$$

is the largest number in the set. (Check this claim by induction on n.)

(f) Let A be a nonempty set, and let $F(A)$ denote the set of all functions from A to A. Then composition of functions is an operation on $F(A)$. (The operation takes each ordered pair of functions (f, g) to the associated composition $f \circ g$.)

(g) If A is a nonempty set, then permutation product is an operation on the symmetric group S_A. (See Section 5.3 for terminology and notation.)

An operation $*$ on a small set $A = \{a_1, a_2, \ldots, a_n\}$ can be specified by a square table with n rows and n columns, in which the entry at the intersection of row i and column j is the element $a_i * a_j$. (*Rows* are horizontal, *columns* are vertical.) We list the elements of A outside the left and top borders of the table as indices for the rows and columns, respectively; and it is customary to adorn the table further with the operation symbol at the upper left corner. For example, the table

$*$	a_1	a_2	a_3
a_1	a_3	a_1	a_2
a_2	a_2	a_3	a_1
a_3	a_1	a_2	a_1

specifies an operation $*$ on the set $A = \{a_1, a_2, a_3\}$. This table tells us that $a_3 * a_1 = a_1$, $a_2 * a_2 = a_3$, and so on. (The table represents a set of nine such equations.)

6.3 DEFINITION. An operation $*$ on a set S is said to be **associative** if for every a, b, $c \in S$ the following equation holds:

(6.4) $$(a * b) * c = a * (b * c).$$

In Equation 6.4 we can see the advantage of using the notation $a * b$ instead of the function notation $*((a, b))$; for in the latter notation, 6.4 becomes

$$*((*((a, b)), c)) = *((a, *((b, c)))),$$

an expression both grotesque and unenlightening.

The reader should verify that in Example 6.2, operations (b) through (g) are associative. In (a), addition and multiplication are associative operations while subtraction and division are not, since

$$(2 - 1) - 1 \neq 2 - (1 - 1) \quad \text{and} \quad (4 \div 3) \div 2 \neq 4 \div (3 \div 2).$$

Notice that to show that a given operation $*$ on a set S is *not* associative, it is enough to exhibit *some* $a, b, c \in S$ such that $(a * b) * c \neq a * (b * c)$. The inequality needn't hold for every choice of $a, b, c \in S$; for example, though $\div$ is not associative, we nevertheless have $(3 \div 5) \div 1 = 3 \div (5 \div 1)$.

6.5 DEFINITION.   Let $*$ be an operation on S. An element $e \in S$ is an **identity element** for $*$ if, for all $a \in S$,

$$e * a = a \quad \text{and} \quad a * e = a.$$

Refer again to Examples 6.2. It is easy to check that 0 is an identity for addition on $\mathbb{R}$ and 1 is an identity for multiplication on $\mathbb{R}$. Now consider whether the *subtraction* operation on $\mathbb{R}$ has an identity element e. If such an element were to exist, then Definition 6.5 would require that

$$e - 1 = 1 \quad \text{and} \quad 1 - e = 1.$$

The first of these equations yields $e = 2$, while the second yields $e = 0$, a contradiction; therefore subtraction on $\mathbb{R}$ has *no* identity element. A similar argument can be used to show that the *division* operation on $\mathbb{R}$ has no identity. In 6.2(b) the empty set is an identity for $\cup$, and the set A is an identity for $\cap$. In 6.2(c) the empty word ϵ is an identity. Checking for identities in 6.2(d) through (g) is left as an exercise.

In each example we have displayed at most one identity. The following result shows that this is not due to a lack of ambition.

6.6 THEOREM.   If the operation $*$ on the set S has an identity, then $*$ has exactly one identity.

PROOF. If e and e_1 are identities for $*$, then

$$e * e_1 = e_1,$$

since e is an identity; but also

$$e * e_1 = e,$$

since e_1 is an identity. Thus $e_1 = e * e_1 = e$. □

6.7 DEFINITION. Suppose the operation $*$ on S has identity element e, and let $a \in S$. If there is an element $b \in S$ satisfying $a * b = b * a = e$, then b is said to be an **inverse** of a with respect to $*$.

6.8 THEOREM. Suppose $*$ is an associative operation on S with identity e. If an element $a \in S$ has an inverse, then it has only one inverse.

PROOF. If b and c are inverses of a, then

$$\begin{aligned}
b &= b * e && \text{(by definition of } e) \\
&= b * (a * c) && \text{(since } c \text{ is an inverse of } a) \\
&= (b * a) * c && \text{(by associativity)} \\
&= e * c && \text{(since } b \text{ is an inverse of } a) \\
&= c. && \text{(by definition of } e) \quad □
\end{aligned}$$

Refer to Examples 6.2. With respect to addition, each $a \in \mathbb{R}$ has inverse $-a$; with respect to multiplication, each $a \neq 0$ has inverse $1/a$. If A is a nonempty set, then $\emptyset$ is the identity with respect to the operation $\cup$ on $P(A)$, but no $X \in P(A) - \{\emptyset\}$ has an inverse, since for every $Y \in P(A)$ we have $X \cup Y \neq \emptyset$. Similarly, though Σ^* has the empty word ϵ as identity, if $w \in \Sigma^* - \{\epsilon\}$ then w has no inverse with respect to concatenation, since concatenation can never decrease word length (so no word z exists satisfying $wz = \epsilon$).

Exercises

1. Suppose the sum and product of integers a, b are denoted $+((a, b))$ and $\cdot((a, b))$, respectively. Use this notation to rewrite the familiar rule $a(b + c) = ab + ac$.

2. Recall from Definition 4.47 that a string of length n over the alphabet Σ is a function $\mathbb{N}_n \to \Sigma$. Use this recollection to give a careful defini-

tion of the concatenation of two strings. (That is, define the concatenation as a *function* instead of using the intuitive phrase "by following one by the other" used in Example 6.2(c).)

3. Let n be a positive integer. How many different binary operations can be defined on a set of n elements?

4. (a) Determine all the real numbers a, b, c for which the equation $(a - b) - c = a - (b - c)$ holds.
   (b) Do the same for the equation $(a \div b) \div c = a \div (b \div c)$.

5. If S is a nonempty set and $n \in \mathbb{N}$, let S^n denote the set of n-tuples of elements of S. An **n-ary operation** on S is a function $S^n \to S$. (In particular, a unary operation is 1-ary and a binary operation is 2-ary.) Suppose $*$ is a binary operation on S. Show that $*$ can be used recursively to define an n-ary operation on S for each $n \geq 2$. (Suggestion: Start by considering the case $n = 3$.)

6. When the subtraction operation is restricted to $\mathbb{N} \times \mathbb{N}$, the result is not an operation on $\mathbb{N}$, according to Definition 6.1, because the condition $a, b \in \mathbb{N}$ does not guarantee that $a - b \in \mathbb{N}$. Determine the smallest set S satisfying $\mathbb{N} \subset S \subseteq \mathbb{R}$ such that the restriction of subtraction to $S \times S$ *is* an operation on S.

7. Show that the division operation on $\dot{\mathbb{R}}$ has no identity element.

8. If an operation $*$ on a finite set S is defined by a table, how can we learn from the table whether there is an identity element for $*$?

9. Suppose $*$ is an operation on the two-element set $\{e, a\}$, and suppose that e is an identity element for $*$. Prove that $*$ is associative.

10. Let $*$ be an operation on a set S. An element $\lambda \in S$ is a **left identity** for $*$ if $\lambda * a = a$ for all $a \in S$; and $\rho \in S$ is a **right identity** for $*$ if $a * \rho = a$ for all $a \in S$.
    (a) Give an example showing that a left identity for $*$ need not be an identity element for $*$.
    (b) Show that if the operation $*$ has both a left identity λ and a right identity ρ, then $*$ has an identity element.

11. Assume S is a finite set with an operation $*$ given by a table. Consider the question, "Does every element of S have an inverse with respect to $*$?" How can we answer this question by studying the table?

12. Show that if $*$ is an associative operation on the set S then

$$((a * b) * c) * d = (a * b) * (c * d) = a * (b * (c * d)), \forall a, b, c, d \in S.$$

13. In each case determine whether the given rule describes an operation on S. If it does, determine whether the operation has an identity element and (if so) whether every element has an inverse.

    (a) S is the set of odd integers, $(a, b) \mapsto ab$.
    (b) S is the set of odd integers, $(a, b) \mapsto a + b$.
    (c) S is the set of even integers, $(a, b) \mapsto a + b$.
    (d) S is the set of even integers, $(a, b) \mapsto ab/2$.
    (e) S is the set of nonzero real numbers, $(a, b) \mapsto a^2 b$.
    (f) S is the set of rational numbers, $(a, b) \mapsto a\sqrt{b^2}$.

14. Suppose $*$ is an associative operation on the three-element set $S = \{e, a, b\}$. Suppose further that e is an identity element for $*$ and that every element of S has an inverse. Prove that there is exactly one possibility for $*$, and exhibit its table.

15. Suppose r and s are fixed integers, and an operation $*$ is defined on $\mathbb{Z}$ by

$$a * b = ra + sb, \qquad \text{for all } a, b \in \mathbb{Z}.$$

Under what conditions on r and s does this operation have an identity element?

6.2 SEMIGROUPS, MONOIDS, AND GROUPS

At this point it is useful to introduce some terminology and notation to streamline our discussion.

6.9 DEFINITION.

(a) A **semigroup** is a nonempty set with an associative operation.

(b) A **monoid** is a semigroup that has an identity element.

(c) A **group** is a monoid in which each element has an inverse. Thus *a group is a nonempty set with an associative operation with respect to which there is an identity element and every element has an inverse.*

From what we have shown in Theorems 6.6 and 6.8, it follows that a group (or monoid) has a *unique* identity, and each element of a group has a *unique* inverse. If G is a group and $*$ is its operation, we usually say "G is a group under $*$."

Semigroups, monoids, and groups are examples of **algebraic structures** (also called **algebraic systems**): sets on which one or more accompanying operations have been defined. If an operation $*$ gives a set G a group structure, we usually say "the group G" instead of "the set G together with its accompanying operation $*$ under which G is a group." Notational brevity has its risks, however: G may be a group (or monoid, or semigroup, or some other kind of algebraic structure that we will discuss later) under one operation but *not* under another operation of interest. So if we say just "the group G," we should be sure that the operation is understood from the context. A reasonable alternative is to define a group as an ordered pair $(G, *)$, where G is a nonempty set and $*$ is an operation on G satisfying the appropriate properties (associativity, etc.). But in practice even the brief notation $(G, *)$ is shortened to G.

6.10 EXAMPLE. From our earlier discussion, it follows that $\mathbb{R}$ is a group under addition, and $\dot{\mathbb{R}} = \mathbb{R} - \{0\}$ is a group under multiplication. Similarly, both $\mathbb{Q}$ and $\mathbb{Z}$ are groups under addition, and $\dot{\mathbb{Q}} = \mathbb{Q} - \{0\}$ is a group under multiplication. But $\mathbb{Z} - \{0\}$ is *not* a group under multiplication, since integers other than 1 and -1 do not have multiplicative inverses in $\mathbb{Z}$. The collection of symmetries of a geometric figure (see Section 5.4) is a group under composition of functions; and the symmetric group S_n and the dihedral group D_n are also groups (under permutation product), as a check of the definitions will show.

Examples of groups have been studied since antiquity, but it was not until the nineteenth century that a general theory of groups was formulated and developed into a major mathematical force. For hundreds of years mathematicians had known the quadratic formula, which gives the roots of polynomials of the form $ax^2 + bx + c$ (where $a, b, c \in \mathbb{R}$), and had sought to extend it to a general formula for the roots of polynomials $a_n x^n + a_{n-1} x^{n-1} + \cdots + a_1 x + a_0$ for any degree $n \in \mathbb{N}$. Like the quadratic formula, the desired formula was required to express a polynomial's roots in terms of its coefficients, using a finite sequence of arithmetic operations (addition, subtraction, multiplication, and division) and radicals. During the period from 1829 to 1832, the French mathematician Evariste Galois laid the foundations of group theory and then used this research to prove that no such formula is possible. Galois associated a group with each polynomial and showed that in order for the desired root formula to exist, the associated group would have to have special properties. He then showed that for each $n \geq 5$, there are polynomials of degree n whose groups do *not* have those properties, hence the roots of those polynomials cannot be determined by such a formula. (Incidentally, Galois, an individual of substantial and varied passions, was shot and killed in a duel at the age of 20, in 1832.)

In this century, group theory has made spectacular contributions to physics and chemistry, notably in connection with quantum mechanics and crystallography. And group theory plays a central role in *algebraic coding theory,* the subject that examines many aspects of the following problem: How can we design our communication networks in such a way that a received message coincides with the message transmitted, despite the presence of interference (or "noise") in the communications medium; and can we do this with reasonable efficiency? There have been wide-ranging applications of group theory to other areas of mathematics, especially through the study of various kinds of symmetry groups, not necessarily of a geometric nature. Of special importance are the symmetric groups S_n; for instance, in Section 5.7 we applied the theory of S_n to a combinatorial problem on the possible configurations of Rubik's Cube, and this is a typical application. A shift from the Cube to quantum mechanics starts when "facelets" are replaced by "atomic particles" in the discussion. The underlying problem—to describe what can go where under the action of a transformation—is then much the same. (But the theory of permutations is just one of many ingredients in the mathematics of quantum mechanics.) Group theory has also made major contributions to *topology,* which is the mathematics of shapes and surfaces and their deformations.

6.11 DEFINITION. An operation $*$ on a set S is **commutative** if $a * b = b * a$ for every $a, b \in S$. We say that a semigroup is commutative or **abelian** if its operation is commutative. (The term *abelian* is after the Norwegian mathematician Niels Abel (1802–1829), whose research on the roots of polynomials was an inspiration for some of Galois's work shortly thereafter.)

6.12 EXAMPLES. The arithmetic operations of addition and multiplication are commutative, while subtraction and division are not. The product operation for permutations in S_n is not commutative if $n \geq 3$ since, for example, $(1 \quad 2 \quad 3)(1 \quad 2) \neq (1 \quad 2)(1 \quad 2 \quad 3)$.

6.13 NOTATION. From now on we will write most group operations as products, using permutation product as a model, so that a new operation symbol needn't be invented each time a new group is discussed. Thus if G is a group under the operation $*$ and $a, b \in G$, we usually write ab instead of $a * b$. We then use the symbol e or 1 to denote the identity; a^{-1} to denote the inverse of a; and, following 5.19 and 5.20, we write

(6.14)
$$a^t = \begin{cases} \underbrace{aa \cdots a}_{t \text{ factors}} & \text{if } t > 0, \\ e & \text{if } t = 0, \\ (a^{-1})^{-t} & \text{if } t < 0. \end{cases}$$

When all of this notation is used, it is customary to call G a **multiplicative** group. (The terminology is a bit loose, since it is only our *notation* that is multiplicative; the group operation may have nothing to do with ordinary multiplication.)

However, if G is known to be abelian, then we often write the group operation additively: $a + b$. Then 0 is used to denote the identity (called the **zero element** of G); $-a$ denotes the inverse of a; and, for $t \in \mathbb{Z}$ and $a \in G$, formula 6.14 becomes

(6.15)
$$ta = \begin{cases} \underbrace{a + a + \cdots + a}_{t \text{ terms}} & \text{if } t > 0, \\ 0 & \text{if } t = 0, \\ (-t)(-a) & \text{if } t < 0. \end{cases}$$

With this notation G is called an **additive** group. Incidentally, in an additive group the expression $a + (-b)$ is usually abbreviated $a - b$.

REMARK. In Theorem 5.22 we observed that laws of exponents (such as $\sigma^s \sigma^t = \sigma^{s+t}$) hold for products in any symmetric group S_n. In fact, the corresponding laws hold in any group G, but we won't show that here.

Exercises

1. Without proof, in each case label the given structure as a semigroup, a monoid, or a group, using the strongest appropriate label. ("Monoid" is stronger than "semigroup," and "group" is stronger than "monoid.") Say "none" if appropriate. If the structure is a monoid, name

its identity element; and if the structure is a group, display the inverse of a typical element.

(a) The nonnegative real numbers under addition.
(b) The integers under multiplication.
(c) The nonzero rational numbers under addition.
(d) The integers under the operation $a * b = a + 2b$.
(e) The even integers under addition.
(f) The nonzero real numbers under the operation $a * b = 2ab$.

2. Let S be a nonempty set, and let $e \in S$. Define an operation $*$ on S by putting $a * b = e$ for all $a, b \in S$.

(a) Show that S is a semigroup under the operation $*$.
(b) Prove that S is a monoid if and only if $S = \{e\}$.

3. Explain why the manipulations that can be done by hand on a shoe-lace constitute a group under the operation "followed by". Then go on to explain why if hands and *scissors* can be used, the manipulations constitute a monoid but *not* a group.

4. Consider Examples 6.2(b) through (e). (In (b) consider each operation separately.)

(a) Show that each of these structures is a semigroup.
(b) Which of these is a monoid? Explain briefly.
(c) Which of these is a group? Explain briefly.

5. Let S and T be semigroups, with each operation written multiplicatively. Now define an operation on the Cartesian product $S \times T$ by

$$(s, t)(s', t') = (ss', tt'), \qquad \forall s, s' \in S, \ t, t' \in T.$$

(a) Show that $S \times T$ is a semigroup under this operation.
(b) Show that if S and T are monoids then so is $S \times T$.
(c) Show that if S and T are groups then so is $S \times T$.

6. Consider the semigroup $S \times T$ defined in Exercise 5. Show that if S is not a monoid then $S \times T$ is not a monoid.

7. For $n \geq 1$, let A_n denote the set of *even* permutations in the symmetric group S_n. Verify that A_n is a group under permutation product. (The group A_n is called the **alternating group on n letters.**)

8. Show that the set of nonzero real numbers of the form $a + b\sqrt{5}$, with $a, b \in \mathbb{Q}$, forms a group under ordinary multiplication.

9. (*Cancellation law*) Prove that if G is a group and $ab = ac$ for some $a, b, c \in G$, then $b = c$. [*Note:* The cancellation law has this important

consequence for finite groups: each row of a group's operation table must contain each group element exactly once. The cancellation law "$ba = ca \Rightarrow b = c$" also holds for groups (by a similar proof), and it follows from this that each *column* of a group's table also contains each group element once.]

10. Let G be a group, and let H be a nonempty subset of G. We say that H is a **subgroup** of G if H is a group under the operation obtained by restricting G's operation to $H \times H$. (*Example:* Exercise 7 shows that A_n is a subgroup of S_n.) In what follows, you may need the cancellation law of Exercise 9.

   (a) Show that if H is a subgroup of G, then the identity element of H is the identity element of G.
   (b) Prove that H is a subgroup of G if and only if $h_1 h_2 \in H$ for all $h_1, h_2 \in H$ and also $h^{-1} \in H$ for all $h \in H$. (Here we are using multiplicative notation, so h^{-1} denotes the inverse of h in G.)
   (c) Prove that H is a subgroup of G if and only if $h_1 h_2^{-1} \in H$ for all $h_1, h_2 \in H$.

11. (Refer to Exercise 10.) Describe a way in which the symmetric group S_m can be viewed as a subgroup of S_n if $m \leqslant n$.

12. Let X be a nonempty set, and let F denote the set of functions from X to X. Prove the following statements.

   (a) The set F is a monoid under composition of functions.
   (b) If X has more than one element then F is not a group.
   (c) The set of bijections from X to X is a group under composition.

13. If a finite semigroup is abelian, how is this apparent from the table for the operation?

14. An archeological dig has uncovered early evidence of algebraic activity, and fragmentary data on three groups has been found in the rubble. In each of the following cases, reconstruct the multiplication table for the group from the given data, and justify your answer. (Incidentally, fossil studies have shown that in ancient days the symbol e always denoted the group identity.) You may freely use the cancellation law stated in Exercise 9, as well as the note at the end of that exercise.

   (a) $G = \{e, a, b\}$.
   (b) $G = \{e, a, b, c\}$ and $a^2 = b$.
   (c) $G = \{e, a, b, c\}$ and $a^2 = b^2 = e$.

15. Show that in any group G the following hold:
    (a) $(a^{-1})^{-1} = a$ for all $a \in G$.
    (b) The mapping $G \to G$ given by $g \mapsto g^{-1}$ is a bijection.
    (c) $(ab)^{-1} = b^{-1}a^{-1}$ for all $a, b \in G$.

16. (In this exercise you may use the results of Exercises 10 and 15.) We define the **center** of a group G to be the set

$$Z(G) = \{x \in G \mid xg = gx \text{ for all } g \in G\}.$$

    (a) Prove that $Z(G)$ is a subgroup of G.
    (b) Find the center of the symmetric group S_2 and of the symmetric group S_3. (Suggestion: Start by exhibiting the multiplication tables of these groups.)
    (c) What can be said about the center of S_n if $n \geq 4$, in view of the result in (b)? Explain.

17. (a) Write the statements in Exercise 10(b) and (c) and Exercise 15 in additive notation.
    (b) The *laws of exponents,* stated for the symmetric group S_n in Theorem 5.22, actually hold in any group. (You need not show this.) State these laws in *additive* notation.

18. Let G be a finite group, with $\#G = n$. (The number of elements in a group is called the **order** of the group.) Show that for each $x \in G$ there is an integer $k \leq n$ such that $x^k = e$. (Suggestion: Start by noticing that two items on the list $x, x^2, \ldots, x^n, x^{n+1}$ must be equal.)

6.3 ISOMORPHISM

If the symbols A and B represent cars, we are likely to write $A = B$ only if A and B represent precisely the same car. But even if $A \neq B$ we will probably regard A and B as "essentially" the same if the two vehicles are the same model and year, that is, if they have the same structure, whether or not they are actually the same object. Now suppose G and H are semigroups. When should we view G and H as being essentially the same? The condition that there be a bijection from G to H is natural, because we would expect "essentially the same" to include a guarantee of equality of size. But note: this condition pays no heed to the operations that constitute the semigroups' algebraic structure; so we need a further condition — an *algebraic* condition.

Consider the semigroups with operations defined by the following four tables:

(6.16)

*	a	b
a	a	a
b	a	a

(1)

+	a	b
a	a	b
b	b	a

(2)

·	y	z
y	y	z
z	z	y

(3)

⊗	r	s
r	s	r
s	r	s

(4)

The first two semigroups have the same underlying set, {a, b}. But (2) has an identity element (namely, a) and hence is a monoid, while (1) has no identity. So we can easily agree that tables (1) and (2) represent essentially different structures. Now compare tables (2) and (3). If we were to transform the symbols a to y, b to z, and + to · , then table (2) would become table (3). So, except for notational differences, semigroups (2) and (3) are the same. Tables (2) and (4) appear at first glance to have more significant differences, but by listing the indexing elements r and s in reverse order at the left and top of table (4), we find that operation ⊗ is also described by

⊗	s	r
s	s	r
r	r	s

Now it becomes clear that semigroup (4) is essentially the same as (2) and (3).

Let's summarize this intuitive discussion. Two semigroups G and H whose operations are given by tables are "essentially the same" (we will say *isomorphic*) if there is a way to transform the elements of G into elements of H so that the table for G is transformed into the table for H. The following definition makes this idea precise, and it also handles semigroups that are too large to be specified by tables.

6.17 DEFINITION. Let G and H be semigroups with respective operations * and ⊗. An **isomorphism** from G to H is a bijection $f: G \to H$ satisfying

$$f(a * b) = f(a) \otimes f(b) \qquad \text{for all } a, b \in G.$$

If an isomorphism exists from G to H, then we write $G \approx H$ and say that G and H are **isomorphic**. (Recall that we used the term informally in Chapter 5; see pages 229 and 232.)

6.18 EXAMPLES.

(a) Let G and H be the semigroups in 6.16, tables (2) and (4), respectively. Define $f: G \to H$ by

$$f(a) = s \quad \text{and} \quad f(b) = r.$$

Then

$$\left. \begin{array}{l} f(a + a) = f(a) = s \\ f(a) \otimes f(a) = s \otimes s = s \end{array} \right\} \quad \text{hence } f(a + a) = f(a) \otimes f(a);$$

$$\left. \begin{array}{l} f(a + b) = f(b) = r \\ f(a) \otimes f(b) = s \otimes r = r \end{array} \right\} \quad \text{hence } f(a + b) = f(a) \otimes f(b).$$

Similarly, $f(b + a) = f(b) \otimes f(a)$ and $f(b + b) = f(b) \otimes f(b)$. (Check these.) Moreover, f is a bijection. Therefore f is an isomorphism from G to H.

(b) Recall the magic of logarithms: they transform multiplication into addition. Let's have a closer look.

The set $\mathbb{R}^+$ of positive real numbers constitutes a group under ordinary multiplication. (Briefly: multiplication is associative; 1 is the identity element; and if $\alpha \in \mathbb{R}^+$ then $\alpha^{-1} \in \mathbb{R}^+$.) Also, we have already observed that the system $\mathbb{R}$ of real numbers is a group under addition. The natural logarithm function

$$\log: \mathbb{R}^+ \to \mathbb{R}$$

can be shown to be a bijection. [For instance, it is increasing (and hence injective) since its derivative is $1/x$, which is positive on $\mathbb{R}^+$.] Moreover, since

$$\log(xy) = \log x + \log y, \quad \text{for all } x, y \in \mathbb{R}^+,$$

it follows that log is an isomorphism from $\mathbb{R}^+$ to $\mathbb{R}$. Our intuition is that multiplication of two numbers in $\mathbb{R}^+$ is transformed into addition of their images in $\mathbb{R}$ by means of the logarithm function.

Later in this chapter we will encounter algebraic structures that come equipped with a *pair* of operations. In that setting, if G has operations $*_1$ and $*_2$ and H has operations $\otimes_1$ and $\otimes_2$, we will say G and H are isomorphic structures if there is a bijection $f: G \rightarrow H$ satisfying

$$f(a *_i b) = f(a) \otimes_i f(b), \qquad \text{for all } a, b \in G \text{ and } i = 1, 2.$$

The following theorem suggests reasons for describing isomorphisms as *structure-preserving* functions.

6.19 THEOREM.   Let $f: G \rightarrow H$ be an isomorphism of semigroups.

(a) If G is a monoid with identity e_G, then H is a monoid with identity $e_H = f(e_G)$.
(b) If G is a group then H is a group, and $(f(a))^{-1} = f(a^{-1})$ for all $a \in G$.

PROOF.   To streamline the notation, we will write the operations in G and H as products, omitting the operation symbols (following 6.13).

(a) To prove the equality $f(e_G) = e_H$, we must show that $f(e_G)h = hf(e_G) = h$ for each $h \in H$. With this goal in mind, pick $h \in H$. Because f is surjective, there is an element $g \in G$ such that $f(g) = h$; then

$$f(e_G)h = f(e_G)f(g) = f(e_G g) = f(g) = h,$$

and similarly $hf(e_G) = h$.
(b) Let $a \in G$. We must check that $f(a^{-1})f(a) = e_H$. (Why will this do the job?) We have

$$f(a^{-1})f(a) = f(a^{-1}a) = f(e_G) = e_H. \qquad \square$$

Exercises

1. Consider the group $\dot{\mathbb{R}}$ of nonzero real numbers under multiplication. In each case determine whether the given mapping f is an isomorphism from $\dot{\mathbb{R}}$ to $\dot{\mathbb{R}}$.

   (a) $x \mapsto \dfrac{1}{x}$

   (b) $x \mapsto x^2$

   (c) $x \mapsto -x$

(d) $x \mapsto \frac{1}{2}x$

(e) $x \mapsto \begin{cases} x^2 & \text{if } x > 0 \\ -x^2 & \text{if } x < 0 \end{cases}$

2. Let G and H be semigroups, and let $f: G \to H$ be an isomorphism. Show that if G is abelian then H is abelian. (Start by picking $h_1, h_2 \in H$. The goal is to show that $h_1 h_2 = h_2 h_1$. Use the fact that f is surjective.)

3. A group G is said to be **cyclic** if there is an element $g \in G$ such that every element in G has the form g^n for some $n \in \mathbb{Z}$. (Such an element g is said to be a **generator** of G.) Prove that if G is cyclic and $G \approx H$ then H is also cyclic.

4. Let $f: G \to H$ be an isomorphism of monoids.

   (a) Use induction to prove that

   $$f(x^n) = (f(x))^n \qquad \text{for all } n \geq 0.$$

   (b) Now suppose G and H are groups. Use Theorem 6.19(b) along with part (a) of this exercise to prove that

   $$f(x^n) = (f(x))^n \qquad \text{for all } n \in \mathbb{Z}.$$

   (If we use additive notation for the group operations, this result takes the form

   $$f(nx) = nf(x) \qquad \text{for all } n \in \mathbb{Z}.)$$

5. An **automorphism** of an algebraic structure A is an isomorphism from A to A. Show that the correspondence $x \mapsto 5x$ describes an automorphism of the group of real numbers under addition, but *not* an automorphism of the group of integers under addition, and *not* an automorphism of the multiplicative group of nonzero real numbers.

6. Prove that if G, H, and K are semigroups and $\varphi: G \to H$ and $\psi: H \to K$ are isomorphisms, then $\psi \circ \varphi: G \to K$ is also an isomorphism.

7. For each $n \in \mathbb{Z}$, let $n\mathbb{Z}$ denote the set of integer multiples of n. That is,

   $$n\mathbb{Z} = \{nk \mid k \in \mathbb{Z}\}.$$

   (a) Show that $n\mathbb{Z}$ is a group under addition.

(b) Verify that if $n \neq 0$, the group $n\mathbb{Z}$ is isomorphic to $\mathbb{Z}$ (also viewed as a group under addition).

8. Let G and H be groups. A function $f: G \rightarrow H$ satisfying

$$f(ab) = f(a)f(b) \qquad \text{for all } a, b \in G$$

is called a **homomorphism** from G to H. (So an isomorphism is just a bijective homomorphism.)

(a) Prove that if $f: G \rightarrow H$ is a homomorphism, then the image $f(G)$ is a subgroup of H. (For an introduction to subgroups, see Exercise 10 of Section 6.2.)

(b) Let e_H denote the identity element of H. We define the **kernel** of the homomorphism f to be the set

$$\ker f = \{x \in G \mid f(x) = e_H\}.$$

Show that $\ker f$ is a subgroup of G. [*Remark.* If f is not injective, then $\ker f$ provides a means of measuring f's failure to be injective. If $\ker f$ has m elements (in other words, if e_H has m preimages in G), it can be shown that f takes exactly m elements of G onto each element of $f(G)$.]

9. Ever since his happy childhood experiences with addition, Stanley had wanted to be a real number, and he especially coveted the role of zero as the additive identity. After a struggle with the government's regulatory agencies, permission was obtained for the construction of the following group G. As a set,

$$G = (\mathbb{R} - \{0\}) \cup \{\text{Stanley}\}.$$

(That is, the real line is punctured at zero, and the hole is plugged with Stanley.) An operation $\oplus$ is defined on G as follows:

$$a \oplus b = \begin{cases} a + b & \text{if } a, b \in \mathbb{R}, \\ a & \text{if } b = \text{Stanley}, \\ b & \text{if } a = \text{Stanley}. \end{cases}$$

Thus, for example,

$$3 \oplus 5 = 8, \qquad -6 \oplus \text{Stanley} = -6,$$

and

$$\text{Stanley} \oplus \text{Stanley} = \text{Stanley}.$$

Prove that G is a group, with Stanley as its identity element, and then prove that G is isomorphic to the additive group of real numbers.

6.4 THE INTEGERS: OPERATIONS AND ORDER

The integers are the focus of *number theory,* one of the most fascinating areas of mathematical activity. The subject is unique in its abundance of appealing problems whose statements are accessible to elementary school students, but whose *solutions* have eluded mathematicians for centuries. Until recently, most of number theory was classified as "pure mathematics," an expression that suggests elegance and beauty to those engaged in its study, and worthless irrelevance to those who insist on immediate practical benefits from the activities of every waking moment. Happily, the computer age has brought a reconciliation of some of the warring camps: computer science has many algebraic and number-theoretic needs, especially in connection with the transmission, coding, and manipulation of numerical data. Pure mathematics has become applied mathematics!

We have already used integers in this book for a variety of purposes, and we have assumed a familiarity with the basic facts of arithmetic. Before proceeding further into number theory, we'll begin by reviewing some of those facts and pointing out some others, mostly without proof. (In fact, in many cases the proofs are surprisingly difficult.)

Addition and multiplication are associative and commutative operations on $\mathbb{Z}$, and each has its own identity: 0 for addition and 1 for multiplication. Additive inverses exist, so $\mathbb{Z}$ is an abelian group under addition. Under multiplication, integers other than 1 and -1 have no inverses, so multiplicatively $\mathbb{Z}$ is an abelian monoid but not a group.

The operations on $\mathbb{Z}$ not only coexist, they *interact* via the **distributive laws:**

(6.20) $\qquad a(b+c) = ab + ac \qquad$ and $\qquad (a+b)c = ac + bc,$

for all $a, b, c \in \mathbb{Z}$. Multiplication and addition are further related by the following implication involving the additive identity: for all $a, b \in \mathbb{Z}$,

(6.21) $\qquad\qquad ab = 0 \quad \Rightarrow \quad a = 0 \ \text{ or } \ b = 0.$

Now we will *prove* several other basic facts about $\mathbb{Z}$ using the assumptions we have made. Resist the temptation to declare the assertions obvious.

6.22 THEOREM.   Let $a, b, c \in \mathbb{Z}$. Then

(a) $0 \cdot a = 0$;
(b) $a \cdot (-b) = -ab$;
(c) (Cancellation law) If $a \neq 0$ and $ab = ac$, then $b = c$.

PROOF.

(a) We have

$$0 \cdot a = (0 + 0) \cdot a = 0 \cdot a + 0 \cdot a.$$

Now add $-(0 \cdot a)$ to both sides of this equation. On the left side we get zero, while on the right side we get

$$
\begin{aligned}
-(0 \cdot a) + (0 \cdot a + 0 \cdot a) &= [-(0 \cdot a) + 0 \cdot a] + 0 \cdot a \\
&\qquad \text{(by associativity of addition)} \\
&= 0 + 0 \cdot a \\
&\qquad \text{(by definition of } -(0 \cdot a)) \\
&= 0 \cdot a. \\
&\qquad \text{(since 0 is the additive identity)}
\end{aligned}
$$

Thus $0 \cdot a = 0$, as desired.

(b) By definition, $-ab$ is the additive inverse of ab, and we know from Theorem 6.8 that there is only one such object. Therefore, to verify the claim we must add $a \cdot (-b)$ to ab and check that the result is zero. But

$$
\begin{aligned}
a \cdot (-b) + ab &= a(-b + b) \\
&= a \cdot 0 \\
&= 0. \quad \text{(by part (a))}
\end{aligned}
$$

(c) Assume the equation $ab = ac$ holds, with $a \neq 0$. Adding $-ac$ to both sides of the assumed equation leads us as follows:

$$
\begin{aligned}
0 &= ab + (-ac) \\
&= ab + a \cdot (-c) \quad \text{(by part (b))} \\
&= a[b + (-c)]. \quad \text{(by the distributive law)}
\end{aligned}
$$

But since $a \neq 0$, we then infer by 6.21 that $b + (-c) = 0$, and therefore that $b = c$. [Note: Another proof can be obtained by multiplying both sides of the given equation ($ab = ac$) by the rational number $1/a$. But our proof has the philosophical advantage of not relying on the existence of any number outside $\mathbb{Z}$.] □

We have the inclusion $\mathbb{N} \subset \mathbb{Z}$; and for every $n \in \mathbb{Z}$ exactly one of the following holds:

(6.23) $\qquad\qquad n \in \mathbb{N}, \qquad n = 0, \qquad -n \in \mathbb{N}.$

Otherwise stated, every integer is either positive, zero, or negative; and the conditions are mutually exclusive. This is called the **trichotomy law**.

We define the order relation "$<$" on $\mathbb{Z}$ by

(6.24) $\qquad a < b \qquad \Leftrightarrow \qquad b = a + n \quad$ for some $n \in \mathbb{N}.$

(This is also written $b > a$.)

6.25 ORDERING PROPERTIES.

(a) For all $a, b \in \mathbb{Z}$, exactly one of the following holds:

$$a < b, \qquad a = b, \qquad a > b.$$

(b) For all $a, b, c \in \mathbb{Z}$,

$$a < b \quad \text{and} \quad b < c \qquad \Rightarrow \qquad a < c. \qquad \textbf{(transitive law)}$$

The **absolute value function** $|\ |$ on $\mathbb{Z}$ is defined by

$$|n| = \begin{cases} n & \text{if } n \geqslant 0, \\ -n & \text{if } n < 0. \end{cases}$$

Here are some of its properties:

$|n| \in \mathbb{N} \cup \{0\}$

$|n| = 0 \quad \Leftrightarrow \quad n = 0 \qquad$ (In particular, if $n \neq 0$ then $|n| \geqslant 1$.)

$|ab| = |a||b|$

$|a + b| \leqslant |a| + |b| \qquad \textbf{(triangle inequality)}$

Exercises

1. Two distributive laws are presented in statement 6.20. Deduce the right-hand law from the left-hand law, using the fact that multiplication is commutative.

2. State 6.22(b) in words, without using any mathematical symbols. (Start like this: "When one integer is multiplied by the additive inverse of . . . ")

3. Prove that the following equations hold for a, b, c, $d \in \mathbb{Z}$.
   (a) $(a + b)(c + d) = ac + bc + ad + bd$
   (b) $(-a)(-b) = ab$

4. The expression $x - y$ is an abbreviation for $x + (-y)$. Use this information to prove the following statement:

$$a - (b + c) = (a - b) - c, \qquad \text{for all } a, b, c \in \mathbb{Z}.$$

5. A direct proof is given in the text for the fact that $0 \cdot a = 0$ for all $a \in \mathbb{Z}$. Now prove this fact by contradiction.

6. Reprove statement 6.22(c) using multiplication by $1/a$, as suggested by the note in the text.

7. Show that statement 6.25(a) is equivalent to the trichotomy law, 6.23.

8. Show that if a, $b \in \mathbb{Z}$, with $b \geqslant 0$, then

$$|a| \leqslant b \qquad \Leftrightarrow \qquad a \leqslant b \text{ and } -a \leqslant b.$$

9. (a) Show that if $a \leqslant b$ and $c \leqslant d$ then $a + c \leqslant b + d$.
   (b) Show that if $0 \leqslant r < b$ and $0 \leqslant r_1 < b$, then

$$0 \leqslant |r - r_1| < b.$$

10. For a, $b \in \mathbb{Z}$, prove the inequality

$$\||a| - |b|\| \leqslant |a - b|.$$

(Suggestion: Start by writing $a = (a - b) + b$, and then apply the triangle inequality.)

6.5 DIVISIBILITY; THE FUNDAMENTAL THEOREM OF ARITHMETIC

6.26 DEFINITION.   Let $a, b \in \mathbb{Z}$, with $a \neq 0$. We say that a **divides** b, b is **divisible** by a, a is a **divisor** or **factor** of b, or b is a **multiple** of a, if $b = ac$ for some $c \in \mathbb{Z}$. The statement "a divides b" is written $a \mid b$, and its negation is written $a \nmid b$.

6.27 EXAMPLES. $3 \mid (-12)$, $6 \mid 18$, $17 \mid 0$, $-5 \mid 20$, $-4 \mid 4$, $6 \nmid 1$.

6.28 THEOREM.   Let $a, b, c \in \mathbb{Z}$.

(a) If $a \mid b$ and $b \neq 0$ then $|a| \leqslant |b|$.

(b) If $a \mid b$ and $a \mid c$ then $a \mid (b + c)$ and $a \mid (b - c)$.

PROOF.

(a) If $a \mid b$ and $b \neq 0$ then $b = ac$ for some $c \in \mathbb{Z}$, with $c \neq 0$. Therefore $|c| \leqslant 1$, and so $|b| = |ac| = |a||c| \geqslant |a|$.

(b) Exercise.   □

We defined the notion of *prime number* in Section 2.10, and we showed in Theorem 2.74 that every integer $n \geqslant 2$ is a product of primes. Given this result, it is reasonable to expect that a thorough understanding of the integers will require an understanding of the primes. To this day, the primes remain rather mysterious creatures, but at least we know this: there are a lot of them. The following theorem is one of the most famous in mathematics, due to the strength of the theorem's statement and the swiftness with which it is proved. To appreciate the theorem fully, its level of sophistication should be compared with that of other realms of human activity over the years. For example, more than 2,000 years after the theorem was proved, it was still widespread medical practice to seek relief for human illness through the application of leeches to the body.

6.29 THEOREM.   (Euclid, ca. 300 B.C.) There are infinitely many prime numbers.

PROOF.   We will use induction to prove the theorem in this form: For every positive integer n, there are at least n prime numbers. It is easy to start a list of primes: 2 is prime. Now suppose $p_1, p_2, \ldots, p_k$ are distinct primes, with $k \geqslant 1$. We must show there is another prime, p_{k+1}. Consider the number $M = (p_1 p_2 \cdots p_k) + 1$; it has a prime divisor p by

Theorem 2.74, say $M = pq$. If $p \in \{p_1, \ldots, p_k\}$, then by relabelling (if necessary) we can suppose $p = p_1$. But then we have

$$1 = M - p_1 p_2 \cdots p_k = p_1 q - p_1 p_2 \cdots p_k = p_1(q - p_2 \cdots p_k).$$

So $p_1 \mid 1$, which is impossible since $p_1 > 1$. Therefore $p \notin \{p_1, \ldots, p_k\}$, and so $\{p_1, \ldots, p_k, p\}$ is a set of $k + 1$ prime numbers. $\square$

The list of prime numbers includes occasional pairs of the form $\{n, n + 2\}$; for instance,

$$\{3, 5\}, \quad \{5, 7\}, \quad \{11, 13\}, \quad \{17, 19\}, \quad \{29, 31\}, \quad \text{and} \quad \{41, 43\}$$

are the first such pairs. These pairs are called **twin primes**. Inspired by Euclid's theorem, we ask:

Are there infinitely many twin primes?

No one knows the answer, though the question has been with us for centuries.

Another famous unsolved problem on primes is known as *Goldbach's conjecture*. In 1742 Goldbach conjectured, but did not prove, the following statement:

Every even integer greater than 2 can be expressed as a sum of two primes. (The primes need not be distinct.)

Evidence in favor of the conjecture:

$$4 = 2 + 2, \quad 6 = 3 + 3, \quad 8 = 3 + 5, \quad 10 = 5 + 5, \quad 12 = 5 + 7,$$

and so on. Write a few more examples. If you can prove the conjecture, your proof will be the first one.

In elementary school we learn an algorithm for dividing one integer by another, obtaining a quotient together with a remainder strictly less than the divisor. For instance, we have

$$\begin{array}{r}
36 \leftarrow \text{quotient} \\
\text{divisor} \rightarrow 16\overline{)581} \\
48 \\
\hline
101 \\
96 \\
\hline
5 \leftarrow \text{remainder}
\end{array}$$

Hence $581 = 16 \cdot 36 + 5$. We also learn (to our sorrow) that there is only one correct solution for each example of this kind. Missing from our early education was a *proof* of the existence and uniqueness of this solution, and we will now provide it.

6.30 THEOREM (DIVISION ALGORITHM).   Let $a, b \in \mathbb{Z}$, with $b > 0$. Then there are integers q and r such that

$$a = bq + r \qquad \text{and} \qquad 0 \leqslant r < b.$$

Moreover, q and r are uniquely determined by these conditions.

PROOF.   First the intuition. The number a is either a multiple of b or strictly between two multiples of b. In the former case we can write $a = bq$. In the latter case we let r denote the distance from a to the nearest multiple of b below a; then r is less than b.

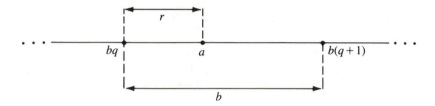

Now the formal proof.

Existence.   Let bq be the largest multiple of b not exceeding a. Then $bq \leqslant a < b(q + 1)$. Define $r = a - bq$. Thus

$$r = a - bq < b(q + 1) - bq = b,$$

as desired.

Uniqueness.   If $a = bq + r$ and also $a = bq_1 + r_1$, with $0 \leqslant r < b$ and $0 \leqslant r_1 < b$, then subtraction of the second equation from the first yields $b(q - q_1) = r_1 - r$; thus $b \mid (r_1 - r)$. But a manipulation of the inequalities gives $b > |r_1 - r|$. (Check this.) Therefore from 6.28(a) we must have $r_1 - r = 0$; so $r_1 = r$ and therefore $q = q_1$.   □

6.31 DEFINITION.   Let a and b be integers. An integer $d \neq 0$ is a **common divisor** of a and b if $d \mid a$ and $d \mid b$. A common divisor d of a and b is said to be the **greatest common divisor** if $d > 0$ and if every common divisor of a and b is also a divisor of d.

6.32 EXAMPLE.

(a) The divisors of 12 are $\pm 1, \pm 2, \pm 3, \pm 4, \pm 6, \pm 12$; the divisors of 16 are $\pm 1, \pm 2, \pm 4, \pm 8, \pm 16$. The common divisors of 12 and 16 are $\pm 1, \pm 2, \pm 4$. The greatest common divisor is 4.

(b) If $a \in \mathbb{Z}$ and $a \neq 0$, then the greatest common divisor of 0 and a is $|a|$.

6.33 THEOREM. If a and b are nonzero integers, than a and b have a unique greatest common divisor.

PROOF. *Existence.* Consider the set

$$L = \{xa + yb \mid x, y \in \mathbb{Z}\}.$$

(The members of L are called the **linear combinations** of a and b.) The set L contains positive integers. For example, we can write

$$|a| = (\pm 1) \cdot a + 0 \cdot b$$

and therefore $|a| \in L$. Let d be the *smallest* positive integer in L; say, $d = x_1 a + y_1 b$, with $x_1, y_1 \in \mathbb{Z}$. If $d \nmid a$ then, by the division algorithm, there are integers q and r such that $a = dq + r$, with $0 < r < d$. But then

$$r = a - dq = a - (x_1 a + y_1 b)q = (1 - x_1 q)a + (-y_1 q)b,$$

which is a positive linear combination of a and b smaller than d. This contradicts the minimality of d in L, therefore $d \mid a$. A similar argument shows that $d \mid b$, so d is a common divisor of a and b.

Now suppose d' is any common divisor of a and b; say, $a = d'a_1$ and $b = d'b_1$. Then

$$d = x_1 a + y_1 b = x_1 d' a_1 + y_1 d' b_1 = d'(x_1 a_1 + y_1 b_1),$$

and hence $d' \mid d$. Thus d is a greatest common divisor of a and b.

Uniqueness. If d and d_1 are both greatest common divisors for a ar d b, then $d_1 \mid d$ (since d is a greatest common divisor) and similarly $d \mid d_1$. Therefore, by 6.28(a), we have $|d| = |d_1|$. But d and d_1 are both positive, so the absolute value signs are redundant, and $d = d_1$.   □

NOTATION. The expression "greatest common divisor" is usually abbreviated **gcd**. Following custom, we will use (a, b) for the gcd of a and *b.* In the mathematical literature, the greatest common divisor is also called the **highest common factor**, and the corresponding abbreviation **hcf** is used.

The fact that (a, b) *is a linear combination of a and b* (see the proof of 6.33) will turn out to be crucial, especially in the proof of Theorem 6.38, which gives a fundamental property of prime numbers.

How can we *compute* the greatest common divisor of two nonzero integers a and b? The proof of Theorem 6.33 is of no use for this purpose, since it requires us to scan an infinite set L for its smallest positive element, and in a finite lifetime we are unlikely to know when we have found it. There is an alternate approach, used in Example 6.32: use trial and error to obtain explicit lists of all the divisors of a and b, and then note the largest entry that appears on both lists. But if a and b are large, this can be a long and extraordinarily tedious process. Fortunately, Euclid developed a faster algorithm for computing the gcd, and the following lemma provides the key to his algorithm.

6.34 LEMMA.   If $a = bq + r$ then $(a, b) = (b, r)$.

PROOF OUTLINE.   Let $d = (a, b)$. Check that every divisor of a and b is also a divisor of r; in particular, $d \,|\, r$ and hence $d \,|\, (b, r)$. [Apply Definition 6.31 to (b, r).] A similar argument shows that $(b, r) \,|\, d$, and the desired equality follows.   □

It is easily checked that $(a, b) = (|a|, |b|)$ for all $a, b \in \mathbb{Z}$. Therefore, for computing greatest common divisors, it suffices to give an algorithm for the gcd of two *positive* integers.

6.35 EUCLID'S ALGORITHM FOR COMPUTATION OF (a, b).

IN WORDS.   Divide b into a, getting a quotient q and **remainder** r_1:

$$a = bq + r_1, \quad \text{with } 0 \leq r_1 < b.$$

If $r_1 = 0$ then $b = (a, b)$; otherwise, divide r_1 into b and get another quotient and remainder r_2 (less than r_1). Continue to divide each new remainder into its predecessor until zero occurs as a remainder. (This must eventually happen since the remainders are successively smaller.) The last nonzero remainder is (a, b). [To see this, observe that $(a, b) = (b, r_1)$ by Lemma 6.34; and $(b, r_1) = (r_1, r_2)$ for the same reason, so $(a, b) = (r_1, r_2)$.

Similar reasoning holds in subsequent steps of the computation. So, if r_k is the last nonzero remainder, we get

$$(a, b) = (r_1, r_2) = (r_2, r_3) = \cdots = (r_k, r_{k+1}) = (r_k, 0) = r_k,$$

as claimed.]

COMPUTER PROGRAM FORMAT.   Here all letters represent nonnegative integers. First observe that if $a = bq + r$, with $0 \leqslant r < b$, then

$$\frac{a}{b} = q + \frac{r}{b}, \qquad \text{with } 0 \leqslant \frac{r}{b} < 1.$$

Therefore $[a/b] = q$, where $[\ ]$ denotes the greatest integer function (see Example 3.8(b)). Now we can write Euclid's algorithm in the format of a program. (If you have had programming experience, convert the material presented here into an actual program.)

```
INPUT A,B
DO UNTIL R = 0:
  → LET R = A − B [A/B]
     IF R = 0 THEN GCD = B
     LET A REPRESENT WHAT HAS BEEN CALLED B UP TO NOW
     LET B REPRESENT WHAT HAS BEEN CALLED R UP TO NOW
     RETURN
END
```

6.36 EXAMPLE.   Compute (2880, 504) and (10374, 2574).

SOLUTION.   We use repeated division, as instructed by Euclid.

$$
\begin{array}{r}
5 \\
b \to 504\overline{)2880} \leftarrow a \\
\underline{2520} \quad 1 \\
r_1 \to 360\overline{)504} \\
\underline{360} \quad 2 \\
r_2 \to 144\overline{)360} \\
\underline{288} \quad 2 \\
r_3 \to 72\overline{)144} \\
\underline{144} \\
r_4 \to \quad 0 \qquad (2880, 504) = 72
\end{array}
$$

$$\begin{array}{r} 4 \\ b \rightarrow 2574\overline{)10374} \leftarrow a \\ \underline{10296} \quad 33 \\ r_1 \rightarrow 78\overline{)2574} \\ \underline{234} \\ 234 \\ \underline{234} \\ r_2 \rightarrow 0 \qquad (10374, 2574) = 78 \end{array}$$

To appreciate the strength of Euclid's algorithm, compare the ease of these computations to the pain of listing all the divisors of each number, as in Example 6.32.

We saw in the proof of Theorem 6.33 that the greatest common divisor (a, b) is equal to $ax + by$ for some integers x and y. If we wish we can *find* x and y by manipulating the data that arises in applying Euclid's algorithm. The following example illustrates the method.

6.37 EXAMPLE.   The first three steps of the computation $(2880, 504) = 72$ in Example 6.36 can be stated as a sequence of equations as follows:

(a) $2880 = 504 \cdot 5 + 360$

(b) $504 = 360 \cdot 1 + 144$

(c) $360 = 144 \cdot 2 + 72$

This yields

$$\begin{aligned} 72 &= 360 + 144 \cdot (-2) &\text{(from (c))} \\ &= 360 + (504 - 360) \cdot (-2) &\text{(from (b))} \\ &= 360 \cdot 3 + 504 \cdot (-2) \\ &= (2880 - 504 \cdot 5) \cdot 3 + 504 \cdot (-2) &\text{(from (a))} \\ &= 2880 \cdot 3 + 504 \cdot (-17). \end{aligned}$$

We proved in Theorem 2.74 that every integer $n \geqslant 2$ can be expressed as a product of prime factors. Our next goal is to prove that this decomposition into primes is *unique*. This is not obvious. For instance, we have

$$24 = 6 \cdot 4 = (3 \cdot 2) \cdot 2^2 = 2^3 \cdot 3 \qquad \text{and} \qquad 24 = 8 \cdot 3 = 2^3 \cdot 3.$$

The final results are the same, but the intermediate factorizations are quite different. A theorem is needed to assure us that we are not witnessing a special feature of the number 24.

Notice this: if a product of two integers is divisible by 5, then at least one of the two integers is divisible by 5. You probably believe this, but can you prove it? The following theorem does the job and, in so doing, it gives a special property of prime numbers that we will need in proving the uniqueness result.

6.38 THEOREM. Let p be a prime number and let a and b be integers. Then the following implication holds:

$$p \mid ab \quad \Rightarrow \quad p \mid a \quad \text{or} \quad p \mid b.$$

PROOF. Assume that $p \mid ab$. Our goal is to deduce the truth of "$p \mid a$ or $p \mid b$." For this purpose it suffices to assume (as we now do) that $p \nmid a$ and deduce that $p \mid b$.

Because $p \nmid a$, we know that $(p, a) = 1$, since the only positive divisors of p are 1 and p. Then from the proof of Theorem 6.33 there are integers x and y such that $xp + ya = 1$. So

$$b = b \cdot 1 = b(xp + ya) = p(xb) + (ab)y.$$

Since $p \mid ab$, both terms on the right side of this equation are divisible by p, and therefore p divides their sum (by 6.28(b)); that is, $p \mid b$. □

6.39 COROLLARY. Let p be prime, and let $a_1, \ldots, a_t \in \mathbb{Z}$. If $p \mid a_1 \cdots a_t$, then $p \mid a_i$ for some i.

PROOF OUTLINE. The case $t = 1$ is trivial, the case $t = 2$ is Theorem 6.38, and induction on t yields the result for all larger t. □

It is easy to see that not every integer has the property stated for primes in Theorem 6.38. For example, $6 \mid 3 \cdot 4$, but $6 \nmid 3$ and $6 \nmid 4$. The next corollary shows that the property in Theorem 6.38 actually completely characterizes the primes.

6.40 COROLLARY. Let m be an integer greater than 1. Then m is prime if and only if the following implication holds for all $a, b \in \mathbb{Z}$:

$$m \mid ab \quad \Rightarrow \quad m \mid a \quad \text{or} \quad m \mid b.$$

PROOF. "Only if" is Theorem 6.38. Conversely, if m is not prime there are integers a and b, with $1 < a < m$ and $1 < b < m$, such that $m = ab$. Then $m \mid ab$, while $m \nmid a$ and $m \nmid b$.   □

Now we can show that the decomposition of an integer $n > 1$ into prime factors is essentially unique.

6.41 FUNDAMENTAL THEOREM OF ARITHMETIC. Let n be an integer greater than 1. Then there are prime numbers $p_1, \ldots, p_r$ such that $n = p_1 \cdots p_r$. Moreover, this factorization of n is unique in the following sense: if $n = q_1 \cdots q_s$ also, with the q's prime, then $r = s$ and the q's are just a rearrangement of the p's. (More technically, there is a permutation $\sigma \in S_r$ such that $p_i = q_{\sigma(i)}$ for $1 \leqslant i \leqslant r$.)

PROOF. The existence of a prime factorization was proved in Theorem 2.74, so only the uniqueness requires proof here.

We will argue by induction on n (in the format of 2.73). Suppose $n = p_1 \cdots p_r = q_1 \cdots q_s$, with the p's and q's prime and $r \leqslant s$. If $n = 2$, then since 2 is prime we have $2 = p_1 = q_1$, and we are done. Now assume $n > 2$, and assume the theorem is known for all integers t satisfying $2 \leqslant t \leqslant n - 1$. Since $p_1 \cdots p_r = q_1 \cdots q_s$, we have $p_1 \mid q_1 \cdots q_s$, so $p_1 \mid q_i$ for some i (by 6.39). By a change of subscripts (if necessary) we can suppose $p_1 \mid q_1$. But q_1 is *prime*, and hence $p_1 = q_1$. It follows from the cancellation law (see 6.22(c)) that $p_2 \cdots p_r = q_2 \cdots q_s$. From the induction hypothesis we then know that $r = s$ and the numbers $q_2, \ldots, q_s$ are a rearrangement of $p_2, \ldots, p_r$.   □

6.42 COROLLARY. Let $n \in \mathbb{Z}$, with $|n| \geqslant 2$. Then n has a unique factorization of the form

$$n = \pm p_1^{\alpha_1} \cdots p_t^{\alpha_t},$$

where $t \geqslant 1$, the p_i are distinct primes satisfying $p_1 \leqslant \cdots \leqslant p_t$, and $\alpha_i \geqslant 1$ for $1 \leqslant i \leqslant t$. (This factorization is called the **standard** or **canonical factorization** of n.)

PROOF OUTLINE. If $n \geqslant 2$ consider the prime decomposition in Theorem 6.41, and gather together all the prime factors equal to p_1, getting $p_1^{\alpha_1}$. Then repeat the process on the remaining prime factors, if there are any.

If n is negative, use the same procedure on $|n|$.   □

The following proposition was known to the Pythagoreans around 400 B.C.

6.43 THEOREM.   The real number $\sqrt{2}$ is irrational.

PROOF.   (By contradiction) If $\sqrt{2}$ is rational we can write

(∗) $$\sqrt{2} = \frac{a}{b}$$

for some integers a and b. By factoring the numerator and denominator into primes and removing any common prime factors, we can assume $(a, b) = 1$. Squaring both sides of equation (∗) yields $2b^2 = a^2$, and so $2 \mid a^2$. Since 2 is prime, it follows from Theorem 6.38 that $2 \mid a$; say $a = 2m$. Thus $2b^2 = 4m^2$, and cancellation gives $b^2 = 2m^2$. Therefore $2 \mid b$, and this contradicts the fact that $(a, b) = 1$. Thus our initial assumption that $\sqrt{2}$ is rational has led to a contradiction, and the proof is complete.   □

We have just shown that there is no rational number whose square is 2. How do we know there is a *real* number whose square is 2? In fact, the existence in $\mathbb{R}$ of square roots of nonnegative real numbers is guaranteed by the way in which $\mathbb{R}$ is constructed. That construction is a complicated matter, and we leave the details for a later course; but we note that Example 3.17 (the divide-and-average method) shows that it is possible to compute a sequence of rational numbers whose squares get closer and closer to 2. Further evidence that in a reasonable world $\sqrt{2}$ must exist was known to the Pythagoreans: in a square with side of length 1, the length d of the diagonal satisfies the equation $d^2 = 1^2 + 1^2 = 2$.

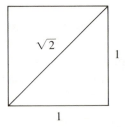

Incidentally, the irrationality of $\sqrt{2}$ has the following intuitive meaning with respect to the picture: given a measuring stick that has been precisely marked to indicate halves, thirds, fourths, fifths, and so on, there is no way to position the stick on the square's diagonal in such a way that the endpoints of the diagonal are both covered by marks on the stick.

If a and b are integers, there are distinct primes $p_1, \ldots, p_t$ and integers $\alpha_i \geq 0$ and $\beta_i \geq 0$, for $1 \leq i \leq t$, such that

(6.44) $a = \pm p_1^{\alpha_1} \cdots p_t^{\alpha_t}$ and $b = \pm p_1^{\beta_1} \cdots p_t^{\beta_t}$.

To see this, start with the standard factorizations of a and b. If there is a prime number occuring in one of these factorizations but not in the other, insert it into the latter factorization with an exponent of zero. For instance, if $a = 1386$ and $b = -1275$, we have the standard factorizations

$$1386 = 2 \cdot 3^2 \cdot 7 \cdot 11 \quad\text{and}\quad -1275 = -3 \cdot 5^2 \cdot 17.$$

In the format of 6.44, these become

$$1386 = 2 \cdot 3^2 \cdot 5^0 \cdot 7 \cdot 11 \cdot 17^0 \text{ and } -1275 = -2^0 \cdot 3 \cdot 5^2 \cdot 7^0 \cdot 11^0 \cdot 17.$$

6.45 EXERCISE.   Given integers a and b with factorizations as in 6.44, show that

$$(a, b) = p_1^{\min(\alpha_1, \beta_1)} \cdots p_i^{\min(\alpha_i, \beta_i)} \cdots p_t^{\min(\alpha_t, \beta_t)},$$

where $\min\{\alpha_i, \beta_i\}$ denotes the minimum of α_i and β_i.

Recall from Definition 5.43 that the **least common multiple** of integers a and b is the smallest nonnegative integer that is a multiple of both a and b. We will write this now as $[a, b]$. With the factorizations in 6.44, we claim that

(6.46) $$[a, b] = p_1^{\max\{\alpha_1, \beta_1\}} \cdots p_t^{\max\{\alpha_t, \beta_t\}}.$$

To see this, first write R for the expression on the right in 6.46, and observe that R is a positive multiple of a and of b. Therefore $[a, b] \leq R$. On the other hand, since $a \mid [a, b]$ and $b \mid [a, b]$, we know that $p_i^{\alpha_i}$ and $p_i^{\beta_i}$ divide $[a, b]$, and hence $p_i^{\max\{\alpha_i, \beta_i\}}$ divides $[a, b]$, for $1 \leq i \leq t$. Therefore, in the standard factorization of $[a, b]$, each prime p_i appears at least to the power $\max\{\alpha_i, \beta_i\}$. Hence $R \mid [a, b]$, and so $R \leq [a, b]$, and the claim follows.

THEOREM 6.47. If a and b are nonzero integers, then

$$[a, b] = \frac{|ab|}{(a, b)}.$$

PROOF OUTLINE. Write a and b as in 6.44. Then write (a, b) as in 6.45, apply the appropriate law of exponents, and compare the result with 6.46. $\square$

6.48 **EXAMPLE.** We compute $[-10140, 2600]$. To use 6.46, we first have to factor the given numbers into primes (this can be tedious!), and then we can read off the answer. We have

$$-10140 = -2^2 \cdot 3 \cdot 5 \cdot 13^2,$$
$$2600 = 2^3 \cdot 5^2 \cdot 13 = 2^3 \cdot 3^0 \cdot 5^2 \cdot 13.$$

This gives $[-10140, 2600] = 2^3 \cdot 3 \cdot 5^2 \cdot 13^2 = 101400$. An alternate approach is to use Euclid's algorithm (6.35) to compute

$$(-10140, 2600) = 260;$$

then by 6.47 we get

$$[-10140, 2600] = \frac{(10140)(2600)}{260} = 101400.$$

Exercises

1. Show that if $a \mid b$ and $a \mid c$ then $a \mid (mb + nc)$ for all $m, n \in \mathbb{Z}$.
2. Label each of the following *true* or *false,* and justify your answer.

   (a) $8 \mid 0$
   (b) $a \mid b$ and $b \mid c \implies a \mid c$
   (c) $a \mid b$ and $a \mid c \implies a \nmid bc$
   (d) $a \mid b \implies -a \mid b$
   (e) $a \mid bc \implies b \mid a$ or $c \mid a$
   (f) $a \mid b$ and $b \mid a \implies a = \pm b$

3. Let a and b be integers, with $a \neq 0$. Prove:

$$a \mid b \iff \frac{b}{a} = \left[\frac{b}{a}\right],$$

where [ ] denotes the greatest integer function.

4. An integer $n \neq 0$ is said to be **composite** if n has a positive divisor other than 1 and n. (For example, -6 and 15 are composite, while -7, 1, and 13 are not composite.) Show that if n is composite, then n has a prime divisor $p \leq \sqrt{|n|}$.

5. (For students with computer programming experience) Write a computer program that accepts an integer $n > 1$ as input and either tells us that n is prime or lists each divisor k of n that satisfies $1 < k < n$. (Suggestion: For each integer k satisfying $2 \leq k \leq \sqrt{n}$, test whether k is a divisor of n by the method of Exercise 3.)

6. The key to Euclid's proof of the assertion that there are infinitely many primes is in showing that the number $M = p_1 \cdots p_k + 1$ has a prime factor distinct from $p_1, p_2, \ldots, p_k$. It is tempting to conjecture that M must actually be prime. Test this conjecture for $1 \leq k \leq 6$.

7. In this exercise we will use an algorithm, called the *sieve of Eratosthenes,* that isolates prime numbers. (Eratosthenes lived from about 276 to 194 B.C. He was also known for his computation of the circumference of the earth and for his treatises on the theater, moral philosophy, and history.) Here is how it works. List the integers from 2 through n. Circle the number 2, then cross off all the higher multiples of 2 (namely, 4, 6, 8, 10, etc.). Circle 3, which is the first remaining number after 2. Then cross off all the multiples of 3 that have not yet been crossed off. Then circle 5. Et cetera. When you are finished with the repeated sieving operations, only the primes will remain on the list, and they will have been circled for emphasis. The exercise: perform the sieve of Eratosthenes with $n = 120$. As you do this, cross off the multiples of different primes by using different kinds of slashes (slashing in different directions, or using different colors), and provide a key for the reader. By this device the reader will immediately know the smallest prime divisor of each composite number on your list. (In view of Exercise 4, after the multiples of 7 have been crossed off, the remaining numbers will all be prime and can be circled to mark this fact.)

8. Define a **prime triple** to be a set of three prime numbers of the form $\{n, n+2, n+4\}$. For instance, $\{3, 5, 7\}$ is a prime triple. Are there any others? Either exhibit another or prove there are none. (Suggestion: First observe what happens when any of the twin primes $\{n, n+2\}$ listed in the text is supplemented with $n+4$.)

9. Suppose $n = k(k+1)(k+2)$ for some $k \in \mathbb{Z}$. Show that n is divisible by 6.

10. Use Euclid's algorithm to compute the following greatest common divisors.

   (a) $(56, 104)$
   (b) $(462, 3003)$
   (c) $(442, -182)$
   (d) $(922, 2161)$

11. Show that when Euclid's algorithm is used to compute the greatest common divisor of two integers, it doesn't matter which integer is labelled a and which is labelled b.

12. Let a and b be integers. Show that if $(a, b) = 1$ then $(a, a + b) = 1$.

13. Show that if a, b, c are integers, not all zero, then

$$((a, b), c) = (a, (b, c)).$$

   Here each ordered pair is understood to represent a greatest common divisor.

14. Write each greatest common divisor computed in Exercise 10 as a linear combination of the given integers.

15. Two nonzero integers are said to be **relatively prime** if their greatest common divisor is 1. Let $a, b, c \in \mathbb{Z}$. Show that if $(a, c) = 1$ and $c \mid ab$, then $c \mid b$. (Hint: Mimic the proof of Theorem 6.38.)

16. In the manner of Example 6.32, list the divisors, the common divisors, and the greatest common divisor of 748 and 1258.

17. Prove that the real number $\sqrt[3]{5}$ is irrational.

18. Suppose $k, n \in \mathbb{N}$. Prove that if n is not the kth power of an integer, then $\sqrt[k]{n}$ is irrational.

19. Give a recursive procedure for constructing a line segment of length $\sqrt{n}$ for each integer $n \geqslant 2$, using only an unmarked straightedge and a compass. Assume that a line segment of length 1 has been given. (Recall that with a straightedge and compass, we can construct a line perpendicular to a given line at a given point.)

20. (a) Suppose d and n are natural numbers, and $d \mid n$. How does the standard factorization of d relate to the standard factorization of n?
    (b) Use part (a) to show how the standard factorization of n can be used to determine the number of positive divisors of n.

21. Compute the least common multiple [462, 1574573] in two ways:
    (a) by factoring both integers into primes and applying formula 6.46;
    (b) by using Euclid's algorithm to compute the greatest common divisor (462, 1574573) and then applying Theorem 6.47.

22. The fact that there are infinitely many prime numbers, together with the apparent abundance of twin primes, might lead the innocent to suspect that the gaps between consecutive primes are of bounded size. Show that this is *not* the case. (Suggestion: Consider sequences of the form

$$n! + 2, \qquad n! + 3, \qquad n! + 4, \ldots$$

where *n* is a large positive integer.)

6.6 CONGRUENCE; DIVISIBILITY TESTS

The *congruence* relation, which we are about to define, is an important vehicle for the discussion of divisibility problems in $\mathbb{Z}$. Moreover, by using congruence, certain computational problems with large integers (whose storage and manipulation may be difficult) can be converted to corresponding problems with small integers. For instance, in this way the hazards of *overflow* in machine calculations can sometimes be eliminated, and without loss of accuracy.

Convention: *throughout this section, all numbers under discussion will be understood to be integers.* For example, the assertion "$m > 0$" will mean "$m \in \mathbb{N}$."

6.49 DEFINITION. Fix $m > 0$. Numbers a and b are said to be **congruent modulo m** if $a - b$ is divisible by m. Symbolically:

$$a \equiv b \pmod{m} \qquad \text{means} \qquad m \mid (a - b).$$

The number m is called the **modulus** of the congruence.

6.50 EXAMPLE. $3 \equiv -5 \pmod{4}$, $0 \equiv 15 \pmod{5}$, $-7 \equiv 5 \pmod{6}$, $5743 \equiv 43 \pmod{100}$.

It is often useful to reformulate the congruence relation as follows:

(6.51) $a \equiv b \pmod{m}$ $\Leftrightarrow$ $a - b = mk$ for some k

                                    $\Leftrightarrow$ $a = b + mk$ for some k.

6.52 EXAMPLES.

(a) Your bicycle's odometer (the mileage gauge) returns to zero after 1000 miles, so after you ride 3472 miles the odometer reads 472. The reason, in our present notation: $3472 \equiv 472 \pmod{1000}$.

(b) Suppose it is now five o'clock. (Never mind whether it is morning or evening.) Consider how to compute what the time will be after 784 hours have elapsed. Since it is now 5 hours past twelve o'clock, after the given time interval it will be 789 hours past twelve o'clock. Division by 12 yields the equation $789 = 12 \cdot 65 + 9$. This equation tells us that in 789 hours, a clock's hour hand will complete 65 full revolutions and will then mark off 9 more hours. The result: in 784 hours from now the time will be nine o'clock. More generally, to compute the time t at h hours after time t_0, we use the division algorithm to write

$$t_0 + h = 12q + r,$$

with $0 \leqslant r < 12$. Then

$$t = \begin{cases} r & \text{if } r \neq 0, \\ 12 & \text{if } r = 0. \end{cases}$$

(The temptation is to write simply "$t = r$", but we do not speak of "zero o'clock".) In congruence notation, the time t is the number satisfying these two conditions:

$$1 \leqslant t \leqslant 12 \quad \text{and} \quad t \equiv t_0 + h \pmod{12}.$$

(Remember: we are assuming that all numbers are integers.)

6.53 THEOREM. Fix $m > 0$. Then congruence modulo m is an equivalence relation on $\mathbb{Z}$.

PARTIAL PROOF.   There are three things we need to verify.

(i) $a \equiv a \pmod{m}$ for all a

(ii) $a \equiv b \pmod{m} \Rightarrow b \equiv a \pmod{m}$

(iii) $a \equiv b \pmod{m}$ and $b \equiv c \pmod{m} \Rightarrow a \equiv c \pmod{m}$

We leave (i) and (ii) as exercises. As for (iii), the hypothesis gives

$$a = b + mk_1 \quad \text{and} \quad b = c + mk_2$$

for some k_1 and k_2 (use 6.51). Then $a = c + m(k_1 + k_2)$, and so $a \equiv c \pmod{m}$. □

We now show that with a fixed modulus, congruences behave in some ways like equations.

6.54 THEOREM. If $a \equiv b \pmod{m}$ and $c \equiv d \pmod{m}$, then

$$a + c \equiv b + d \pmod{m} \quad \text{and} \quad ac \equiv bd \pmod{m}.$$

PROOF. The hypothesis gives $a = b + mk_1$ and $c = d + mk_2$ for some k_1 and k_2. Then

$$a + c = (b + d) + m(k_1 + k_2),$$

so $a + c \equiv b + d \pmod{m}$. Also,

$$ac = bd + m(k_1 d + k_2 b + mk_1 k_2),$$

so $ac \equiv bd \pmod{m}$. □

6.55 COROLLARY. Suppose there are congruences $a_1 \equiv b_1 \pmod{m}$, $a_2 \equiv b_2 \pmod{m}$, ..., $a_n \equiv b_n \pmod{m}$. Then

$$\sum_{i=1}^{n} a_i \equiv \sum_{i=1}^{n} b_i \pmod{m} \quad \text{and} \quad \prod_{i=1}^{n} a_i \equiv \prod_{i=1}^{n} b_i \pmod{m}.$$

In particular, if $a \equiv b \pmod{m}$ then $a^n \equiv b^n \pmod{m}$ for all $n \geq 1$.

PROOF IDEA. Argue by induction on n, using Theorem 6.54 to establish the result when $n = 2$. □

6.56 MAGIC TRICK. Take the first four digits of your telephone number, add your weight to the nearest pound, and multiply the result by 9.

Then write down the sum of the digits of the answer. If your sum has one digit, stop; if your sum has two digits write down *their* sum. Prediction: Your answer is 9. Sample:

$$
\begin{array}{rr}
\text{First four digits of telephone number} & 3785 \\
\text{weight} & +\ 197 \\
\hline
 & 3982 \\
\times\quad 9 & \\
\hline
 & 35838
\end{array}
$$

$$
3 + 5 + 8 + 3 + 8 = 27
$$
$$
2 + 7 = \ 9
$$

An explanation of 6.56 requires a close look at decimal notation. A nonnegative integer n written in base 10 has the form $a_t a_{t-1} \cdots a_1 a_0$, where each a_i satisfies $0 \le a_i \le 9$; each a_i is called a **decimal digit** of n. Here the juxtaposition of digits does not mean multiplication; instead, it means this:

$$
n = a_t \cdot 10^t + a_{t-1} \cdot 10^{t-1} + \cdots + a_1 \cdot 10 + a_0 = \sum_{i=0}^{t} a_i \cdot 10^i.
$$

For example, $4708 = 4 \cdot 10^3 + 7 \cdot 10^2 + 8$. The following result is the key to the "magic" displayed in 6.56.

6.57 THEOREM. Every nonnegative integer is congruent modulo 9 to the sum of its decimal digits. Symbolically, if $0 \le a_i \le 9$ for $0 \le i \le t$, then

$$
\sum_{i=0}^{t} a_i \cdot 10^i \equiv \sum_{i=0}^{t} a_i \pmod 9.
$$

PROOF. Clearly $10 \equiv 1 \pmod 9$. Then three applications of Corollary 6.55 yield the following three statements:

$$
\begin{array}{ll}
10^i \equiv 1 \pmod 9 & \text{for all } i \ge 0 \\
a_i \cdot 10^i \equiv a_i \pmod 9 & \text{for } 0 \le i \le t \\
\sum_{i=0}^{t} a_i \cdot 10^i \equiv \sum_{i=0}^{t} a_i \pmod 9. & \square
\end{array}
$$

6.58 COROLLARY (TEST FOR DIVISIBILITY BY 9). An integer is a multiple of 9 if and only if the sum of its decimal digits is a multiple of 9.

PROOF OUTLINE.   First notice that it suffices to check the assertion for *nonnegative* integers. Then verify the following statement:

(6.59)   If $a \equiv b \pmod{m}$, then $m \mid a \Leftrightarrow m \mid b$.

Finally, apply 6.59 to the congruence in the statement of Theorem 6.57, with $m = 9$.   □

Now we can understand the idea underlying 6.56. Once we somehow produce a multiple of 9, we know that the sum of its digits is also a multiple of 9, as is each subsequent digital sum. Eventually we reach a one-digit multiple of 9, and that's 9.

It is a straightforward matter to extend these results on decimal representation of integers to results in other bases, as follows. If $b \geqslant 2$, writing an integer n **in the base b** means writing $n = a_t a_{t-1} \ldots a_0$ as an abbreviation for

$$n = \sum_{i=0}^{t} a_i b^i, \qquad \text{with } 0 \leqslant a_i \leqslant b - 1.$$

Then the congruence in Theorem 6.57 generalizes to

$$\sum_{i=0}^{t} a_i b^i \equiv \sum_{i=0}^{t} a_i \pmod{b - 1}$$

by essentially the same argument. Corollary 6.58 extends to this: An integer is a multiple of $b - 1$ if and only if the sum of its digits in the base b representation is a multiple of $b - 1$.

6.60 THEOREM (TEST FOR DIVISIBILITY BY 11).   An integer n with decimal representation $n = a_t a_{t-1} \ldots a_0$ is divisible by 11 if and only if the number $a_t - a_{t-1} + a_{t-2} - \cdots \pm a_0$ is divisible by 11.

PROOF OUTLINE.   We are given that

$$n = \sum_{i=0}^{t} a_i \cdot 10^i.$$

Start with the congruence $10 \equiv -1 \pmod{11}$, then proceed as in the proof of Theorem 6.57, but this time deduce the congruence

(*) $$\sum_{i=0}^{t} a_i \cdot 10^i \equiv \sum_{i=0}^{t} (-1)^i a_i \quad (\text{mod } 11).$$

Apply 6.59 (with $m = 11$) to (*), and observe that the number $\sum_{i=0}^{t}(-1)^i a_i$ is equal to $\pm(a_t - a_{t-1} + a_{t-2} - \cdots \pm a_0)$. $\square$

   6.61 EXAMPLE.   The number 319,245,386,597,518,260 is divisible by 11, because the number

$$3 - 1 + 9 - 2 + 4 - 5 + 3 - 8 + 6 - 5 + 9 - 7 + 5 - 1 + 8 - 2 + 6 - 0 = 22$$

is divisible by 11.

   We will return to congruence considerations toward the end of the next section, after we have developed the terminology of *ring theory.*

Exercises

1. In each part answer *true* or *false,* without proof.
   (a) $-297 \equiv 533613 \quad (\text{mod } 5)$
   (b) $a \equiv b \;(\text{mod } m)$ and $n \mid m \;\Rightarrow\; a \equiv b \;(\text{mod } n)$
   (c) $a \equiv b \;(\text{mod } m)$ and $m \mid n \;\Rightarrow\; a \equiv b \;(\text{mod } n)$
   (d) $a \equiv b \;(\text{mod } m) \;\Rightarrow\; a^2 \equiv b^2 \;(\text{mod } m^2)$
   (e) $a \equiv b \;(\text{mod } m)$ and $a \equiv b \;(\text{mod } n) \;\Rightarrow\; a \equiv b \;(\text{mod}(m, n))$
   (f) $a \equiv b \;(\text{mod } m)$ and $a \equiv b \;(\text{mod } n) \;\Rightarrow\; a \equiv b \;(\text{mod}[m, n])$
   (g) $a \equiv b \;(\text{mod } m)$ and $a \equiv b \;(\text{mod } n) \;\Rightarrow\; a \equiv b \;(\text{mod } mn)$
   (h) $(10^{50} - 1)^{12} \equiv 6 \quad (\text{mod } 3)$

2. For which integers a and b is the statement $a \equiv b \;(\text{mod } 1)$ correct?

3. Prove that if n is an odd integer then $n^2 \equiv 1 \;(\text{mod } 8)$.

4. Prove that every integer n satisfies the congruence $n^3 \equiv n \;(\text{mod } 6)$.

5. Consider the integer $n = 123456789101112 13. \,.\,.100$. (The number is written by listing the positive integers from 1 to 100 without spaces between them.)

   (a) How many digits does n have?
   (b) Is n divisible by 9?

6. Prove statement 6.59.

7. In the country of Zzyzzx the clocks have eight-hour faces. Suppose it is now five o'clock in Zzyzzx. What time will it be in 4725 hours?

8. Suppose it is now 8 A.M. in Boston. What time of day will it be in Boston in 6538 hours? (Be sure to indicate whether the hour will be A.M. or P.M.)

9. An assortment of numbered tiles is assembled on a rack, forming an integer n, and the rack is placed on view in a local gallery. A clumsy visitor overturns the rack. He hurriedly gathers up the tiles and repositions them on the rack, producing an integer n'. Show that the difference between n and n' is divisible by 9.

10. (a) Show that if $(k, m) = 1$ then the following cancellation law holds:

$$ka \equiv kb \pmod{m} \quad \Rightarrow \quad a \equiv b \pmod{m}.$$

   (b) Is the condition $(k, m) = 1$ in part (a) essential?

11. Let m be a positive integer.
   (a) Show that every integer is congruent modulo m to exactly one of the integers $0, 1, \ldots, m - 1$.
   (b) Suppose $(k, m) = 1$. Show that every integer is congruent modulo m to exactly one of the integers $0, k, 2k, \ldots, (m - 1)k$.

12. In each part, fill the blank with one or more numbers from the set $\{0, 1, \ldots, 7\}$, using the smallest number of elements that yields a true statement. Then prove the statement.
   (a) If n is the square of an even integer, then $n \equiv$ ____ $\pmod{8}$.
   (b) If n is the square of an odd integer, then $n \equiv$ ____ $\pmod{8}$.

13. (a) Use the results of Exercise 12 to show that the equation $98765432 = x^2 + y^2$ is not solvable in $\mathbb{Z}$.
   (b) Is the equation $78135 = x^2 + y^2 + z^2$ solvable in $\mathbb{Z}$?
   [*Historical note* (not needed for the solution of this exercise). In the eighteenth century it was shown by J. L. Lagrange that every positive integer can be expressed in the form $x^2 + y^2 + z^2 + w^2$ for some $x, y, z, w \in \mathbb{Z}$.]

14. Suppose n is a positive integer whose digits add to 24. Name a positive divisor of n other than 1 and n, and prove your claim.

15. (a) Let n be a positive integer. Prove that if $2^n - 1$ is prime, then n is prime.

(b) Rewrite the statement proved in (a), but with $2^n - 1$ expressed as a number in binary (base 2) form.

16. Write the integer 35481 in base 7 notation. (Suggestion: Begin by finding the largest integer t such that $7^t \leqslant 35481$. Then find $a_t \in \mathbb{N}$ such that $35481 = a_t \cdot 7^t + r$, with $0 \leqslant r < 7^t$. The number a_t is the leftmost digit of the answer.)

17. (a) Describe a simple procedure for converting a number given in base 2 notation to base 4 notation. Your procedure should go directly from base 2 to base 4, without using decimal notation as an intermediary.
   (b) Illustrate your method by converting the binary number

$$11101011011001101$$

to base 4 notation.

18. Consider the following list of numbers:

   1
   12
   123
   $\vdots$
   123456789
   1234567891
   12345678912
   123456789123
   $\vdots$

   What is the first number on this list that is divisible by 11?

19. Design a simple procedure for testing whether an integer given in binary notation is divisible by 3, and justify your claim.

20. For a positive integer a, let $D(a)$ denote the integer obtained by taking the sum of a's digits, then taking the sum of the digits of *that* number, and continuing this process until a one-digit number is obtained. (We did this sort of thing in 6.56.) Show that

$$D(a + b) = D(D(a) + D(b)) \qquad \text{and} \qquad D(ab) = D(D(a) \cdot D(b)).$$

(Hint: Use Theorem 6.57.)
   [The results of this exercise can be used to check computations with large integers. For instance, if computation seems to show that

$ab = c$, then according to the exercise we must have $D(D(a) \cdot D(b)) = D(c)$. Checking that this is the case will catch eight errors out of nine. (Think about why this is so.) This procedure of checking computations by replacing each integer by the sum of its digits is called **casting out nines**.]

6.7 RINGS AND FIELDS

In this section we begin our study of algebraic structures that have two interacting operations. The need for such a study will come as no surprise to anyone who has multiplied and added numbers on the same day.

6.62 DEFINITION. A **ring** is a set R with two operations, here denoted by addition and multiplication, such that

1. R is an abelian group with respect to addition;
2. R is a semigroup with respect to multiplication;
3. addition and multiplication are linked by the **distributive laws**:

$$a(b + c) = ab + ac \quad \text{and} \quad (a + b)c = ac + bc, \qquad \text{for all } a, b, c \in R.$$

If multiplication is commutative then R is called a **commutative ring**. If R has a multiplicative identity, that element is denoted 1 and is called the **unit, unity,** or **identity** element of R, and R is called a **ring with unit**. (As in any additive group, the **additive identity** in R is called **zero**, written 0.)

Our childhoods provide us with our first examples of algebraic structures: $\mathbb{Z}$, $\mathbb{Q}$, and $\mathbb{R}$ are all rings with respect to the usual numerical operations of addition and multiplication. In fact, they are all commutative rings with the number 1 serving as unit element in each case. Now we will consider some less familiar examples of rings.

6.63 EXAMPLES.

(a) Let E be the set of even integers, with standard addition and multiplication. Then E is a commutative ring with no unit element.
(b) Let G by any abelian group, with its operation denoted as addition; define a product operation on G by setting $ab = 0$ for all $a, b \in G$. This makes G into a commutative ring, with an admit-

tedly uninteresting product. (A bit of work will show that G has a unit element if and only if $G = \{0\}$.)

(c) Let $\mathbb{Z}$ (mod 3) denote the set $\{0, 1, 2\}$, with the sum ($+$) and product ($\cdot$) of two elements defined to be the remainders after the usual sum and product (respectively) are divided by 3. Explicitly, the operations $+$ and $\cdot$ are given by the following tables:

$+$	0	1	2		$\cdot$	0	1	2
0	0	1	2		0	0	0	0
1	1	2	0		1	0	1	2
2	2	0	1		2	0	2	1

This algebraic system is a commutative ring with unit element 1. (We postpone the details; see page 337.)

(d) Let F be the set of all functions from $\mathbb{R}$ to $\mathbb{R}$. If $f, g \in F$, define $f + g \in F$ by

$$(f + g)(r) = f(r) + g(r), \qquad \text{for all } r \in \mathbb{R};$$

and define the product $fg \in F$ by

$$(fg)(r) = f(r) \cdot g(r), \qquad \text{for all } r \in \mathbb{R}.$$
$$\uparrow$$
(multiplication of real numbers)

(*Do not confuse fg with the composition f $\circ$ g.*) Then F is a commutative ring with unit. Its zero element is the constant function $z: \mathbb{R} \rightarrow \mathbb{R}$ given by $z(r) = 0$ for all $r \in \mathbb{R}$. The unit element in F is the constant function $I: \mathbb{R} \rightarrow \mathbb{R}$ given by $I(r) = 1$ for all $r \in \mathbb{R}$.

(e) Let A be a set. Define operations on the power set $P(A)$ as follows. "Addition," denoted $\oplus$, is given by

$$X \oplus Y = (X - Y) \cup (Y - X), \qquad \text{for all } X, Y \in P(A).$$

The picture:

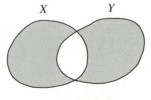

$X \oplus Y$ (shaded)

(The set $X \oplus Y$ is usually called the **symmetric difference** of X and Y.) For "multiplication" we use set intersection, $\cap$. We assert that with these operations, $P(A)$ is a commutative ring with unit. The proof is based largely on Theorem 2.26 and it requires some extensive set-theoretic flagellation; we will sketch only part of it here. It is easy to check that both $\oplus$ and $\cap$ are commutative, and we know (from 2.26(a)) that $\cap$ is associative. Verifying that $\oplus$ is associative requires more effort, and we leave the details to the reader. However, it may be helpful first to observe that $X \oplus (Y \oplus Z)$ and $(X \oplus Y) \oplus Z$ are both represented by the shaded region in the following Venn diagram:

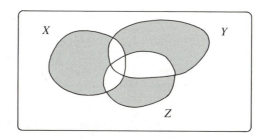

(Some other properties of the symmetric difference operation are given in Exercise 8 on page 68.) The zero element is $\emptyset$, since

$$X \oplus \emptyset = (X - \emptyset) \cup (\emptyset - X) = X, \quad \text{for all } X \in P(A).$$

(Also $\emptyset \oplus X = X$, by commutativity of $\oplus$.) Each $X \in P(A)$ is its own additive inverse, since $X \oplus X = (X - X) \cup (X - X) = \emptyset$. Distributive laws hold; for instance, we have

$$
\begin{aligned}
X \cap (Y \oplus Z) &= X \cap [(Y - Z) \cup (Z - Y)] \\
&= [X \cap (Y - Z)] \cup [X \cap (Z - Y)] \\
&= [(X \cap Y) - (X \cap Z)] \cup [(X \cap Z) - (X \cap Y)] \\
&= (X \cap Y) \oplus (X \cap Z).
\end{aligned}
$$

We note that A is the unit element for the ring, since $A \cap X = X \cap A = X$ for every $X \in P(A)$.

(f) A **two-by-two** (written 2×2) **matrix** over $\mathbb{R}$ is a square array of four real numbers. Some samples:

$$
\begin{pmatrix} 0 & 3 \\ -1 & 8 \end{pmatrix}, \quad
\begin{pmatrix} 1 & \pi \\ 0 & 1 \end{pmatrix}, \quad
\begin{pmatrix} \sqrt{2} & 7 \\ 0 & 0 \end{pmatrix}, \quad
\begin{pmatrix} 0 & 0 \\ 0 & 0 \end{pmatrix}, \quad
\begin{pmatrix} -3 & -4 \\ -5 & -2 \end{pmatrix}.
$$

The numbers positioned inside a matrix are its **entries**. Let $M_2(\mathbb{R})$ denote the set of all 2×2 matrices over $\mathbb{R}$. Define the **sum** of two matrices A and B to be the matrix whose entries are each the sum of the like-positioned entries in A and B. Symbolically:

(6.64)
$$\begin{pmatrix} a_{11} & a_{12} \\ a_{21} & a_{22} \end{pmatrix} + \begin{pmatrix} b_{11} & b_{12} \\ b_{21} & b_{22} \end{pmatrix} = \begin{pmatrix} a_{11} + b_{11} & a_{12} + b_{12} \\ a_{21} + b_{21} & a_{22} + b_{22} \end{pmatrix}.$$

$$\qquad\quad A \qquad\qquad\quad B \qquad\qquad\qquad A + B$$

Here a subscript ij on an entry indicates that the entry is in the ith row (horizontal) and jth column (vertical). Matrix **product** is defined as follows:

(6.65)
$$\begin{pmatrix} a_{11} & a_{12} \\ a_{21} & a_{22} \end{pmatrix}\begin{pmatrix} b_{11} & b_{12} \\ b_{21} & b_{22} \end{pmatrix} = \begin{pmatrix} a_{11}b_{11} + a_{12}b_{21} & a_{11}b_{12} + a_{12}b_{22} \\ a_{21}b_{11} + a_{22}b_{21} & a_{21}b_{12} + a_{22}b_{22} \end{pmatrix}.$$

$$\qquad\quad A \qquad\quad B \qquad\qquad\qquad AB$$

This definition is grossly nonintuitive, yet it turns out to be the key to a vast array of applications of matrices to the description of natural phenomena.

According to these definitions of sum and product we have

$$\begin{pmatrix} 1 & 2 \\ 5 & -1 \end{pmatrix} + \begin{pmatrix} 3 & 1 \\ -7 & 8 \end{pmatrix} = \begin{pmatrix} 4 & 3 \\ -2 & 7 \end{pmatrix}$$

and

$$\begin{pmatrix} 6 & 2 \\ -1 & 3 \end{pmatrix}\begin{pmatrix} 1 & -2 \\ 3 & 4 \end{pmatrix} = \begin{pmatrix} 12 & -4 \\ 8 & 14 \end{pmatrix}.$$

It can be checked that with these operations, $M_2(\mathbb{R})$ is a ring with unit, with

$$\begin{pmatrix} 0 & 0 \\ 0 & 0 \end{pmatrix}$$

as the zero element and

$$\begin{pmatrix} 1 & 0 \\ 0 & 1 \end{pmatrix},$$

the **identity matrix**, as the unit element. Because of the complexity of matrix multiplication, checking some of the ring axioms (such as associativity of multiplication) requires the patient comparison of rather complicated expressions. Other axioms follow easily from the ring properties of $\mathbb{R}$. For instance, the proof that matrix addition is commutative goes as follows. Let

$$A = \begin{pmatrix} a_{11} & a_{12} \\ a_{21} & a_{22} \end{pmatrix} \quad \text{and} \quad B = \begin{pmatrix} b_{11} & b_{12} \\ b_{21} & b_{22} \end{pmatrix}$$

be in $M_2(\mathbb{R})$. Then

$$A + B = \begin{pmatrix} a_{11} + b_{11} & a_{12} + b_{12} \\ a_{21} + b_{21} & a_{22} + b_{22} \end{pmatrix} \quad \text{(by definition of matrix addition)}$$

$$= \begin{pmatrix} b_{11} + a_{11} & b_{12} + a_{12} \\ b_{21} + a_{21} & b_{22} + a_{22} \end{pmatrix} \quad \text{(by commutativity of addition in } \mathbb{R}\text{)}$$

$$= \begin{pmatrix} b_{11} & b_{12} \\ b_{21} & b_{22} \end{pmatrix} + \begin{pmatrix} a_{11} & a_{12} \\ a_{21} & a_{22} \end{pmatrix}$$

$$= B + A.$$

The reader should check some of the other ring axioms, such as associativity of addition, and the fact that the zero and unit elements of $M_2(\mathbb{R})$ are as claimed. Notice that *matrix multiplication is not commutative,* and hence the ring $M_2(\mathbb{R})$ *is not a commutative ring.* For example,

$$\begin{pmatrix} 0 & 1 \\ 2 & 3 \end{pmatrix} \begin{pmatrix} 1 & 2 \\ 3 & 4 \end{pmatrix} = \begin{pmatrix} 3 & 4 \\ 11 & 16 \end{pmatrix},$$

while

$$\begin{pmatrix} 1 & 2 \\ 3 & 4 \end{pmatrix} \begin{pmatrix} 0 & 1 \\ 2 & 3 \end{pmatrix} = \begin{pmatrix} 4 & 7 \\ 8 & 15 \end{pmatrix}.$$

The following theorem shows that two properties demonstrated earlier for $\mathbb{Z}$ actually hold in any ring. The proof, which we omit here, is a virtual copy of the proof of Theorem 6.22(a) and (b). Notice that

Theorem 6.66 does *not* include a claim that the cancellation law of 6.22(c) holds in all rings. For instance, in $M_2(\mathbb{R})$ we have

$$\begin{pmatrix} 1 & 2 \\ 0 & 0 \end{pmatrix}\begin{pmatrix} 5 & 0 \\ 1 & 0 \end{pmatrix} = \begin{pmatrix} 1 & 2 \\ 0 & 0 \end{pmatrix}\begin{pmatrix} 3 & -2 \\ 2 & 1 \end{pmatrix},$$

but we cannot cancel

$$\begin{pmatrix} 1 & 2 \\ 0 & 0 \end{pmatrix}$$

from both sides of the equation.

6.66 THEOREM. Let R be a ring. Then

(a) $0 \cdot a = 0$　for all $a \in R$;
(b) $a \cdot (-b) = -(ab) = (-a) \cdot b$　for all $a, b \in R$.

6.67 DEFINITION. Let R be a ring with unit element 1. An element in R is said to be **invertible in R** if it has an inverse in R with respect to the multiplication operation. That is, $x \in R$ is invertible if there is an element $y \in R$ such that $xy = yx = 1$. From Theorem 6.8 we know that an invertible element x has a *unique* multiplicative inverse, and we denote it by x^{-1}.

6.68 THEOREM. Let R be a ring with unit, and let G be the set of invertible elements in R. Then G is a multiplicative group.

PROOF. Clearly $1 \in G$, since $1 \cdot 1 = 1$, so G is nonempty and has a multiplicative identity. If $a \in G$ then $a^{-1} \in R$; and a^{-1} has inverse a, since $aa^{-1} = a^{-1}a = 1$, so $a^{-1} \in G$. Multiplication in G is associative since R is a multiplicative semigroup. Finally, we must check that multiplication is really an operation on G, that is, if $a, b \in G$ then $ab \in G$. (Popular jargon: we are checking that G is "closed" under multiplication.) But if $a, b \in G$ then $a^{-1}, b^{-1} \in G$ and

$$(ab)(b^{-1}a^{-1}) = (b^{-1}a^{-1})(ab) = 1.$$

Thus ab is invertible and $(ab)^{-1} = b^{-1}a^{-1}$; that is, $ab \in G$. This completes the proof that G is a group.　$\square$

Invertible elements in a ring R are also called **units** of R. Of course, we are already using the word "unit" to mean the multiplicative identity of R, and there is at most one of those. (Moreover, if there is no multiplicative identity then "invertible" has no meaning.) So it is customary to refer to "*the* unit" to mean the multiplicative identity 1, and "*a* unit" for an invertible element that may or may not equal 1. We can now safely label the group G in Theorem 6.68 as *the group of units* of R, without fear that we are decreeing G to have 1 as its only member.

6.69 DEFINITION.   A **field** is a commutative ring with unit element $1 \neq 0$, in which *every* nonzero element is invertible. Otherwise put, a field is a ring in which the nonzero elements constitute an abelian group under multiplication.

6.70 EXAMPLES.

(a) The rings $\mathbb{R}$ and $\mathbb{Q}$ are fields; but $\mathbb{Z}$ is not a field since, except for ± 1, nonzero elements of $\mathbb{Z}$ have no multiplicative inverses in $\mathbb{Z}$.

(b) In the ring $\mathbb{Z}$ (mod 3), defined in 6.63(c), the group of units is $\{1, 2\}$, making $\mathbb{Z}$ (mod 3) a field.

(c) In Example 6.63(d), a function $f \in F$ is a unit in F if and only if $f(r) \neq 0$ for all $r \in \mathbb{R}$; then the inverse of f in F is the function g defined by $g(r) = 1/f(r)$ for all $r \in \mathbb{R}$. Notice that g is *not* usually the function we have previously denoted by f^{-1}. Indeed, for a function $f \in F$ to be a unit in F, it need not be invertible as defined in Section 3.3.

(d) In Example 6.63(e), the unit element is the whole set A, since $A \cap B = B$ for every $B \in P(A)$. In fact, A is the *only* unit in the ring $P(A)$. For if $B \in P(A)$ and $B \neq A$, then $B \subset A$, so there is no $C \in P(A)$ such that $B \cap C = A$. (Remember: $\cap$ is the "product" operation in $P(A)$.)

(e) Now consider the matrix ring $M_2(\mathbb{R})$ defined in 6.63(f). Since multiplication in this ring is not commutative, $M_2(\mathbb{R})$ is automatically disqualified from being a field. But $M_2(\mathbb{R})$ does have a multiplicative identity, so it makes sense to look for units: a matrix A is a unit in $M_2(\mathbb{R})$ only if there exists a matrix $X \in M_2(\mathbb{R})$ such that

(*)
$$AX = XA = \begin{pmatrix} 1 & 0 \\ 0 & 1 \end{pmatrix}.$$

Some matrices are easily ruled out as candidates for unit-hood. For instance, if

$$A = \begin{pmatrix} 0 & 0 \\ 1 & 0 \end{pmatrix}$$

and if

$$X = \begin{pmatrix} w & x \\ y & z \end{pmatrix}$$

is any matrix in $M_2(\mathbb{R})$, then

$$AX = \begin{pmatrix} 0 & 0 \\ 1 & 0 \end{pmatrix} \begin{pmatrix} w & x \\ y & z \end{pmatrix} = \begin{pmatrix} 0 & 0 \\ w & x \end{pmatrix} \neq \begin{pmatrix} 1 & 0 \\ 0 & 1 \end{pmatrix},$$

and hence A is not a unit of $M_2(\mathbb{R})$. On the other hand, if a, b, c, d are real numbers satisfying the condition $ad - bc \neq 0$, then it is readily checked that

$$\begin{pmatrix} a & b \\ c & d \end{pmatrix}$$

is invertible, with inverse

$$\begin{pmatrix} \dfrac{d}{ad - bc} & \dfrac{-b}{ad - bc} \\ \dfrac{-c}{ad - bc} & \dfrac{a}{ad - bc} \end{pmatrix}.$$

(Check this by seeing that the two matrices fit the roles of A and X in (∗).) This shows that the condition $ad - bc \neq 0$ is *sufficient* in order for

$$\begin{pmatrix} a & b \\ c & d \end{pmatrix}$$

to be invertible. It turns out that the condition is *necessary* as well, but we won't go into the details here.

6.71 DEFINITION.   Let R be a commutative ring. A nonzero element $x \in R$ is said to be a **zero divisor** if there is a nonzero element $y \in R$ such that $xy = 0$.

6.72 EXAMPLES.

(a) Let $A = \{1, 2\}$; then $P(A) = \{\emptyset, \{1\}, \{2\}, A\}$ can be made into a commutative ring by using symmetric difference for addition and $\cap$ for multiplication, as in Examples 6.63(e) and 6.70(d). Then $\emptyset$ is the zero element, and $\{1\} \cap \{2\} = \emptyset$; so $\{1\}$ and $\{2\}$ are zero divisors.

(b) Let D denote the ring of all 2×2 **diagonal matrices** over $\mathbb{R}$; that is,

$$D = \left\{ \begin{pmatrix} a & 0 \\ 0 & b \end{pmatrix} \,\middle|\, a, b \in \mathbb{R} \right\},$$

with the usual matrix operations of addition and multiplication. Then D is a ring with unit (check this claim); moreover, D is *commutative*, because

$$\begin{pmatrix} a & 0 \\ 0 & b \end{pmatrix} \begin{pmatrix} c & 0 \\ 0 & d \end{pmatrix} = \begin{pmatrix} c & 0 \\ 0 & d \end{pmatrix} \begin{pmatrix} a & 0 \\ 0 & b \end{pmatrix} = \begin{pmatrix} ac & 0 \\ 0 & bd \end{pmatrix}, \quad \forall a, b, c, d \in \mathbb{R}.$$

This ring D has zero divisors, as the following equation demonstrates:

$$\begin{pmatrix} 3 & 0 \\ 0 & 0 \end{pmatrix} \begin{pmatrix} 0 & 0 \\ 0 & 4 \end{pmatrix} = \begin{pmatrix} 0 & 0 \\ 0 & 0 \end{pmatrix}.$$

(c) The ring $\mathbb{Z}$ has *no* zero divisors. This is the content of statement 6.21 in Section 6.4.

6.73 THEOREM.   Let R be a commutative ring. Then the following conditions are equivalent:

(i) R has no zero divisors;

(ii) For all $a, b \in R$, if $a \neq 0$ and $ab = ac$ then $b = c$.
     (This is called the **cancellation law** in R.)

PROOF.   (i) $\Rightarrow$ (ii): Assume (i) is true, and suppose that $a \neq 0$ and

$ab = ac$. Then

$$0 = ab - ac = ab + (-ac) = ab + a(-c) = a(b + (-c)) = a(b - c).$$

But since $a \neq 0$ and R has no zero divisors, $b - c$ must equal 0; that is, $b = c$.

(ii) $\Rightarrow$ (i): Assume (ii) is true, and suppose that $x \in R$ is nonzero. We must show that x is not a zero divisor. If for some $y \in R$ we have $xy = 0$, then $xy = x \cdot 0$; so by condition (ii) it follows that $y = 0$. Hence x is not a zero divisor. (The logic: the statement "$xy = 0 \Rightarrow y = 0$" is equivalent to "$y \neq 0 \Rightarrow xy \neq 0$," and therefore x fails the criterion of Definition 6.71.) $\square$

6.74 DEFINITION. Let R be a commutative ring with unit element $1 \neq 0$. If R has no zero divisors, then R is called an **integral domain**.

The ring $\mathbb{Z}$ is the world's most popular integral domain. Also, we claim that if F is any field, then F is an integral domain. To see this, by Theorem 6.73 it suffices to check the cancellation law. With this in mind, let $a, b, c \in F$. Then

$$\begin{aligned}
a \neq 0 \quad \text{and} \quad ab = ac \quad &\Rightarrow \quad a^{-1}(ab) = a^{-1}(ac) \\
&\Rightarrow \quad (a^{-1}a)b = (a^{-1}a)c \\
&\Rightarrow \quad 1 \cdot b = 1 \cdot c \\
&\Rightarrow \quad b = c.
\end{aligned}$$

This proves the cancellation law and hence the claim.

Up to now we have seen only three examples of fields: $\mathbb{Q}$, $\mathbb{R}$, and $\mathbb{Z}$ (mod 3). We will conclude this section by generalizing the construction of $\mathbb{Z}$ (mod 3) in such a way as to obtain an infinite family of rings. In particular, the discussion will yield *an infinite family of finite fields,* one for each prime number p.

Fix an integer $m \geqslant 2$. Then for any $n \in \mathbb{Z}$, the division algorithm gives unique integers q and r satisfying

$$n = mq + r \quad \text{and} \quad 0 \leqslant r \leqslant m - 1.$$

The number r here is called the **residue of n modulo m**. Thus the residue of n modulo m is just the remainder after n is divided by m.

Now let $\mathbb{Z}$ (mod m) denote the set $\{0, 1, \ldots, m - 1\}$; in our present terminology, $\mathbb{Z}$ (mod m) is the set of all residues modulo m. We make

$\mathbb{Z}$ (mod m) into a ring by defining operations $\oplus$ and $\odot$ as follows: for $a, b \in \mathbb{Z}$ (mod m),

(6.75)
$$a \oplus b \quad \text{is the residue of} \quad a + b \quad \text{modulo } m,$$
$$a \odot b \quad \text{is the residue of} \quad ab \quad \text{modulo } m.$$

For instance, the operation tables for $\mathbb{Z}$ (mod 3) were given in Example 6.63(c), and $\mathbb{Z}$ (mod 6) has operations $\oplus$ and $\odot$ given by the following tables:

(6.76)

$\oplus$	0	1	2	3	4	5
0	0	1	2	3	4	5
1	1	2	3	4	5	0
2	2	3	4	5	0	1
3	3	4	5	0	1	2
4	4	5	0	1	2	3
5	5	0	1	2	3	4

$\odot$	0	1	2	3	4	5
0	0	0	0	0	0	0
1	0	1	2	3	4	5
2	0	2	4	0	2	4
3	0	3	0	3	0	3
4	0	4	2	0	4	2
5	0	5	4	3	2	1

We assert that $\mathbb{Z}$ (mod m), equipped with $\oplus$ and $\odot$, is a commutative ring with unit element 1. Since $a + b = b + a$ and $ab = ba$, it follows from 6.75 that $\oplus$ and $\odot$ are commutative operations. However, associativity requires more work. First observe the following implications:

(6.77)
$$(a \oplus b) \oplus c = r \implies (a + b) + c \equiv r \pmod{m},$$
$$\text{and } 0 \leqslant r \leqslant m - 1;$$
$$a \oplus (b \oplus c) = s \implies a + (b + c) \equiv s \pmod{m},$$
$$\text{and } 0 \leqslant s \leqslant m - 1.$$

But $(a + b) + c = a + (b + c)$, and each integer has a *unique* residue modulo m. Therefore $r = s$; that is,

$$(a \oplus b) \oplus c = a \oplus (b \oplus c).$$

Associativity of $\odot$ is proved similarly. The remaining properties of commutative rings with unit (see Definition 6.62) are left for the reader to verify.

6.78 THEOREM. If p is prime then $\mathbb{Z}$ (mod p) is a field. Thus there are infinitely many finite fields.

PROOF. By the preceding discussion, $\mathbb{Z}$ (mod p) is a commutative ring with unit. To show that $\mathbb{Z}$ (mod p) is a field (see Definition 6.69), it remains to check that every nonzero element has a multiplicative inverse. Now let $a \in \mathbb{Z}$ (mod p), with $a \neq 0$. Then $1 \leqslant a \leqslant p - 1$, and so $(a, p) = 1$. Therefore, by the proof of Theorem 6.33 we know there exist integers x and y such that $ax + py = 1$, and this gives $ax \equiv 1$ (mod p). Now let x_1 be the residue of x modulo p. Then $x_1 \equiv x$ (mod p), and therefore $ax_1 \equiv ax \equiv 1$ (mod p). That is, in $\mathbb{Z}$ (mod p) we have $a \odot x_1 = 1$, and so $a^{-1} = x_1$. $\square$

REMARK. Suppose that $m \geqslant 2$ and m is *not* prime; say $m = ab$, with $1 < a < m$ and $1 < b < m$. Then $\mathbb{Z}$ (mod m) is not an integral domain, because $a \odot b = 0$ in $\mathbb{Z}$ (mod m). [For example, a check of Table 6.76 shows that 2, 3, and 4 are zero divisors in $\mathbb{Z}$ (mod 6).] Hence $\mathbb{Z}$ (mod m) is not a field.

Exercises

1. Let $A = \{1, 2\}$. Exhibit tables for the ring operations $\oplus$ and $\cap$ on $P(A)$, as defined in Example 6.63(e).

2. Define

$$T_2(\mathbb{R}) = \left\{ \begin{pmatrix} a & 0 \\ b & c \end{pmatrix} \middle| a, b, c \in \mathbb{R} \right\}.$$

   The members of $T_2(\mathbb{R})$ are called 2×2 **lower triangular matrices** over $\mathbb{R}$.

   (a) Show that $T_2(\mathbb{R})$ is a ring with unit under the operations of matrix addition and multiplication.
   (b) Is $T_2(\mathbb{R})$ a commutative ring?

3. Let

$$R = \left\{ \begin{pmatrix} a & a \\ 0 & 0 \end{pmatrix} \middle| a \in \mathbb{Z} \right\}.$$

   (a) Show that R is a ring under matrix addition and multiplication.
   (b) Does R have a unit element?

4. (a) Let R be a commutative ring with unit, and let $\alpha \in R$. Use mathematical induction to prove that for all $n \geqslant 1$,

$$\alpha^n - 1 = (\alpha - 1)(\alpha^{n-1} + \alpha^{n-2} + \cdots + \alpha + 1).$$

(b) Deduce that if n is a positive integer and $2^n - 1$ is prime, then n is prime.

5. Determine the invertible elements of the ring F given in Example 6.63(d).

6. A ring R in which $r^2 = r$ for all $r \in R$ is said to be a **Boolean ring**.

   (a) Show that a Boolean ring is commutative.
   (b) Show that if R is a Boolean ring then $r + r = 0$ for all $r \in R$.
   (c) Give an example of a two-element Boolean ring.

7. Consider the ring D of 2×2 diagonal matrices over $\mathbb{R}$. (See Example 6.72(b).) Show that if $r \in D$ then *exactly one* of the following holds: $r = 0$, r is a zero divisor, r is a unit of D.

8. Let R be a commutative ring, and let $r, s \in R$. Prove:

   (a) If r is invertible then r is not a zero divisor.
   (b) If r is a zero divisor and $rs \neq 0$, then rs is a zero divisor.

9. Give an example of two zero divisors (in the same ring) whose sum is neither zero nor a zero divisor.

10. Show that the set $R = \{a + b\sqrt{2} \mid a, b \in \mathbb{Q}\}$ is a field under ordinary addition and multiplication.

11. Stories abound of prison camp inmates playing chess with pieces fashioned from bread crumbs, thereby imposing the structure of one set (chess pieces) on another set (bread crumbs). Now let's perform a similar maneuver in an algebraic context. Suppose R is a ring, S is a set disjoint from R, and $f: R \to S$ is a bijection. Use these raw materials to define operations on S making S into a ring and making f into an isomorphism from R onto S.

12. Let R be a commutative ring. A subset $A \subseteq R$ is an **ideal** of R if A is an additive subgroup of R (see Exercise 10 in Section 6.2) and the following statement holds:

$$r \in R \text{ and } a \in A \implies ra \in A.$$

(We say A is **closed under multiplication from R**.)

   (a) Show that $\{0\}$ and R are ideals of R.
   (b) Show that if A is an ideal of R, and A contains an invertible element of R, then $A = R$.
   (c) Show that if R is a field then the only ideals of R are $\{0\}$ and R.

13. (Refer to Exercise 12.) Let m be an integer, and let (m) denote the set of all multiples of m; that is,

$$(m) = \{mr \mid r \in \mathbb{Z}\}.$$

Show that (m) is an ideal of the ring $\mathbb{Z}$.

14. (Refer to Exercise 13.) Let a and b be integers, and let $d = (a, b)$. Show that (d) is the smallest ideal in $\mathbb{Z}$ that contains the ideals (a) and (b). That is, show that $(d) \supseteq (a) \cup (b)$, and then show that if A is *any* ideal containing $(a) \cup (b)$ then $A \supseteq (d)$.

15. (Refer to Exercise 13.)

   (a) Show that if A and B are ideals in a given commutative ring R, then $A \cap B$ is also an ideal of R.
   (b) Let $a, b \in \mathbb{Z}$, and let $m = [a, b]$. Show that $(m) = (a) \cap (b)$.

16. (Refer to Exercise 13.) Show that every ideal A in $\mathbb{Z}$ has the form (m) for some integer m. Do this by filling in the details of the following outline.

   (a) First consider the case $A = \{0\}$.
   (b) Suppose $A \neq \{0\}$. Show that A contains a positive integer.
   (c) Let m denote the smallest positive integer in A. Use the division algorithm to show that every element of A is a multiple of m.

17. (Refer to Exercise 16.)

   (a) Describe an infinite descending chain of ideals in $\mathbb{Z}$, that is, a family of ideals of the form $A_1 \supset A_2 \supset A_3 \supset \cdots$.
   (b) In $\mathbb{Z}$, is there an infinite *ascending* chain of ideals, $A_1 \subset A_2 \subset A_3 \subset \cdots$?

18. (Refer to Exercise 12.) Let $J = \mathbb{R} \times \mathbb{R}$. Make J into a ring by means of coordinatewise operations as follows:

$$\left. \begin{aligned} (a, b) + (c, d) &= (a + c, b + d) \\ (a, b) \cdot (c, d) &= (ac, bd) \end{aligned} \right\} \quad \forall a, b, c \in \mathbb{R}.$$

With these operations, J is a commutative ring with unit. (You needn't verify this.)

   (a) Let $I_1 = \{(r, 0) \mid r \in \mathbb{R}\}$. Show that I_1 is an ideal of J.
   (b) Define $I_2 = \{(0, r) \mid r \in \mathbb{R}\}$. A proof like the one in part (a) will show that I_2 is an ideal of J. (You needn't check this.) Show that $J, \{0\}, I_1$, and I_2 are the *only* ideals of J.

19. Inspection of Example 6.63(f) shows that the only property required of $\mathbb{R}$ to make $M_2(\mathbb{R})$ into a ring is that $\mathbb{R}$ itself be a ring. Accordingly, if K is any ring, we can follow the scheme of 6.63(f) and make $M_2(K)$ into a ring, with operations given by 6.64 and 6.65.

    (a) Show that $M_2(K)$ is not commutative if K has at least two elements.
    (b) What does a typical element of $M_2(M_2(\mathbb{R}))$ look like?
    (c) Show that the ring $M_2(M_2(\mathbb{R}))$ has an identity element.
    (d) Show that if A and B are invertible elements of $M_2(\mathbb{R})$, then

$$\begin{pmatrix} A & \begin{pmatrix} 0 & 0 \\ 0 & 0 \end{pmatrix} \\ \begin{pmatrix} 0 & 0 \\ 0 & 0 \end{pmatrix} & B \end{pmatrix} \quad \text{and} \quad \begin{pmatrix} \begin{pmatrix} 0 & 0 \\ 0 & 0 \end{pmatrix} & A \\ B & \begin{pmatrix} 0 & 0 \\ 0 & 0 \end{pmatrix} \end{pmatrix}$$

    are invertible elements of $M_2(M_2(\mathbb{R}))$.

20. We have seen several reasons why $M_2(\mathbb{R})$ is not a field. (For instance, multiplication is not commutative, two nonzero elements can have zero product, and not every element is invertible.) Show that $M_2(\mathbb{R})$ *contains* a field.

21. (a) Display the addition and multiplication tables for $\mathbb{Z}$ (mod 7).
    (b) What is the multiplicative inverse of 5 in $\mathbb{Z}$ (mod 7)?

22. If E and F are fields, with $E \supseteq F$, and the operations on F are restrictions of the operations on E, then we say F is a **subfield** of E, or E is an **extension** of F.

    (a) Show that if E is an extension of F, then the zero element and multiplicative identity of E are those of F.
    (b) Show that if E is an extension of the field $\mathbb{Z}$ (mod 2), then $a + a = 0$ for all $a \in E$.
    (c) There is a four-element field $E = \{0, 1, a, b\}$ that is an extension of $\mathbb{Z}$ (mod 2). (Assume this without proof.) Write the addition and multiplication tables for E.

    [*Remark.* Part (c) postulates the existence of an extension of $\mathbb{Z}$ (mod 2) having 2^2 elements. More generally, for every prime p and every natural number n, there is an extension of $\mathbb{Z}$ (mod p) that has p^n elements. Furthermore, there is essentially only one such field for a given p and n, in the sense that any two are isomorphic. Proving these assertions is beyond the scope of our treatment here.]

6.8 POLYNOMIALS

We all remember playing with polynomials in high school algebra. For example, we can easily compute results like

$$(x^2 + 2)(x^2 + 3x - 5) = x^4 + 3x^3 - 3x^2 + 6x - 10.$$

Once we have computed this, we will then instantly agree that

$$(y^2 + 2)(y^2 + 3y - 5) = y^4 + 3y^3 - 3y^2 + 6y - 10$$

without actually checking the computation. *Why* do we agree so readily, and what *are* these things we are manipulating, anyway? For instance, are they members of some set? Our memories also include some concern with finding *roots* of polynomials, and we recall that the *quadratic formula* was a highlight of the festivities.

The goal of this section is to develop a perspective for the systematic study of polynomials, mostly by viewing the subject through ring-theoretic glasses. We will see that polynomials behave much like integers and, after we have weathered an initial storm of definitions, we will find many analogies with number theory to explore in some detail.

Throughout this section F will denote a field. In a first reading it may be helpful to think of F as the field of real numbers.

For the purposes of this section, a **sequence** in F will mean a function from the nonnegative integers into F. Thus with each integer $i \geq 0$ we associate an element $a_i \in F$. Intuitively, a sequence is just a list $(a_0, a_1, a_2, \dots)$ of elements of F (called **terms** of the sequence), denoted $\{a_i\}_{i \geq 0}$ for short. From our work on functions (see Equation 3.5), it follows that two sequences $\{a_i\}_{i \geq 0}$ and $\{b_i\}_{i \geq 0}$ are equal if and only if $a_i = b_i$ for all $i \geq 0$.

6.79 DEFINITION. A sequence $\{a_i\}_{i \geq 0}$ in F is a **polynomial** if it has only finitely many nonzero terms.

For computational purposes, we introduce the following notation and terminology to describe a polynomial $f = \{a_i\}_{i \geq 0}$. If for some n we have $a_n \neq 0$ and $a_i = 0$ for all $i \geq n + 1$, we represent f by the expression

(6.80) $f = a_n x^n + a_{n-1} x^{n-1} + \cdots + a_1 x + a_0$

or, more briefly, by

$$\sum_{i=0}^{n} a_i x^i \quad \text{or} \quad \sum_{i \geqslant 0} a_i x^i;$$

and we say that f has **degree** n, written $\deg f = n$. The elements $a_0, a_1, \ldots$ are called the **coefficients** of f. The coefficient a_0 is called the **constant term** and, if $\deg f = n$, the coefficient a_n is the **leading coefficient**. If $a_i = 0$ for all $i \geqslant 0$, we call f the **zero polynomial** and write $f = 0$ (or, more fancily, $f = \Sigma_{i \geqslant 0} 0 x^i$). The zero polynomial is not assigned a numerical degree.

If f and g are nonzero polynomials, say, $f = \Sigma_{i \geqslant 0} a_i x^i$ and $g = \Sigma_{i \geqslant 0} b_i x^i$, then it follows from the preceding discussion that $f = g$ if and only if $\deg f = \deg g$ and $a_i = b_i$ for all i satisfying $0 \leqslant i \leqslant \deg f$.

6.81 EXAMPLE. The symbol x represents the sequence $(0, 1, 0, 0, \ldots)$, and it is a polynomial of degree 1. For traditional reasons, the symbol x is sometimes called an **indeterminate** or a **variable**. The expression $x^4 - 2x + 8$ represents a polynomial of degree 4, namely, the sequence

$$(8, -2, 0, 0, 1, 0, 0, \ldots).$$

Notice that we have followed the custom of streamlining the notation by purging inessential zeros: we wrote $x^4 - 2x + 8$ instead of the more complete $x^4 + 0x^3 + 0x^2 + (-2)x + 8$. No information is lost, because the exponents on the x's indicate which sequence term is being represented by each coefficient.

Now let $F[x]$ denote the set of all polynomials with coefficients in the field F. We will make $F[x]$ into a ring by defining addition and multiplication of polynomials according to the rules of high school algebra. Explicitly, we define addition by

$$(6.82) \qquad \sum_{i \geqslant 0} a_i x^i + \sum_{i \geqslant 0} b_i x^i = \sum_{i \geqslant 0} (a_i + b_i) x^i.$$

In words: we add polynomials by adding corresponding coefficients. In sequence notation, this definition takes the form

$$(6.83) \quad (a_0, a_1, a_2, \ldots) + (b_0, b_1, b_2, \ldots)$$
$$= (a_0 + b_0, a_1 + b_1, a_2 + b_2, \ldots).$$

The advantage of 6.82 over 6.83 is a visual one: using 6.82 we can instantly write down the equation

$$(x^9 - 5x^7 + 2x^5 - x^3 + x - 1) + (x^8 + x^5 - x + 4)$$
$$= x^9 + x^8 - 5x^7 + 3x^5 - x^3 + 3.$$

But in format 6.83 it becomes more difficult to see what gets added to what; we have no exponents (or subscripts) to guide us, and it is easy to lose our place:

$$(-1, 1, 0, -1, 0, 2, 0, -5, 0, 1, 0, 0, \ldots)$$
$$+ (4, -1, 0, 0, 0, 1, 0, 0, 1, 0, 0, \ldots) = ?$$

It follows immediately from 6.82 or 6.83 that polynomial addition is a commutative and associative operation, with 0 as the zero element and $\sum_{i=0}^{n}(-a_i)x^i$ as the additive inverse of $\sum_{i=0}^{n}a_i x^i$. Let's rephrase that last statement:

$$-\left(\sum_{i=0}^{n} a_i x^i\right) = \sum_{i=0}^{n} (-a_i)x^i.$$

Thus $F[x]$ is an abelian group under addition. Incidentally, now that we have observed that polynomial addition is commutative, we are free to vary our symbol order. For example,

$$x^3 + 2x - 5 = x^3 - 5 + 2x = -5 + 2x + x^3,$$

and so on.

Polynomial multiplication is more complicated; it is defined by the formula

$$(6.84) \quad \left(\sum_{i \geqslant 0} a_i x^i\right)\left(\sum_{j \geqslant 0} b_j x^j\right) = \sum_{k \geqslant 0} c_k x^k, \quad \text{with } c_k = \sum_{i+j=k} a_i b_j.$$

Thus $c_0 = a_0 b_0$, $c_1 = a_0 b_1 + a_1 b_0$, $c_2 = a_0 b_2 + a_1 b_1 + a_2 b_0$, and so on. According to formula 6.84, the product $(1 + x + x^2 + x^4)(-5 + 3x + x^2)$ is equal to

$$(*) \quad [1 \cdot (-5)] + [1 \cdot 3 + 1 \cdot (-5)]x + [1 \cdot 1 + 1 \cdot 3 + 1 \cdot (-5)]x^2$$
$$+ [1 \cdot 1 + 1 \cdot 3]x^3 + [1 \cdot 1 + 1 \cdot (-5)]x^4 + 3x^5 + x^6,$$

which condenses to

$$x^6 - 4x^4 + 4x^3 - x^2 - 2x - 5.$$

When doing computations by hand, we are more likely to use a standard multiplication format for the job:

$$
\begin{array}{l}
\quad\quad\quad x^4 + \ x^2 + \ x + 1 \\
\quad\quad\quad\quad\ x^2 + 3x - 5 \\
\hline
\quad -5x^4 \quad\quad\quad - 5x^2 - 5x - 5 \quad \leftarrow \quad -5 \cdot (x^4 + x^2 + x + 1) \\
\quad\ \ 3x^5 \quad\quad + 3x^3 + 3x^2 + 3x \quad \leftarrow \quad 3x \cdot (x^4 + x^2 + x + 1) \\
x^6 \quad\quad + \ x^4 + \ x^3 + \ x^2 \quad\quad\quad \leftarrow \quad x^2 \cdot (x^4 + x^2 + x + 1) \\
\hline
x^6 + 3x^5 - 4x^4 + 4x^3 - \ x^2 - 2x - 5 \quad \leftarrow \quad \text{TOTAL}
\end{array}
$$

We claim: under the operations of addition and multiplication we have just defined, $F[x]$ is a commutative ring with unit element 1. The proof is largely a matter of careful subscript management, and we'll skip most of it. But to hint at the flavor, here is a proof that multiplication in $F[x]$ is commutative:

$$\left(\sum_{i \geqslant 0} a_i x^i\right)\left(\sum_{j \geqslant 0} b_j x^j\right) = \sum_{k \geqslant 0}\left(\sum_{i+j=k} a_i b_j\right) x^k \qquad \text{(by 6.84)}$$

$$= \sum_{k \geqslant 0}\left(\sum_{i+j=k} b_j a_i\right) x^k \qquad \text{(since multiplication in } F \text{ is commutative)}$$

$$= \left(\sum_{j \geqslant 0} b_j x^j\right)\left(\sum_{i \geqslant 0} a_i x^i\right). \qquad \text{(by 6.84)}$$

Our custom of omitting any polynomial term that has coefficient zero makes each field element $a \in F$ indistinguishable from the polynomial $a + 0 \cdot x + 0 \cdot x^2 + \cdots$, which in turn represents the sequence $(a, 0, 0, \ldots)$ in F. In fact, from now on we will consider them the same object. (This convention is mathematically acceptable, but we won't justify it here.) Thus we are viewing F as a subset of $F[x]$ from now on, namely, F consists of the polynomials of degree zero (along with the element 0).

The ring $F[x]$ is an integral domain. To prove this, we need only show that the product of two nonzero polynomials f and g is nonzero. In brief, suppose

$$f = a_m x^m + \text{(lower-degree terms)} \quad \text{and} \quad g = b_n x^n + \text{(lower-degree terms)},$$

with a_m, $b_n \in \dot{F} = F - \{0\}$. Then

(6.85) $fg = a_m b_n x^{m+n} +$ (lower-degree terms),

and $a_m b_n \neq 0$ because F is an integral domain. Therefore $fg \neq 0$, which is what we needed to prove. Also notice that 6.85 shows that if f and g are nonzero polynomials, then

(6.86) $\deg(fg) = \deg f + \deg g$.

The ring $F[x]$ shares many properties with the ring $\mathbb{Z}$ of integers. As we have seen, $F[x]$ is an integral domain, but there is more to the resemblance than that. As in $\mathbb{Z}$, there is a concept of *divisibility*: if $f, h \in F[x]$, we say f is a **divisor** of h (f **divides** h, h is a **multiple** of f) if there is a polynomial $q \in F[x]$ such that $h = fq$; we write $f | h$.

There is also a *division algorithm*: if $f, h \in F[x]$, with $f \neq 0$, then there exist polynomials $q, r \in F[x]$ such that

(6.87) $h = fq + r$,   with $r = 0$ or $0 \leqslant \deg r < \deg f$.

Moreover, the polynomials q (the **quotient**) and r (the **remainder**) that satisfy 6.87 are *unique*. Here the degree function gives a measure of "size" for polynomials, just as the absolute value function does for integers. In both contexts the division algorithm dictates a unique quotient and remainder, with the remainder a smaller size than the divisor (which is f here).

The actual determination of q and r proceeds as follows. If $\deg h < \deg f$, then choose $q = 0$ and $r = h$. Otherwise, suppose

$f = a_m x^m +$ (lower-degree terms)   and $h = b_n x^n +$ (lower-degree terms),

with $m \leqslant n$. Then $(b_n/a_m)x^{n-m}$ is the leading term of the quotient q. To obtain the next term of q, replace h by

$$h - \left(\frac{b_n}{a_m} x^{n-m} \right) \cdot f$$

and repeat the procedure given in the previous two sentences. The leading term of the quotient at this second stage will be the second term of q. Continue this process to get all the lower-degree terms of q. Eventually a remainder r is achieved that is either zero or has degree less than m. This method will be easier to understand with an explicit example, so here it is.

6.88 EXAMPLE.   Let $f = x^2 + 2$ and $h = 5x^4 + x^3 + 11x^2 + 3x + \frac{5}{2}$. Find $q, r \in \mathbb{Q}[x]$ such that $h = fq + r$, with $r = 0$ or $0 \leqslant \deg r < 2$.

SOLUTION.   We will carry out and display the procedure just given, using a long-division format. (We have labelled the relevant items.)

$$
\begin{array}{r}
\dfrac{b_n}{a_m}x^{n-m} \searrow \\
\overline{5x^2 + x + 1} \quad \leftarrow q \\
f \rightarrow x^2 + 2\,\overline{)\,5x^4 + x^3 + 11x^2 + 3x + \tfrac{5}{2}}\quad \leftarrow h \\
\overset{a_m x^m}{\nearrow}\;\underline{5x^4 \qquad\quad + 10x^2}\qquad \leftarrow \left(\dfrac{b_n}{a_m}x^{n-m}\right)\cdot f \\
x^3 + \quad x^2 + 3x + \tfrac{5}{2} \\
b_n x^n \qquad \underline{x^3 \qquad\qquad + 2x} \qquad h - \left(\dfrac{b_n}{a_m}x^{n-m}\right)\cdot f \\
x^2 + \; x + \tfrac{5}{2} \\
\underline{x^2 \qquad\quad + 2} \\
x + \tfrac{1}{2} \quad \leftarrow r
\end{array}
$$

This gives

$$
\underbrace{5x^4 + x^3 + 11x^2 + 3x + \tfrac{5}{2}}_{h} = \underbrace{(x^2 + 2)}_{f}\cdot \underbrace{(5x^2 + x + 1)}_{q} + \underbrace{x + \tfrac{1}{2}}_{r}.
$$

A polynomial $f = \sum_{i \geqslant 0} a_i x^i$ in $F[x]$ can be viewed as a function from F to F by defining $f(\alpha) = \sum_{i \geqslant 0} a_i \alpha^i$ for all $\alpha \in F$. It follows from this definition and the definitions of polynomial addition and multiplication that if f and g are polynomials and $\alpha \in F$, then

(6.89) $(f + g)(\alpha) = f(\alpha) + g(\alpha)$ and $(fg)(\alpha) = f(\alpha) \cdot g(\alpha)$.

We say that an element $\alpha \in F$ is a **root** of f if $f(\alpha) = 0$.

(In the special case in which F is the field $\mathbb{R}$ of real numbers, there are standard techniques available for graphing polynomial functions in the xy-plane. Then the roots of a polynomial $f \in \mathbb{R}[x]$ are the x-coordinates of the points where the graph of f crosses the x-axis.)

Our next objective is to use the division algorithm to get a bound on the number of roots of a polynomial.

6.90 LEMMA.   Let $f \in F[x]$ be a polynomial of degree $n \geqslant 1$, and let $\alpha \in F$. Then α is a root of f if and only if $x - \alpha$ is a divisor of f.

PROOF.   By the division algorithm for polynomials, there are polynomials q and r such that

$$
f = (x - \alpha)q + r, \qquad \text{where } r = 0 \quad \text{or} \quad \deg r = 0
$$

(that is, $r \in F$). Now view f as a function from F to F; then $f(\alpha) = (\alpha - \alpha)q + r = r$. In particular, α is a root of f if and only if $r = 0$, and the conclusion follows.   □

6.91 THEOREM.   A polynomial in $F[x]$ of degree $n \geqslant 1$ has at most n roots.

PROOF.   Let f be a polynomial of degree n. We use induction on n.

If $n = 1$ then $f = ax + b$ for some $a, b \in F$, with $a \neq 0$; then $-b/a$ is the only root of f.

Now suppose $n > 1$, and suppose further that the result has been checked for polynomials of lower degree. Let α be a root of f. (If f has *no* roots we are done.) By Lemma 6.90 there is a factorization $f = (x - \alpha) \cdot g$, where g is a polynomial of degree $n - 1$. If β is a root of f different from α, then

$$0 = f(\beta) = (\beta - \alpha) \cdot g(\beta).$$

But $\beta - \alpha \neq 0$, and therefore we must have $g(\beta) = 0$; that is, β is a root of g. But g has at most $n - 1$ roots, by the induction hypothesis, and therefore f has at most n roots altogether.   □

6.92 DEFINITION.   A field F is **algebraically closed** if every polynomial in $F[x]$ of positive degree has a root in F.

It is not obvious that any algebraically closed fields exist. For instance, neither $\mathbb{Q}$ nor $\mathbb{R}$ is algebraically closed, because the polynomial $x^2 + 1$ has no roots in either field. (Why?) Moreover, we claim that *no finite field is algebraically closed.* For if F is a finite field, say $F = \{\alpha_1, \ldots, \alpha_n\}$, then consider the polynomial

$$g = (x - \alpha_1)(x - \alpha_2) \cdots (x - \alpha_n) + 1.$$

Check that $g(\alpha) = 1$ for all $\alpha \in F$; therefore g has no roots in F, proving our claim. The field of *complex numbers,* to be defined in the next section, turns out to be algebraically closed.

We saw earlier that the ring $F[x]$, like $\mathbb{Z}$, is an integral domain and has a division algorithm for finding quotients and remainders. We conclude this section by pushing our analysis of $F[x]$ a bit further, with our eyes open for "$\mathbb{Z}$-ish" behavior.

Since 1 is the unit element of $F[x]$, the only invertible elements of the ring $F[x]$ are the nonzero elements of F. For if $f \in F[x]$ is invertible, then $fg = 1$ for some $g \in F[x]$. Thus by 6.86,

$$\deg f + \deg g = \deg 1 = 0,$$

which implies that $\deg f = 0$, and hence $f \in \dot{F}$.

Now consider a polynomial in $F[x]$ that has positive degree. Such a polynomial f is said to be **reducible** over F if there exist $g,\, h \in F[x]$, with $\deg g > 0$ and $\deg h > 0$, such that $f = gh$; otherwise f is said to be **irreducible** over F. For instance, it follows from 6.86 that every polynomial of degree 1 is irreducible.

There is a network of ideas on factorization in $F[x]$ that closely parallels the theory in $\mathbb{Z}$. (So much so, in fact, that we won't retrace the proofs in the present context.) *Greatest common divisors* exist. (Use the definition of gcd in 6.31, except now all elements are polynomials, and omit the condition "$d > 0$," which has no meaning in $F[x]$.) Computation of gcd's proceeds by means of Euclid's algorithm (6.35), except that the condition "$0 \leqslant r_1 < b$" is now replaced by "either $r_1 = 0$ or $0 \leqslant \deg r_1 < \deg b$." There is also a theorem on unique factorization. For this purpose, we need the concept of a **prime** in $F[x]$, which is an irreducible polynomial whose leading coefficient is 1. Then Theorem 6.41 has the following analogue in $F[x]$:

6.93 FUNDAMENTAL THEOREM OF ARITHMETIC IN $F[x]$. Every polynomial $f \in F[x]$ with $\deg f \geqslant 1$ has a factorization

$$f = \lambda p_1^{\alpha_1} \cdots p_t^{\alpha_t},$$

where $\lambda \in F - \{0\}$, the p_i are distinct prime polynomials, and $\alpha_i \geqslant 0$ for $1 \leqslant i \leqslant t$. Moreover, this factorization is unique, except for the order in which the primes are listed. (The factorization in this theorem is called the **standard factorization** of f.)

Exercises

In these exercises F will always denote a field. Any further assumptions about F in an exercise will be stated explicitly.

1. Compute the sum and the product of

$$f = 2x^3 + 4x^2 + 3 \quad \text{and} \quad g = 3x^3 + x + 2$$

   viewed as polynomials with coefficients in the finite field $\mathbb{Z}$ (mod 5).

2. Show that the element x does not have an inverse in the polynomial ring $F[x]$, and therefore $F[x]$ is not a field.

3. Use mathematical induction to prove the following law in $F[x]$:

$$(x - 1)(x^{n-1} + x^{n-2} + \cdots + x + 1) = x^n - 1, \quad \text{for all } n \geqslant 1.$$

4. Let $F = \mathbb{Z}$ (mod 2). Write $(x^2 + x + 1)^2 \in F[x]$ as a polynomial with exactly three nonzero coefficients.

5. Let f and g be nonzero polynomials in $F[x]$. Prove that

$$f \mid g \text{ and } g \mid f \quad \Leftrightarrow \quad g = \alpha f \text{ for some } \alpha \in F - \{0\}.$$

6. Show that $F[x]$ contains infinitely many prime polynomials. Do this by imitating the proof of Euclid's theorem (6.29), which asserts that there are infinitely many prime numbers. Make the appropriate adjustments for the shift from $\mathbb{Z}$ to $F[x]$.

7. Let g be a polynomial in $F[x]$, with deg $g = 2$ or 3. Show that g has a root in F if and only if g is reducible over F.

8. Use the results of Exercise 7 to produce an irreducible polynomial of degree 2 and an irreducible polynomial of degree 3 with coefficients in $\mathbb{Z}$ (mod 2).

9. Let $F = \mathbb{Z}$ (mod 2). List all the polynomials of degree 3 in $F[x]$, and write each as a product of irreducible polynomials in $F[x]$.

10. Show that the assertion proved in Exercise 7 does not hold without some restriction on deg g. Do this by giving an example of a reducible polynomial of degree 4 in $\mathbb{Q}[x]$ that has no roots in $\mathbb{Q}$.

11. In analytic geometry we learn that the equation $x^2 + y^2 = 1$ describes a circle of radius 1 centered at the origin in the xy-plane. That circle has infinitely many points, so if we write $f(x, y) = x^2 + y^2 - 1$, we therefore know that there are infinitely many points (a, b) such that $f(a, b) = 0$. Does this contradict Theorem 6.91?

12. Let $F = \mathbb{Z}$ (mod 5). Exhibit a nonzero polynomial $g \in F[x]$ such that $g(\alpha) = 0$ for all $\alpha \in F$.

13. Let F be a finite field, with $\#F = q$.

    (a) Let $\alpha \in F - \{0\}$. Show that $\alpha^i = \alpha^j$ for some integers i and j satisfying $1 \leqslant i < j \leqslant q$. Deduce that $\alpha^k = 1$ for some k satisfying $1 \leqslant k \leqslant q - 1$.

    (*Remark:* In fact, $\alpha^{q-1} = 1$ for all $\alpha \in F - \{0\}$; but you need not prove this.)

    (b) Suppose $F = \{\alpha_1, \alpha_2, \ldots, \alpha_q\}$. Give a factorization of $x^q - x$ into irreducible polynomials in $F[x]$.

14. Let $h = x^4 + x^2 + 2$ and $f = x^2 + 2x + 2$. Find $q, r \in F[x]$ such that $h = fq + r$, with $r = 0$ or $0 \leqslant \deg r < 2$, under the following circumstances:

    (a) $F = \mathbb{Z} \pmod 3$;
    (b) $F = \mathbb{R}$.

15. (a) Show that the ring $F[x]$ has infinitely many elements.

    (b) Show that if F is a finite field, there are only finitely many functions from F to F.

    (c) The paragraph containing statement 6.89 shows that a polynomial in $F[x]$ can act as a function from F to F. What can you now conclude in view of (a) and (b)?

16. Let F be an infinite field, and let f and g be polynomials in $F[x]$ of positive degree. Show that if $f(\alpha) = g(\alpha)$ for all $\alpha \in F$, then $f = g$.

17. In adapting 6.31 to define a notion of "greatest common divisor" in $F[x]$, replace the condition "$d > 0$" (which has no meaning for polynomials in general) by the requirement "d has leading coefficient 1." This guarantees uniqueness of greatest common divisors in $F[x]$, and gcd's can be found by Euclid's algorithm.

    (a) Use Euclid's algorithm to compute the greatest common divisor of the polynomials $x^4 + x^3 - x^2 + x - 2$ and $x^2 + 5x + 6$ over the field $\mathbb{Q}$ of rational numbers.

    (b) Now, with the result from part (a), can you think of another method (involving trial and error and luck) that might have yielded the same result?

18. Let $f = a_n x^n + a_{n-1} x^{n-1} + \cdots + a_1 x + a_0 \in F[x]$, and let $b \in F$. The computation $f(b)$ was discussed in Example 3.3(f). Here is another perspective on that computation.

    (a) Show that there exist a polynomial $q \in F[x]$ and a field element $c \in F$ such that $f = (x - b)q + c$.

    (b) Deduce that $f(b) = c$.

19. (For students with programming experience) Write a computer program that accepts two polynomials (in some form) as input and computes their product.

6.9 COMPLEX NUMBERS

The square of a real number is nonnegative, and therefore the polynomial $x^2 + 1$ has no real roots. Centuries ago, mathematicians found it convenient to pretend that $x^2 + 1$ had an "imaginary" root, which they labelled i. Once they adopted this fantasy and made the further assumption that i could interact with real numbers via addition and multiplication, mathematicians were then able to obtain formulas for roots of polynomials without being forced to invent still more imaginary quantities. For instance, using the assertion that $i^2 = -1$, it is easy to check that $1 + i$ is a root of $x^2 - 2x + 2$, a polynomial that has no real roots. More generally, the quadratic formula can be used to find roots of any polynomial of the form $ax^2 + bx + c$ in which $a, b, c \in \mathbb{R}$ and $a \neq 0$. The name *complex number* was given to any expression of the form $a + bi$ in which $a, b \in \mathbb{R}$.

Nearly 200 years ago, the complex number system was given a firm mathematical foundation, so that it was no longer necessary to view complex numbers as convenient fictions that magically produced credible results. In this section we will consider the system $\mathbb{C}$ of complex numbers from this fantasy-free perspective. We will observe that $\mathbb{C}$ is a field and give geometric interpretations to its operations of addition and multiplication, and we will make some observations about the roots of polynomials with complex coefficients.

We begin our treatment of complex numbers by endowing the ordinary xy-plane with a field structure. We do this by defining two suitable operations, the second of which may seem somewhat bizarre. Explicitly, we define *addition* of points in $\mathbb{R} \times \mathbb{R}$ by

$$(a, b) + (c, d) = (a + c, b + d);$$

that is, we add points by adding their respective coordinates. Some examples:

$$(3, -17) + (-1, \pi) = (2, -17 + \pi)$$
$$\left.\begin{array}{l} (a, 0) + (b, 0) = (a + b, 0) \\ (0, a) + (0, b) = (0, a + b) \\ (0, 0) + (a, b) = (a, b) \end{array}\right\} \quad \text{for all } a, b \in \mathbb{R}.$$

If (a, b), (c, d), and $(0, 0)$ are all different, then they occupy three vertices of a parallelogram whose fourth vertex is at $(a, b) + (c, d) = (a + c, b + d)$. (See the picture.)

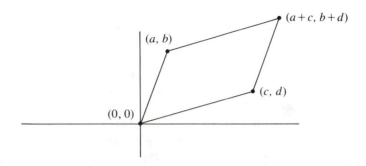

Now define a *product* operation on $\mathbb{R} \times \mathbb{R}$ by the rule

$$(a, b)(c, d) = (ac - bd, ad + bc).$$

For example,

$$(1, 2)(3, -4) = (11, 2)$$
$$(4, 0)(0, -3) = (0, -12)$$
$$\left.\begin{array}{l}(a, 0)(b, 0) = (ab, 0)\\(1, 0)(a, b) = (a, b)\end{array}\right\} \quad \text{for all } a, b \in \mathbb{R}.$$

It is a tedious but straightforward matter to verify that, with these operations of addition and multiplication, $\mathbb{R} \times \mathbb{R}$ is a field. It is called the field of **complex numbers** and denoted $\mathbb{C}$. The zero element of $\mathbb{C}$ is $(0, 0)$ and the multiplicative identity is $(1, 0)$. The additive inverse of (a, b) is $(-a, -b)$ and, if $(a, b) \neq (0, 0)$, the multiplicative inverse of (a, b) is

$$\left(\frac{a}{a^2 + b^2}, \frac{b}{a^2 + b^2}\right).$$

Checking that the product operation is associative can be done as follows. Let $z_1, z_2, z_3 \in \mathbb{C}$; say, $z_1 = (a, b)$, $z_2 = (c, d)$, and $z_3 = (e, f)$, with $a, b, c, d, e, f \in \mathbb{R}$. Then

$$(z_1 z_2)z_3 = (ac - bd, ad + bc)(e, f)$$
$$= (ace - bde - adf - bcf, acf - bdf + ade + bce),$$

and computation of $z_1(z_2 z_3)$ gives the same result. We leave the other field properties to the reader. Incidentally, when we represent the field of complex numbers as a plane, we call it the **complex plane**.

It is our usual custom to view the elements of the field $\mathbb{R}$ of real numbers as the points on a marked line, like this:

On the other hand, in analytic geometry we usually view $\mathbb{R}$ as embedded in the plane as the horizontal axis. That is, *we regard an ordered pair $(r, 0)$ as indistinguishable from the real number r.* (The jargon: we *identify* $(r, 0)$ and r.) We will adopt this convention here. That is, a complex number of the form $(r, 0)$ with $r \in \mathbb{R}$ will now be written more simply as r. Once this is done we have the set inclusion $\mathbb{R} \subseteq \mathbb{C}$. (We've seen this kind of "identification" process before: in the previous section we found it convenient to consider the elements of a field F as elements of the polynomial ring $F[x]$.) From now on we will refer to the horizontal axis in the complex plane as the **real axis**. Since the definition of product in $\mathbb{C}$ gives

$$(r, 0)(a, b) = (ra, rb), \qquad \text{for all } a, b, r \in \mathbb{R},$$

with our new notational convention this becomes

$$r(a, b) = (ra, rb), \qquad \text{for all } a, b, r \in \mathbb{R}.$$

Now let $i = (0, 1)$. Then

$$i^2 = (0, 1)(0, 1) = (-1, 0) = -1.$$

Thus the complex number $i = (0, 1)$ is a root of the polynomial $x^2 + 1$. Also notice that if $(a, b) \in \mathbb{C}$, then

(6.94) $(a, b) = (a, 0)(1, 0) + (b, 0)(0, 1) = a + bi.$

Moreover, it follows from this that

$$a + bi = c + di \qquad \Leftrightarrow \qquad a = c \quad \text{and} \quad b = d.$$

We have therefore proved the following result.

6.95 THEOREM. Every complex number z has a unique expression of the form $z = a + bi$, with $a, b \in \mathbb{R}$.

The numbers $a, b \in \mathbb{R}$ obtained in Theorem 6.95 are called the **real** and **imaginary parts** of z, respectively; we write $a = \mathrm{Re}\, z$ and $b = \mathrm{Im}\, z$. Thus $z \in \mathbb{R}$ if and only if $\mathrm{Im}\, z = 0$. If $z \neq 0$ and $\mathrm{Re}\, z = 0$, then z is said to be a **pure imaginary** number. So when $\mathbb{C}$ is pictured as the usual xy-plane, the pure imaginary numbers occupy the vertical axis (except for the origin), while the real numbers constitute the horizontal axis. Incidentally, this usage of the words "real" and "imaginary" has historical origins. The fact is, *there is nothing imaginary about complex numbers;* they are just ordered pairs of real numbers. It is their operations that give them their special properties.

It is fruitful to represent complex numbers in terms of *polar coordinates,* as we will now describe. Define the **absolute value** of a complex number z, denoted $|z|$, to be the distance of z from the origin. Thus if $z = a + bi$, with $a, b \in \mathbb{R}$, then

$$|z| = \sqrt{a^2 + b^2},$$

and the line segment $0z$ drawn from the origin to z has length $|z|$.

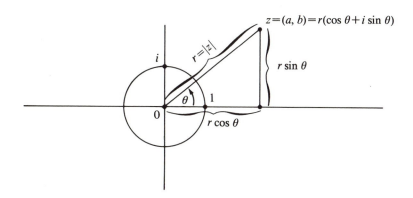

Now if $|z| = r > 0$, it follows from trigonometry that

(6.96)
$$\begin{aligned} z = (a, b) &= (r \cos \theta, r \sin \theta) \\ &= r(\cos \theta, \sin \theta), \end{aligned}$$

where θ is the angle made by the line segment $0z$ with the positive real axis. Now, using Equation 6.94, we see that 6.96 can be rewritten, and every nonzero complex number can be expressed in the form

(6.97)
$$z = r(\cos \theta + i \sin \theta),$$

with $0 < r = |z|$. This is called the **polar form** of z. An angle θ for which 6.97 holds is called an **argument** of z, denoted arg z. Angles are measured in *radians,* and a positive θ corresponds to a counterclockwise rotation through an angle θ measured from the positive real axis.

6.98 EXAMPLE. Write the complex numbers $z = 1 + \sqrt{3}\,i$ and $w = -3 + 3i$ in polar form.

SOLUTION. We have $|z| = \sqrt{1^2 + (\sqrt{3})^2} = 2$. To find arg z, it is helpful to consider this diagram:

We have arg $z = \sin^{-1}(\sqrt{3}/2) = \pi/3$. Therefore

$$z = 2\left(\cos\frac{\pi}{3} + i\sin\frac{\pi}{3}\right).$$

Similarly, $|w| = \sqrt{(-3)^2 + 3^2} = 3\sqrt{2}$, and from the diagram

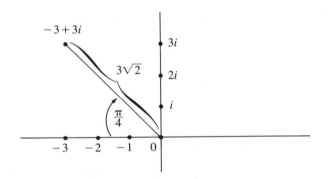

we deduce that

$$\arg w = \pi - \frac{\pi}{4} = \frac{3\pi}{4}.$$

Therefore

$$w = 3\sqrt{2}\left(\cos\frac{3\pi}{4} + i\sin\frac{3\pi}{4}\right).$$

Now consider two nonzero complex numbers in polar form, say

$$z = r(\cos \theta + i \sin \theta) \quad \text{and} \quad z_1 = r_1(\cos \theta_1 + i \sin \theta_1).$$

Then

$$zz_1 = rr_1[(\cos \theta \cos \theta_1 - \sin \theta \sin \theta_1) + i(\sin \theta \cos \theta_1 + \cos \theta \sin \theta_1)]$$
$$= rr_1[\cos(\theta + \theta_1) + i \sin(\theta + \theta_1)],$$

by the miracles of trigonometric identities. Therefore we have

$$(6.99) \quad |zz_1| = rr_1 = |z||z_1| \quad \text{and} \quad \arg(zz_1) = \theta + \theta_1 = \arg z + \arg z_1.$$

In words: to multiply two nonzero complex numbers we multiply their absolute values and add their arguments. Thus, geometrically, multiplication of a complex number w by the nonzero complex number $z = r(\cos \theta + i \sin \theta)$ has the effect of stretching the line segment $0w$ by a factor of r and rotating that segment about the origin by the angle θ. (If $|z| < 1$ then we speak of "contracting" instead of "stretching.") Thus the use of polar form has led us to a geometrical interpretation of multiplication in $\mathbb{C}$.

When $z = z_1 = r(\cos \theta + i \sin \theta)$, 6.99 yields

$$z^2 = r^2(\cos 2\theta + i \sin 2\theta).$$

More generally, it follows by induction on n that

$$(6.100) \qquad [r(\cos \theta + i \sin \theta)]^n = r^n(\cos n\theta + i \sin n\theta)$$

for all integers $n \geq 0$. Formula 6.100 is known as **De Moivre's formula**.

Let n be a positive integer, and for each integer k satisfying $0 \leq k \leq n - 1$ consider the complex number

$$(6.101) \qquad \qquad \zeta_k = \cos \frac{2\pi k}{n} + i \sin \frac{2\pi k}{n}.$$

(Note: ζ is the Greek letter zeta.) In the complex plane, the ζ_k's lie on the unit circle, since $|\zeta_k| = 1$ for each k. In fact, they occupy the vertices of a regular n-gon centered at the origin, with one vertex at $(1, 0)$. Now observe:

$$(\zeta_k)^n = \cos\left(n \cdot \frac{2\pi k}{n}\right) + i \sin\left(n \cdot \frac{2\pi k}{n}\right) \qquad \text{(by De Moivre's formula)}$$

$$= 1. \qquad \text{(since } \cos 2\pi k = 1 \text{ and } \sin 2\pi k = 0)$$

Therefore the n complex numbers $\zeta_0, \zeta_1, \ldots, \zeta_{n-1}$ are roots of the polynomial $x^n - 1$. Moreover, they are the *only* roots of $x^n - 1$, since by 6.91 a polynomial of degree n has at most n roots. The numbers $\zeta_0, \zeta_1, \ldots, \zeta_{n-1}$ defined by Equation 6.101 are called the **complex nth roots of unity**.

    6.102 EXAMPLE.   The complex sixth roots of unity are $\zeta_0, \zeta_1, \ldots, \zeta_5$, where

$$\zeta_0 = \cos 0 + i \sin 0 = 1,$$

$$\zeta_1 = \cos\frac{\pi}{3} + i \sin\frac{\pi}{3} = \frac{1}{2} + i\frac{\sqrt{3}}{2},$$

$$\zeta_2 = \cos\frac{2\pi}{3} + i \sin\frac{2\pi}{3} = -\frac{1}{2} + i\frac{\sqrt{3}}{2},$$

$$\zeta_3 = \cos\pi + i \sin\pi = -1,$$

$$\zeta_4 = \cos\frac{4}{3}\pi + i \sin\frac{4}{3}\pi = -\frac{1}{2} - i\frac{\sqrt{3}}{2},$$

$$\zeta_5 = \cos\frac{5}{3}\pi + i \sin\frac{5}{3}\pi = \frac{1}{2} - i\frac{\sqrt{3}}{2}.$$

Therefore, although the polynomial $x^6 - 1$ has only two roots in $\mathbb{R}$, the six numbers $\zeta_0, \ldots, \zeta_5$ are its roots in $\mathbb{C}$. The picture:

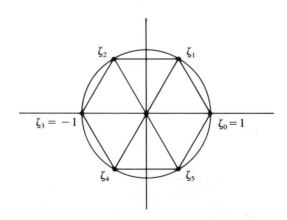

As we have seen, some polynomials in $\mathbb{R}[x]$ have no roots in $\mathbb{R}$. On the other hand, our discussion of roots of unity shows that a polynomial can have roots in $\mathbb{C}$ that are outside $\mathbb{R}$. In fact, the following remarkable fact holds for the field of complex numbers: *Every polynomial with complex coefficients has a complex root.* This assertion is called the **Fundamental Theorem of Algebra.** In the terminology of field theory (see Definition 6.92):

(6.103)   The field of complex numbers is algebraically closed.

The proof is beyond the scope of our treatment here.

Now define a mapping from $\mathbb{C}$ to $\mathbb{C}$, denoted $z \mapsto \bar{z}$ and called **complex conjugation,** by

$$\overline{a + bi} = a - bi.$$

For example, $\overline{2 + 3i} = 2 - 3i$, $\bar{0} = 0$, $\bar{i} = -i$, $\bar{\pi} = \pi$, $\overline{-3 - 4i} = -3 + 4i$, and in Example 6.102, $\bar{\zeta_1} = \zeta_5$ and $\bar{\zeta_4} = \zeta_2$. The following properties are easily checked: for all $w, z \in \mathbb{C}$,

(6.104)

(a) $\overline{w + z} = \bar{w} + \bar{z}$ and $\overline{wz} = \bar{w}\bar{z}$,

(b) $\bar{\bar{z}} = z$,

(c) $\bar{z} = z \iff z \in \mathbb{R}$.

Geometrically, the complex conjugate $\bar{z}$ is the reflection of z in the real axis, as pictured.

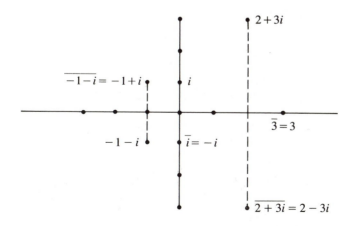

We will conclude this section by using complex conjugation and the fact that $\mathbb{C}$ is algebraically closed to obtain a fundamental property of polynomials with *real* coefficients.

6.105 THEOREM. If $f \in \mathbb{R}[x]$ and $\deg f \geq 1$, then f can be expressed as a product of polynomials in $\mathbb{R}[x]$ of degree at most 2. An irreducible (over $\mathbb{R}$) quadratic factor will occur in this product if and only if f has a nonreal root in $\mathbb{C}$.

PROOF. Let $f = a_n x^n + a_{n-1} x^{n-1} + \cdots + a_0$, with $a_i \in \mathbb{R}$ for each i. We have $\mathbb{R}[x] \subseteq \mathbb{C}[x]$, so by the Fundamental Theorem of Algebra (6.103) we know that f has a root $\lambda_1 \in \mathbb{C}$. Therefore $f = (x - \lambda_1) \cdot g$ for some $g \in \mathbb{C}[x]$, by 6.90. If $\deg g \geq 1$, repeat this argument on g. Eventually we obtain

(*) $$f = a_n(x - \lambda_1) \cdots (x - \lambda_n),$$

where $\lambda_1, \ldots, \lambda_n$ are the roots of f in $\mathbb{C}$. (Incidentally, the λ_i's needn't be distinct.)

Suppose one of the roots, call it β, belongs to $\mathbb{C} - \mathbb{R}$. We have

(**) $$0 = a_n \beta^n + a_{n-1} \beta^{n-1} + \cdots + a_1 \beta + a_0,$$

and application of complex conjugation to both sides of (**) yields

$$0 = \overline{a_n \beta^n + a_{n-1} \beta^{n-1} + \cdots + a_0} = \overline{a_n \beta^n} + \overline{a_{n-1} \beta^{n-1}} + \cdots + \overline{a_0}$$
$$= a_n \bar{\beta}^n + a_{n-1} \bar{\beta}^{n-1} + \cdots + a_0 = f(\bar{\beta}).$$

Therefore $\bar{\beta}$ is also a root of f. Thus the nonreal roots of f come in pairs of the form $\{\beta, \bar{\beta}\}$; so we can rewrite (*) in the form

$$f = a_n(x - \alpha_1) \cdots (x - \alpha_r)(x - \beta_1)(x - \bar{\beta}_1) \cdots (x - \beta_t)(x - \bar{\beta}_t),$$

where $\alpha_1, \ldots, \alpha_r$ are the real roots of f and $\beta_1, \bar{\beta}_1, \ldots, \beta_t, \bar{\beta}_t$ are the nonreal roots.

Now let $\beta = a + bi$ be one of the nonreal roots, with $a, b \in \mathbb{R}$. Observe that

$$(x - \beta)(x - \bar{\beta}) = [x - (a + bi)][x - (a - bi)]$$
$$= \underbrace{x^2 - 2ax + (a^2 + b^2)}_{p} \in \mathbb{R}[x].$$

The quadratic polynomial p is irreducible over $\mathbb{R}$ because it has no real roots. This completes the proof of the theorem's first statement, and it also proves the "if" part of the theorem's second statement.

Conversely, if f has a quadratic factor $q \in \mathbb{R}[x]$ that is irreducible over $\mathbb{R}$, then q cannot have a real root, by Lemma 6.90; but the quadratic formula provides a complex root, which is also a root of f. $\quad\square$

6.106 EXAMPLES.

(a) Consider the polynomial $f = x^4 + x^3 + x^2 - 3x \in \mathbb{R}[x]$. Clearly 0 is a root and x is a divisor of f, so $f = x(x^3 + x^2 + x - 3)$. Next, 1 is a root of $x^3 + x^2 + x - 3$, making $x - 1$ a divisor. The division algorithm gives the explicit factorization

$$x^3 + x^2 + x - 3 = (x - 1)(x^2 + 2x + 3).$$

Finally, the quadratic polynomial $x^2 + 2x + 3$ is irreducible over $\mathbb{R}$ because its roots are not real: the quadratic formula gives them as $-1 + \sqrt{2}i$ and $-1 - \sqrt{2}i$. Thus f has the prime factorization

$$f = x(x - 1)(x^2 + 2x + 3)$$

over $\mathbb{R}$, and over $\mathbb{C}$ we have the more complete factorization

$$\begin{aligned} f &= x(x - 1)[x - (-1 + \sqrt{2}i)][x - (-1 - \sqrt{2}i)] \\ &= x(x - 1)(x + 1 - \sqrt{2}i)(x + 1 + \sqrt{2}i). \end{aligned}$$

(b) Let $f = x^4 + 5x^2 - 36$. High school algebra provides the factorization over $\mathbb{R}$:

$$f = (x^2 - 4)(x^2 + 9) = (x - 2)(x + 2)(x^2 + 9).$$

The polynomial $x^2 + 9$ is irreducible over $\mathbb{R}$, but its complex roots are $3i$ and $-3i$. Therefore over $\mathbb{C}$ we have

$$f = (x - 2)(x + 2)(x - 3i)(x + 3i).$$

Exercises

1. Write each of the following complex numbers in the form $a + bi$, with $a, b \in \mathbb{R}$.

   (a) $(3 + 2i)^2$ (b) i^{857} (c) i^{-903}

(d) $\dfrac{1+i}{1-i}$ (e) $5 \cdot \left(\dfrac{-1+\sqrt{3}i}{2}\right)^3$

(f) $(1+i)^{20}$ (You can use the binomial theorem with $n = 20$, but there is a simpler way.)

(g) $\dfrac{2+5i}{7i}$ (h) $\dfrac{2+i}{3i-4} + \dfrac{3-i}{1+i}$

2. Prove the distributive law $w(y+z) = wy + wz$, for $w, y, z \in \mathbb{C}$.

3. Use the definition of multiplication in $\mathbb{C}$ to prove that if $w, z \in \mathbb{C}$, then $wz = 0$ if and only if $w = 0$ or $z = 0$.

4. (Use a hand-held calculator for this one.) Let $z = 1 + .1417745i$.
   (a) Compute $|z|$.
   (b) What integer n makes $|z^n|$ approximately equal to 2?

5. Suppose $|z| = \frac{1}{2}$ and $\arg z = \pi/4$. Write z in the form $a + bi$, with $a, b \in \mathbb{R}$. (Do not use a calculator.)

6. Suppose $\arg z = \pi/4$ and $|z| = .9$. Draw a curve (freehand) on which all the numbers z^n are located, for $n \in \mathbb{Z}$ and $1 \leqslant n \leqslant 24$, and mark the points corresponding to those numbers. (A calculator will help.)

7. Repeat Exercise 6, but with $|z| = 1.1$.

8. Let M denote set of all 2×2 matrices of the form

$$\begin{pmatrix} a & b \\ -b & a \end{pmatrix},$$

with $a, b \in \mathbb{R}$.
   (a) Show that M is a ring under the operations of matrix addition and multiplication.
   (b) Prove that there is an isomorphism $f: \mathbb{C} \to M$. (Suggestion: See the paragraph following Examples 6.18.)

9. List all the 18th roots of unity that satisfy $0 \leqslant \arg z \leqslant \pi/2$.

10. Determine the polar form of each of the following complex numbers. In each case choose $\arg z$ to satisfy the inequality $-\pi < \arg z \leqslant \pi$. (This choice for $\arg z$ is called the **principal value** of $\arg z$, usually denoted Arg z.) Draw a picture showing the location of each number in the complex plane.
   (a) -1
   (b) $1-i$

(c) $-\dfrac{1}{2}+\dfrac{\sqrt{3}}{2}i$

(d) $-\sqrt{6}+\sqrt{2}i$

11. Define the **distance** between complex numbers z and w to be the number $|z - w|$.

(a) Show that this value is equal to the usual distance between z and w viewed as points in the xy-plane.

(b) Find all the complex numbers whose distance from i is 5 and whose distance from $1 - i$ is 4. (There are exactly two such numbers.)

12. Draw pictures of the following sets of complex numbers.

(a) $\{z \mid 1 \leqslant |z| < 2 \quad \text{and} \quad \tfrac{3}{4}\pi < \arg z \leqslant \pi\}$

(b) $\{z \mid |z - 1| \leqslant 1 \quad \text{and} \quad |z - 1 - i| < 1\}$

13. What complex number is located at the midpoint of the line segment connecting $2 - 3i$ to $1 + 5i$? Draw a picture as a check on your answer.

14. A polynomial in $\mathbb{C}[x]$ of the form $x^n + x^{n-1} + \cdots + x + 1$ is called a **cyclotomic polynomial**. The dictionary tells us that *cyclotomy* is the "mathematical theory of the division of the circle into equal parts." Why is "cyclotomic" an appropriate adjective for such a polynomial?

15. (a) Suppose α and β are both roots of the polynomial $x^8 - 16$ in $\mathbb{C}$. Show that α/β is a root of unity.

(b) Draw a picture showing all the roots of $x^8 - 16$ in the complex plane.

16. Show that $1/\bar{z} = \overline{1/z}$ for all $z \in \mathbb{C} - \{0\}$.

17. Let z be a complex number.

(a) Verify that $z + \bar{z}$ and $z\bar{z}$ are real numbers.

(b) Express Re z, Im z, and $|z|^2$ in terms of z and $\bar{z}$, using only addition, subtraction, multiplication, and division.

(c) Compute the real number

$$\frac{(z + \bar{z})^2 + \left(\dfrac{1}{i}(z - \bar{z})\right)^2}{z\bar{z}}, \qquad \text{for } z \neq 0.$$

18. Verify statements 6.104(a) and (b). Then do whatever else is needed to show that complex conjugation is an isomorphism from the field $\mathbb{C}$ to itself.

19. Do *one* of the following.
    (a) Give an example of a polynomial $f \in \mathbb{C}[x]$ of degree 3 having a complex root z for which the conjugate $\bar{z}$ is *not* a root of f.
    (b) Explain why there are no examples of the kind described in (a).

20. Use Theorem 6.105 to show that every polynomial of odd degree in $\mathbb{R}[x]$ has a real root. (Another proof can be given using the intermediate value theorem from differential calculus, but that is not our approach here.)

21. In each of the following cases, write the given polynomial as a product of irreducible polynomials in $\mathbb{R}[x]$. Then go on to factor the polynomial into *linear* factors (factors of degree 1) in $\mathbb{C}[x]$.
    (a) $x^6 - 1$
    (b) $x^4 + 2x^2 - 35$
    (c) $x^6 - x^4 + x^2 - 1$
    (d) $x^4 - 2x^3 - 8x + 16$

6.10 BOOLEAN ALGEBRAS AND RELATED ORDER STRUCTURES

Is the power set of a set S a ring with respect to union and intersection? We know from set theory that union and intersection are commutative and associative operations on $P(S)$, and that each of these operations has an identity element: $\emptyset$ is the identity for union and S for intersection. Moreover, we have the distributive law $A \cap (B \cup C) = (A \cap B) \cup (A \cap C)$. So $P(S)$ is well on its way to being a commutative ring with unit, with $\cup$ as "addition" and $\cap$ as "multiplication." [Or vice versa, perhaps, for there is also the distributive law $A \cup (B \cap C) = (A \cup B) \cap (A \cup C)$.] But there is an obstruction along the road to ringville: with respect to each of the operations $\cup$ and $\cap$, most members of $P(S)$ have no inverse. Specifically, if $A \neq \emptyset$ there is no B such that $A \cup B = \emptyset$; and if $A \neq S$ there is no B such that $A \cap B = S$.

Now let A' denote the complement of A in S. Observe that A' serves as a sort of "twisted inverse" for A with respect to both operations: $A \cup A' = S$, the identity for $\cap$, and $A \cap A' = \emptyset$, the identity for $\cup$. So, while we cannot call $P(S)$ a *ring* with respect to the operations $\cup$ and $\cap$, there is enough algebraic structure here to warrant further study, and even a name: $P(S)$ is an example of a *Boolean algebra*.

Before defining Boolean algebras, let's recall that many of the properties of sets have counterparts in the mechanics of statements, with disjunction ($\vee$), conjunction ($\wedge$), and negation ($\sim$) corresponding to the set operations of union, intersection, and complimentation, respectively. For instance, compare the results on set equality given by Theorem 2.26 with the results on logical equivalence given by 1.28 and 1.30–1.35. The results deal with different topics but they are clearly parallel. We saw in Chapter 1 that results on logical equivalence can be applied to the design of electrical switching circuits, and especially to the problem of minimizing the complexity of circuitry. We will return to that problem later in this section.

We will prove only a few results on Boolean algebras in this section. Nevertheless, we will do enough to establish a clear connection between these abstract algebraic structures and some concrete applications.

6.107 DEFINITION.   A **Boolean algebra** is a set B with two operations, presently denoted $\wedge$ and $\vee$ (verbalized "wedge" and "vee," respectively), satisfying the following axioms:

(i) Both $\wedge$ and $\vee$ are commutative and associative operations, and each has an identity element. The identity for $\wedge$ is denoted 1, and that for $\vee$ is denoted 0. Moreover, $1 \neq 0$ (and therefore $\wedge$ and $\vee$ are different operations). As with rings, the element 1 is called the **unit element** of B, and 0 is called **zero**.

(ii) **Distributive laws** hold:

$$a \wedge (b \vee c) = (a \wedge b) \vee (a \wedge c),$$
$$a \vee (b \wedge c) = (a \vee b) \wedge (a \vee c),$$

for all $a, b, c \in B$.

(iii) For each $a \in B$,

$$a \wedge a = a \quad \text{and} \quad a \vee a = a.$$

These are the **idempotency laws**.

(iv) For each $a \in B$ there is an element $a' \in B$, called the **complement** of a, satisfying

$$a \vee a' = 1 \quad \text{and} \quad a \wedge a' = 0.$$

6.108 EXAMPLES.

(a) Let S be any nonempty set. Then the power set $P(S)$ is a Boolean algebra under the operations $\cap$ and $\cup$, with respective identities S and $\emptyset$. The Boolean algebra complement is the usual set complement. (We checked most of the details in Chapter 2; see especially Theorem 2.26.)

(b) We know from Definition 6.107 that a Boolean algebra must have at least two elements. Here is a two-element Boolean algebra. Let $\mathbb{B} = \{0, 1\}$, and define operations $\vee$ and $\wedge$ in $\mathbb{B}$ by the following tables:

$\vee$	0	1		$\wedge$	0	1
0	0	1		0	0	0
1	1	1		1	0	1

These operations are called **Boolean addition** and **Boolean multiplication**, respectively. Notice in particular that $1 \vee 1 = 1$ [in contrast to the equality $1 + 1 = 0$ in the ring $\mathbb{Z}$ (mod 2)]. These operations may seem strange if we base our view solely on our experience with numbers. But if we replace 0 with *false* and 1 with *true*, then the operation tables describe the truth-value behavior of propositions under the logical operations of disjunction and conjunction. Finally, define complements by $0' = 1$ and $1' = 0$. It is now straightforward to check that with these definitions of $\vee$, $\wedge$, and $'$, the set $\mathbb{B}$ is a Boolean algebra.

(c) Let $\mathbb{B}^n$ denote the set of n-tuples of 0s and 1s, and define $\vee$, $\wedge$, and $'$ on $\mathbb{B}^n$ coordinatewise, using the operations just defined in $\mathbb{B}$:

$$(a_1, \ldots, a_n) \vee (b_1, \ldots, b_n) = (a_1 \vee b_1, \ldots, a_n \vee b_n),$$
$$(a_1, \ldots, a_n) \wedge (b_1, \ldots, b_n) = (a_1 \wedge b_1, \ldots, a_n \wedge b_n),$$
$$(a_1, \ldots, a_n)' = (a_1', \ldots, a_n'),$$

where the symbols $\wedge$, $\vee$, and $'$ on the right denote operations in $\mathbb{B}$. For instance, in $\mathbb{B}^4$ we have

$$(1, 0, 1, 0) \vee (1, 0, 0, 1) = (1, 0, 1, 1),$$
$$(1, 0, 1, 0) \wedge (1, 0, 0, 1) = (1, 0, 0, 0),$$
$$(1, 0, 1, 0)' = (0, 1, 0, 1),$$
$$(1, 1, 1, 1)' = (0, 0, 0, 0).$$

It can be checked that $\mathbb{B}^n$ is a Boolean algebra under these operations, with $1 = (1, \ldots, 1)$ and $0 = (0, \ldots, 0)$. The elements of $\mathbb{B}^n$ are called **n-bit strings**.

(d) A function $\mathbb{B}^n \to \mathbb{B}$ is called a **Boolean function** (or, more completely, a **Boolean function in n variables over $\mathbb{B}$**). If f and g are Boolean functions, define new Boolean functions $f \vee g$, $f \wedge g$, and f' by the formulas

$$\left. \begin{aligned} (f \vee g)(x) &= f(x) \vee g(x) \\ (f \wedge g)(x) &= f(x) \wedge g(x) \\ f'(x) &= (f(x))' \end{aligned} \right\} \quad \text{for all } x \in \mathbb{B}^n.$$

With these operations, the family of Boolean functions becomes a Boolean algebra. The unit element is the function with action

$$x \mapsto 1, \quad \forall x \in \mathbb{B}^n,$$

and the zero element is the function with action

$$x \mapsto 0, \quad \forall x \in \mathbb{B}^n.$$

The verification of these assertions is left to the reader.

The Boolean algebras in 6.108(b), (c), and (d) are all finite, and (a) is also finite if S is a finite set. In each case observe that the cardinality (the number of elements) is a power of 2. (In (d), each of the 2^n elements of $\mathbb{B}^n$ has two possible images under a Boolean function; hence there are $2^{(2^n)}$ such functions. See Exercise 19(a) on page 172.) We are compelled to make a conjecture: The cardinality of a finite Boolean algebra is of the form 2^k for some positive integer k. We will sketch a confirmation of this conjecture later in the section.

The following theorem gives a more complete description of the complementation process.

6.109 THEOREM. Let B be a Boolean algebra. Then the following statements hold:

(a) Every element of B has a *unique* complement, and therefore complementation is a function from B to B.
(b) For each $x \in B$ the equation $(x')' = x$ holds.
(c) Complementation is a bijection from B to B.

PROOF.

(a) If x' and $\bar{x}$ are complements of x in B, then

$$x' = x' \vee 1 = x' \vee (x \vee \bar{x}) = (x' \vee x) \vee \bar{x} = 1 \vee \bar{x} = \bar{x}.$$

(b) The fact that $(x')' = x$ follows from the equations

$$x \vee x' = 1 \quad \text{and} \quad x \wedge x' = 0.$$

(c) Surjectivity follows immediately from (b), because complementation has the action $x' \mapsto x$, for each $x \in B$. Injectivity follows from the implications

$$x' = y' \quad \Rightarrow \quad (x')' = (y')' \quad \Rightarrow \quad x = y. \qquad \square$$

Recall that in any *ring* we have $a \cdot 0 = 0$ for each element a. (See Theorem 6.66.) The following two Boolean algebra results have a similar flavor, though the proof here is rather different from the ring argument.

6.110 THEOREM.   Let B be a Boolean algebra. Then, for all $a \in B$,

$$\text{(i)} \quad a \vee 1 = 1 \quad \text{and} \quad \text{(ii)} \quad a \wedge 0 = 0.$$

PROOF.   (i) $\begin{aligned}[t] a \vee 1 &= (a \vee 1) \wedge 1 \\ &= (a \vee 1) \wedge (a \vee a') \\ &= a \vee (1 \wedge a') \quad \text{(distributive law)} \\ &= a \vee a' = 1 \end{aligned}$

To prove statement (ii), first observe that statement (i) transforms into statement (ii) when we substitute $\wedge$ for $\vee$ and 0 for 1. Then go on to observe that the proof of (i) transforms into a proof of (ii) if each $\vee$ is replaced by $\wedge$, each $\wedge$ by $\vee$, and each 1 by 0. $\square$

6.111 DUALITY PRINCIPLE.   Each of the Boolean algebra axioms (listed in 6.107) remains an axiom when the symbols $\wedge$ and $\vee$ are substituted for each other, and a simultaneous mutual substitution is performed with the symbols 0 and 1. This replacement procedure is called **dualizing**, and it transforms a statement into its **dual**. If P is any proposition about Boolean algebras that can be proved using only the axioms in 6.107, then the dual of P is also true, and its proof is the dual of the proof of P. Once again: *the proof of the dual is the dual of the proof.*

As an instance of the duality principle in action, we note that having proved 6.110(i), the proof of 6.110(ii) can be summarized as follows: "Dualize." We cannot resist requoting Alfred North Whitehead (from page 56):

> This example shows that, by the aid of symbolism, we can make transitions in reasoning almost mechanically by the eye, which otherwise would call into play the higher faculties of the brain.

With the help of Theorem 6.110, we now obtain a result suggesting that we are witnessing a reincarnation of our set theory chapter. Later on we'll see that this is no hoax.

6.112 THEOREM (DE MORGAN'S LAWS).   Let B be a Boolean algebra. Then for all $a, b \in B$,

(i) $(a \wedge b)' = a' \vee b'$ and   (ii) $(a \vee b)' = a' \wedge b'.$

PROOF.   We need only consider (i), because (ii) is its dual. To show that $a' \vee b'$ is the complement of $a \wedge b$, we must verify that

$$(a' \vee b') \vee (a \wedge b) = 1 \quad \text{and} \quad (a' \vee b') \wedge (a \wedge b) = 0.$$

For the first statement, we have

$$(a' \vee b') \vee (a \wedge b) = [(a' \vee b') \vee a] \wedge [(a' \vee b') \vee b]$$
$$\text{(distributive law)}$$
$$= [b' \vee (a' \vee a)] \wedge [a' \vee (b' \vee b)]$$
$$\text{(associative and commutative laws)}$$
$$= (b' \vee 1) \wedge (a' \vee 1)$$
$$= 1 \wedge 1 = 1.$$

For the second statement, we start by dualizing the first, obtaining

$$(a' \wedge b') \wedge (a \vee b) = 0,$$

for all $a, b \in B$. In particular, this holds with a' and b' in the roles of a

and b, respectively; that is,

$$[(a')' \wedge (b')'] \wedge (a' \vee b') = 0.$$

This is clearly equivalent to the statement we are after.   □

Let's restate De Morgan's laws in the terminology of earlier sections of this chapter. A Boolean algebra B is a monoid with respect to each of the operations $\wedge$ and $\vee$. Let B_1 and B_2 denote the set B viewed as a monoid under $\wedge$ and $\vee$, respectively. De Morgan's laws then assert that the complementation mapping (which is a bijection from B to B, by 6.109) is simultaneously an isomorphism from B_1 to B_2 and an isomorphism from B_2 to B_1. Remarkable!

Our next goal is to use the operations on a Boolean algebra B to define a notion of relative size for elements of B. This will lead us to some geometric intuition that has been lacking up to now, intuition that will be reinforced by a diagram that captures the essence of the algebraic structure on B.

6.113 DEFINITION. A relation $\leq$ on a set X is said to be a **partial ordering** (or, more briefly, an **ordering**) on X if the following three properties hold:

(i) $a \leq a$ for all $a \in X$ (*reflexive* property)

(ii) For all $a, b \in X$,

$$a \leq b \quad \text{and} \quad b \leq a \;\Rightarrow\; a = b. \quad (\textit{antisymmetric} \text{ property})$$

(iii) For all $a, b, c \in X$,

$$a \leq b \quad \text{and} \quad b \leq c \;\Rightarrow\; a \leq c. \quad (\textit{transitive} \text{ property})$$

A set X with a partial ordering is said to be a **partially ordered set** (sometimes shortened to **poset**). It is convenient to abbreviate the two-part statement "$a \leq b$ and $a \neq b$" by "$a < b$."

6.114 EXAMPLES.

(a) The usual numerical less-than-or-equal-to relation is a partial ordering on the set $\mathbb{R}$ of real numbers.

(b) For any set S, the power set $P(S)$ is partially ordered by the set-inclusion relation $\subseteq$.

(c) The set $\mathbb{N}$ of natural numbers is partially ordered by the "divides" relation (see Definition 6.26).

A finite partially ordered set can be represented by a graph, called a **Hasse diagram**, as follows. Let X and $\leqslant$ be as in Definition 6.113. Each element of X corresponds to a vertex of the graph. If $a, b \in X$ and $a < b$, then vertex b is located higher than vertex a in the diagram *and* there is an ascending edge-path that starts at a and ends at b.

6.115 EXAMPLES.

- (a) The inclusion relation on the power set $P(\{1, 2\})$ has the following graph:

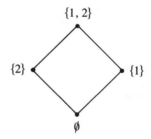

(b) The Hasse diagram for $P(\{1, 2, 3\})$ under set inclusion is more elaborate:

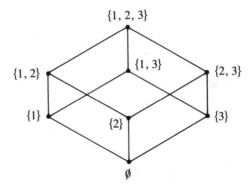

(c) George and Sally marry and have two children, Joan and Bob. Some years later George dies mysteriously, and Sally marries Jim shortly thereafter. Soon their delightful twins Clarissa and Henry are born. Meanwhile, Joan marries Archie, and their baby Lucy arrives within a year. Bob becomes a hermit and never has a family. Eventually Clarissa marries Oliver, after a protracted engagement, and the birth of their little Wolfgang brings Sally a second grandchild at last. On this set of people, the partial ordering "is the same person or is an ancestor of" has the following Hasse diagram:

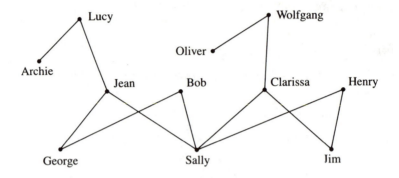

(d) The "divides" relation on the set {1, 2, 3, . . . , 12} has this diagram:

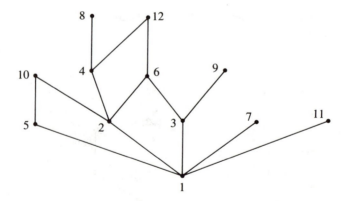

(e) Examples (a) through (d) use a partially ordered set to produce a graph. Now let's carry out an opposite procedure; namely, we will start with a suitable graph and use it to define a partial ordering. For instance, consider this piece of backyard latticework used for training roses:

Label its vertices with the elements of your favorite 25-element set S. Then for $a, b \in S$, define $a \leq b$ to mean that either $a = b$ or there is an ascending path from a to b in the diagram. Then "$\leq$" is a partial ordering on S.

6.116 DEFINITION. Let S be a partially ordered set, and let A be a subset of S. An element $b \in S$ is an **upper bound** for A if the inequality $x \leq b$ holds for all $x \in A$; and an upper bound b for A is a **least upper bound** for A if every upper bound b' for A satisfies the inequality $b \leq b'$. Similarly, $c \in S$ is a **lower bound** for A if $c \leq x$ for all $x \in A$; and a lower bound c for A is a **greatest lower bound** for A if every lower bound c' for A satisfies the inequality $c' \leq c$.

Notice that a subset of a partially ordered set can have at most one least upper bound. For if b and b' are both least upper bounds for A, then $b \leq b'$ and $b' \leq b$, and hence $b = b'$ by the antisymmetric property. Similarly, such a set can have at most one greatest lower bound. A least upper bound for A is denoted **lub A**, and a greatest lower bound for A is denoted **glb A**. Observe that such bounds need not always exist. For instance, in Example 6.115(c) the set {Lucy, Wolfgang} has no upper bound, hence no least upper bound. On the other hand, in that example Sally is a lower bound, indeed the *only* lower bound, for the set {Lucy, Wolfgang}; therefore Sally = glb {Lucy, Wolfgang}.

6.117 EXAMPLE. Consider a partially ordered set S with this tree-like Hasse diagram:

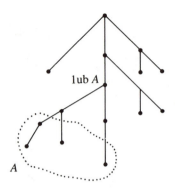

A subset of this set S has a lower bound if and only if its vertices are all on the same ascending path, in which case its greatest lower bound is the lowest vertex on that path. But every subset has an upper bound, namely, the element of S represented by the top vertex in the diagram. And the *least* upper bound of a subset is the lowest vertex of S that can be reached by ascent from each vertex of the subset.

Let B be a Boolean algebra. Define a relation $\leq$ on B by

$$(6.118) \qquad\qquad a \leq b \;\Leftrightarrow\; a \vee b = b.$$

We claim that this relation is a partial ordering on B. To prove it we must verify that the relation is reflexive, antisymmetric, and transitive. For each $a \in B$ we have $a \leq a$, because $a \vee a = a$ by idempotency. Now let $a, b \in B$. Then

$$a \leq b \;\text{ and }\; b \leq a \quad\Rightarrow\quad a \vee b = b \;\text{ and }\; b \vee a = a \quad \text{(from 6.118)}$$
$$\Rightarrow\quad a = b. \qquad\qquad \text{(because } a \vee b = b \vee a\text{)}$$

So $\leq$ is antisymmetric. Finally, suppose $a \leq b$ and $b \leq c$. We must show that $a \leq c$; that is, we must show that $a \vee c = c$. We have

$$a \leq b \;\text{ and }\; b \leq c \quad\Rightarrow\quad a \vee b = b \;\text{ and }\; b \vee c = c$$
$$\Rightarrow\quad a \vee c = a \vee (b \vee c)$$
$$= (a \vee b) \vee c = b \vee c = c$$
$$\Rightarrow\quad a \leq c.$$

Therefore $\leq$ is transitive, and our claim is proved.

The following theorem reveals the linkage between a Boolean algebra's algebraic structure and its structure as a partially ordered set.

6.119 THEOREM.   Let B be a Boolean algebra, and let $\leq$ be the partial ordering on B defined by 6.118. Then

(i) $0 \leq a \leq 1$ for all $a \in B$;

(ii) $a \wedge b = \mathrm{glb}\{a, b\}$ for all $a, b \in B$;

(iii) $a \vee b = \mathrm{lub}\{a, b\}$ for all $a, b \in B$.

PROOF.

(i) Because 0 is the identity for $\vee$, we have $0 \vee a = a$ and therefore $0 \leq a$. We observed in 6.110 that $a \vee 1 = 1$, therefore $a \leq 1$.

(ii) We first show that $a \wedge b$ is a lower bound for $\{a, b\}$; that is,

$$a \wedge b \leq a \quad \text{and} \quad a \wedge b \leq b.$$

This is equivalent to showing

(6.120) $(a \wedge b) \vee a = a \quad \text{and} \quad (a \wedge b) \vee b = b.$

But

$$(a \wedge b) \vee a = (a \wedge b) \vee (a \wedge 1) = a \wedge (b \vee 1) = a \wedge 1 = a,$$

and the proof of $(a \wedge b) \vee b = b$ is similar. Now we have to prove that $a \wedge b$ is the *greatest* lower bound. For this purpose, suppose $c \leq a$ and $c \leq b$; we must show that $c \leq a \wedge b$. Our hypothesis on c can be restated like this:

$$c \vee a = a \quad \text{and} \quad c \vee b = b.$$

Then

$$c \vee (a \wedge b) = \underbrace{(c \vee a)}_{a} \wedge \underbrace{(c \vee b)}_{b} = a \wedge b,$$

and therefore $c \leq a \wedge b$, as desired.

(iii) The outline of the argument is similar to that for (ii), but the details are simpler. They are left for the reader to supply. $\square$

Equations 6.120 are called the **absorption laws** in B.

Let's put Theorem 6.119 in perspective. We have seen that the algebraic structure on a Boolean algebra B induces a partial ordering on B; and Theorem 6.119 shows that, with respect to that ordering, every pair of elements has a greatest lower bound and a least upper bound. The theorem also shows that if the explicit definitions of the operations $\wedge$ and $\vee$ were somehow lost and we were left with only the induced partial ordering on B, then $\wedge$ and $\vee$ could be reconstructed using 6.119(ii) and (iii). Thus the algebraic structure and the ordering structure on B completely determine each other, and they are therefore equivalent in a sense. The geometric interpretation of Theorem 6.119 is that on the Hasse diagram for B, the element 1 is on top, 0 is on the bottom, $a \vee b$ is the lowest vertex reachable by ascent from both a and b, and $a \wedge b$ is the highest vertex reachable by descent from both a and b.

6.121 EXAMPLE. Let S be a nonempty set. If we view the power set $P(S)$ as a Boolean algebra, with operations $\cap$ and $\cup$ for $\wedge$ and $\vee$, respectively (see Example 6.108(a)), then the resulting partial ordering $\leqslant$ is defined by

$$A \leqslant B \quad \Leftrightarrow \quad A \cup B = B.$$

It is an easy exercise in set theory to check that

$$A \subseteq B \quad \Leftrightarrow \quad A \cup B = B.$$

Therefore the partial ordering induced by the algebraic operations on $P(S)$ is just set inclusion; that is, the relations $\leqslant$ and $\subseteq$ on $P(S)$ are the same.

A partially ordered set L in which every pair of elements has both a greatest lower bound and a least upper bound is called a **lattice**. (The partially ordered set diagrammed in 6.115(e) is an example of a lattice, and it suggests the origin of the terminology.) A lattice need not be a Boolean algebra under the operations defined by 6.119(ii) and (iii), even if there exist elements 0 and 1 such that 6.119(i) holds. For example, suppose L is the set of all positive divisors of 24, under the partial ordering "divides." (That is, $a \leqslant b$ means $a \mid b$, as defined in 6.26.) This ordered set is a lattice, and it has the following Hasse diagram:

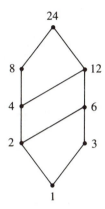

Now we can use 6.119(ii) and (iii) to define operations on L. Then 24 is the unit element, 1 is the zero element, and we have $4 \vee 6 = 12$, $4 \wedge 6 = 2$, and so on. However, these operations do *not* make L into a Boolean algebra. For if L were a Boolean algebra, we would have $4 \vee 4' = 24$ and $4 \wedge 4' = 1$. To satisfy the first of these conditions, we must have $4' = 24$. But $4 \wedge 24 = 4$, and hence $4' \neq 24$, a contradiction.

Now we can state the main structure theorem for finite Boolean algebras, which shows that the power sets of finite sets are essentially the *only* finite Boolean algebras. We provide a brief sketch of the proof; the details will be left for more advanced courses.

6.122 THEOREM. Let B be a finite Boolean algebra. Then B is isomorphic to the Boolean algebra $P(S)$ for some finite set S.

PROOF SKETCH. We first define an **atom** of B to be a minimal non-zero element of B; that is, an element $a \neq 0$ such that whenever $0 \leq x \leq a$, it follows that $x = 0$ or $x = a$. On the Hasse diagram for B the atoms are represented by the vertices closest to 0.

The main thrust of the argument is to prove that every element $x \in B$ decomposes uniquely into atoms. More explicitly, the proof shows that if $0 \neq x \in B$ and $a_1, \ldots, a_t$ are the distinct atoms that are less than or equal to x, then $x = a_1 \vee \cdots \vee a_t$. And this decomposition is unique in the sense that if also $x = b_1 \vee \cdots \vee b_k$, where the b's are distinct atoms, then $k = t$ and $\{b_1, \ldots, b_t\} = \{a_1, \ldots, a_t\}$.

Once that part of the proof has been completed, it follows that if S is the set of atoms in B, then there is a bijection $P(S) \rightarrow B$ given by the correspondence

$$\{a_1, \ldots, a_t\} \mapsto a_1 \vee \cdots \vee a_t$$

for each nonempty set $\{a_1, \ldots, a_t\} \subseteq S$; and $\emptyset \mapsto 0$.

It turns out that this bijection is actually an isomorphism of the Boolean algebra $P(S)$ onto B.   □

6.123 EXAMPLE.   Consider the Boolean algebra $\mathbb{B}^3$ defined by Example 6.108(c). The associated partial ordering on $\mathbb{B}^3$ (given by 6.118) has the following Hasse diagram:

(*)

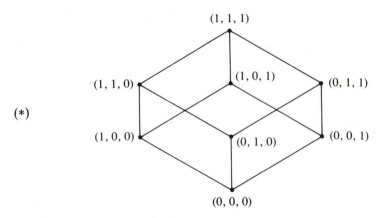

The atoms of $\mathbb{B}^3$ are therefore $a_1 = (1, 0, 0)$, $a_2 = (0, 1, 0)$, and $a_3 = (0, 0, 1)$. Now let $S = \{a_1, a_2, a_3\}$. We saw in Example 6.121 that the partial ordering determined by the Boolean algebra structure on $P(S)$ is just set inclusion, so as a partially ordered set, $P(S)$ has the following Hasse diagram:

(**)

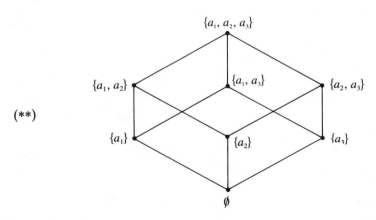

The isomorphism $P(S) \to \mathbb{B}^3$ given by Theorem 6.122 can be viewed as a mapping of diagram (**) onto diagram (*) that takes each vertex of (**) onto the like-positioned vertex of (*).

6.124 COROLLARY. If B is a finite Boolean algebra, then $\#B = 2^n$ for some integer $n \geqslant 1$.

PROOF.   Let S denote the set of atoms in B; then S has n elements for some $n \geqslant 1$, and therefore $\#P(S) = 2^n$. The proof of Theorem 6.122 provides a bijection $P(S) \to B$, and we are done.   □

6.125 COROLLARY. If B is a finite Boolean algebra, then there is an isomorphism $B \approx \mathbb{B}^n$ for some $n \geqslant 1$; here $\mathbb{B}$ denotes the two-element Boolean algebra whose elements 0 and 1 are the zero element and unit element, respectively, of B. (Thus every finite Boolean algebra is essentially the Boolean algebra of n-bit strings, for a suitable n.)

PROOF OUTLINE.   Let $\{a_1, \ldots, a_n\}$ be the set of atoms in B. It follows from the proof of Theorem 6.122 that a typical element in B can be expressed in the form

$$(\lambda_1 \wedge a_1) \vee (\lambda_2 \wedge a_2) \vee \cdots \vee (\lambda_n \wedge a_n)$$

for exactly one n-tuple $(\lambda_1, \ldots, \lambda_n) \in \mathbb{B}^n$. The mapping $B \to \mathbb{B}^n$ given by

$$(\lambda_1 \wedge a_1) \vee \cdots \vee (\lambda_n \wedge a_n) \mapsto (\lambda_1, \ldots, \lambda_n)$$

is the desired isomorphism.   □

We conclude this section by briefly indicating some relationships between Boolean algebras and the theory of switching circuits. We first discussed switching circuits in connection with the concept of logical equivalence, in Section 1.6.

Suppose we want to represent a given Boolean function $f: \mathbb{B}^n \to \mathbb{B}$ by a switching circuit. More explicitly, given n input wires, each of which can have current or not (corresponding to value assignment 1 or 0, respectively), we want to construct a circuit whose output, 1 or 0, corresponds to the value of f on each member of $\mathbb{B}^n$.

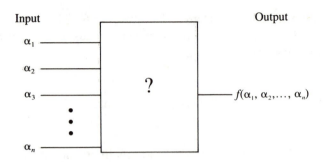

Can we do it? And if we can, can we do it with a circuit that is in some sense minimal?

First we informally introduce the term "Boolean expression." A **Boolean expression generated by n distinct symbols $x_1, \ldots, x_n$** (none of which is in $\mathbb{B}$) is an expression formed from elements of $\mathbb{B} \cup \{x_1, \ldots, x_n\}$ using the symbols $\vee, \wedge$, and $'$ a finite number of times (along with parentheses as needed for clarity), as if we were operating in a Boolean algebra. Thus, the elements of $\mathbb{B} \cup \{x_1, \ldots, x_n\}$ are themselves Boolean expressions; and if f and g are Boolean expressions, then so are $f \vee g$, $f \wedge g$, and f'. For example, here are some Boolean expressions generated by x_1, x_2, x_3, and x_4:

$$((x_1 \vee x_2) \wedge x_3)' \vee (x_1' \vee x_2), \qquad 1 \wedge x_4,$$

$$x_1 \wedge x_2 \wedge x_3 \wedge x_4, \qquad x_1 \vee x_1 \vee x_1.$$

We let **Bool($x_1, \ldots, x_n$)** denote the collection of Boolean expressions generated by $x_1, \ldots, x_n$.

Each $f \in \text{Bool}(x_1, \ldots, x_n)$ can be viewed as a function from $\mathbb{B}^n$ to $\mathbb{B}$ as follows: for each n-tuple $(\alpha_1, \ldots, \alpha_n) \in \mathbb{B}^n$, define $f(\alpha_1, \ldots, \alpha_n)$ to be the element of $\mathbb{B}$ obtained by replacing each x_i in the formula for f by α_i and then computing the result.

6.126 EXAMPLE.   The functional action of the Boolean expression $f = 1 \wedge (x_1 \wedge x_2')'$ is given by the following table.

α_1	α_2	$f(\alpha_1, \alpha_2)$
0	0	1
0	1	1
1	0	0
1	1	1

There are infinitely many Boolean expressions, but only finitely many (in fact, $2^{(2^n)}$) functions $\mathbb{B}^n \to \mathbb{B}$, so we can expect many Boolean expressions to represent the same function. For instance, all the Boolean expressions $x_1, x_1 \wedge x_1, x_1 \wedge x_1 \wedge x_1, \ldots$ represent the same function. From now on we will call two Boolean expressions *equal* if they represent the same function. So, with this convention in effect, we have

$$x_1 \wedge (x_2 \vee x_3) = (x_1 \wedge x_2) \vee (x_1 \wedge x_3) \quad \text{and} \quad x_1 \vee x_1' = 1.$$

(Here 1 denotes the constant function with action $\alpha \mapsto 1$ for all $\alpha \in \mathbb{B}^n$.)

It is true, but not obvious, that every Boolean function can be represented by a Boolean expression. We will demonstrate one way to produce such an expression, but first here is some notation. If x is one of the symbols generating our Boolean expressions, we define $x^1 = x$ and $x^0 = x'$. With this notation, $x_1^{\delta_1} \wedge \cdots \wedge x_n^{\delta_n}$ is a Boolean expression for each $(\delta_1, \ldots, \delta_n) \in \mathbb{B}^n$. Now if f is any Boolean function, we have $f(\delta_1, \ldots, \delta_n) \in \mathbb{B}$ for each $(\delta_1, \ldots, \delta_n) \in \mathbb{B}^n$, and therefore

$$(6.127) \qquad f(\delta_1, \ldots, \delta_n) \wedge x_1^{\delta_1} \wedge \cdots \wedge x_n^{\delta_n}$$

is a Boolean expression. The expression

$$[f(0, \ldots, 0) \wedge x_1^0 \wedge \cdots \wedge x_n^0] \vee [f(0, \ldots, 0, 1) \wedge x_1^0 \wedge \cdots \wedge x_n^1]$$
$$(6.128) \qquad \vee \cdots \vee [f(1, \ldots, 1) \wedge x_1^1 \wedge \cdots \wedge x_n^1]$$

is obtained by joining together (using $\vee$) all the expressions of the form 6.127 for which $(\delta_1, \ldots, \delta_n) \in \mathbb{B}^n$. It can be shown that 6.128 represents the Boolean function f. This representation of f is called the **disjunctive normal form** for f.

6.129 EXAMPLE. We can use the disjunctive normal form to determine a Boolean expression that represents the function $f: \mathbb{B}^3 \to \mathbb{B}$ given by the two left-hand sections of the following table.

δ_1	δ_2	δ_3	$f(\delta_1, \delta_2, \delta_3)$	$x_1^{\delta_1} \wedge x_2^{\delta_2} \wedge x_3^{\delta_3}$
0	0	0	0	$x_1' \wedge x_2' \wedge x_3'$
0	0	1	1	$x_1' \wedge x_2' \wedge x_3$
0	1	0	0	$x_1' \wedge x_2 \wedge x_3'$
0	1	1	1	$x_1' \wedge x_2 \wedge x_3$
1	0	0	0	$x_1 \wedge x_2' \wedge x_3'$
1	0	1	1	$x_1 \wedge x_2' \wedge x_3$
1	1	0	1	$x_1 \wedge x_2 \wedge x_3'$
1	1	1	1	$x_1 \wedge x_2 \wedge x_3$

We fill in the right-hand section of the table, then we use 6.128 to construct the disjunctive normal form for f; we omit the terms with coefficient $f(\delta_1, \delta_2, \delta_3) = 0$. This yields the Boolean expression

$$(x_1' \wedge x_2' \wedge x_3) \vee (x_1' \wedge x_2 \wedge x_3) \vee (x_1 \wedge x_2' \wedge x_3)$$
$$\vee (x_1 \wedge x_2 \wedge x_3') \vee (x_1 \wedge x_2 \wedge x_3).$$

Once a Boolean expression has been obtained that represents a given Boolean function, a switching circuit can be designed to simulate it, using AND, OR, and NOT gates, as in Section 1.6.

The trouble with the disjunctive normal form is that it generally gives a representation that is far from minimal, and the resulting circuit is likely to be expensive. (For instance, it is easy to check that the function in Example 6.129 is also represented by the expression $(x_1 \wedge x_2) \vee x_3$, which yields a far simpler circuit.) Fortunately, in view of our agreement to equate Boolean expressions that represent the same function, $\text{Bool}(x_1, \ldots, x_n)$ is now a Boolean algebra; therefore, ideas and techniques from the theory of Boolean algebras can be invoked in pursuit of representations that are minimal in some sense. We refer the interested reader to Elliott Mendelson, *Boolean Algebra and Switching Circuits* (New York: McGraw-Hill, 1970) and to the references provided in that text. See especially the *Quine–McCluskey method* for a procedure for pruning the disjunctive normal form.

Exercises

1. In the mathematical literature on Boolean algebras, the notation $a + b$ is sometimes used instead of $a \vee b$, and ab is used instead of $a \wedge b$.

   (a) Write the distributive laws and idempotency laws in this notation.
   (b) Show that $\mathbb{Z}$ is not a Boolean algebra under the usual arithmetic operations of addition and multiplication.

2. Is the ring $\mathbb{Z} \pmod 2$ a Boolean algebra?

3. Give a more detailed proof of the equation $(x')' = x$ stated in Theorem 6.109(b).

4. (a) Verify that the following cancellation law holds in a Boolean algebra:

$$a \wedge b = a \wedge c \quad \text{and} \quad a' \wedge b = a' \wedge c \quad \Rightarrow \quad b = c.$$

(b) If the phrase "and $a' \wedge b = a' \wedge c$" were omitted from the statement proved in part (a), would the result still be valid?

5. Let B be a Boolean algebra with operations $\vee$ and $\wedge$. Define operations of addition and multiplication on B by

$$a + b = (a \wedge b') \vee (a' \wedge b) \quad \text{and} \quad ab = a \wedge b.$$

(a) Verify that with these new operations, B is a commutative ring with unit element.
(b) Show that in the ring B we have $a^2 = a$ for all a, and use this to prove that $a + a = 0$ for all $a \in B$.
(*Definition.* A commutative ring R with unit in which $a^2 = a$ for all $a \in R$ is called a **Boolean ring**.)

6. Let B be a Boolean algebra.
(a) Prove that $a \wedge (a' \vee b) = a \wedge b$ for all $a, b \in B$.
(b) Write the dual of the equation in part (a), and illustrate the duality principle by showing that the dual of the proof of (a) is the proof of the dual of (a).

7. Prove that the idempotency laws (Definition 6.107(iii)) actually follow from the other Boolean algebra axioms.

8. Let B be a Boolean algebra. Show that the following holds in B:

$$a \wedge b' = 0 \quad \Leftrightarrow \quad a \wedge b = a.$$

9. In each part, say whether the given relation on the given set is a partial ordering, and justify your answer.
(a) The relation "weighs the same as" on the set of human beings.
(b) Let $S = \{x \in \mathbb{Z} \mid x \geqslant 2\}$. For $a, b \in S$, define $a \sim b$ to mean $(a, b) \geqslant 2$, where (a, b) is the greatest common divisor of a and b.
(c) Let $J = \{x \in \mathbb{R} \mid x \geqslant 1\}$. Define $a \approx b$ to mean $a \leqslant b^3$.
(d) Let S be the set of all propositions (that is, the set of all truth-valued statements; see Chapter 1). Consider the relation "$\Rightarrow$" on S. For the purposes of this exercise, view two propositions as *equal* if they are equivalent (see page 15).

10. Exhibit a Hasse diagram for each of the following partially ordered sets:
(a) the "divides" relation on $\{x \in \mathbb{Z} \mid 3 \leqslant x \leqslant 25\}$;
(b) the "equals" relation on $\{1, 2, 3, 4\}$;
(c) the inclusion relation on the collection of all members of $P(\{1, 2, 3, 4\})$ that have an odd number of elements.

11. (Refer to Section 6.5 for background material.) Let B be the set of all positive divisors of 30.

   (a) Show that B is a Boolean algebra under the operations

   $$a \vee b = [a, b] \quad \text{and} \quad a \wedge b = (a, b).$$

   (The challenging step is the statement and proof of the distributive laws.)

   (b) Draw the Hasse diagram for B.

12. A partial ordering $\leq$ on a set X is said to be a **total ordering** if for every $a, b \in X$, either $a \leq b$ or $b \leq a$. Now consider the partial ordering $\subseteq$ on $P(\{1, 2, 3\})$.

   (a) Why is this not a total ordering?
   (b) Name a totally ordered subset of $P(\{1, 2, 3\})$ that has maximum size.
   (c) Consider the set $\{x \in \mathbb{Z} \mid 1 \leq x \leq 70\}$ partially ordered by "divides." Without attempting to draw the massive Hasse diagram corresponding to this partially ordered set, exhibit a totally ordered subset that is as big as possible, and justify briefly.

13. (Refer to Exercise 12.) Given a partially ordered set S that is not totally ordered, it may be necessary somehow to extend the original ordering to a total ordering. For example, suppose we have a number of tasks to perform, and some must precede others, and we can do only one task at a time. (Here "is equal to, or must precede" is a partial ordering of the tasks.) We then need to produce a total ordering of the tasks consistent with the precedence requirements. That is, when we view the relations as sets of ordered pairs, the new relation must contain the old one.

   (a) Characterize the Hasse diagrams of totally ordered sets.
   (b) Produce a total ordering of the power set $P(\{1, 2, 3\})$ that extends the set-inclusion relation, and exhibit the associated Hasse diagram.

   (Intuitively, the extension of a partial ordering to a total ordering is achieved by distorting the original Hasse diagram into a straight line, taking care that each ascending path in the original diagram is transformed into an ascending segment of the final linear graph. The transition from a partial ordering to total ordering is called *topological sorting*.)

14. Let A and B be Boolean algebras, with unit elements 1_A, 1_B and zero elements 0_A, 0_B, respectively. Let $\varphi: A \rightarrow B$ be an isomorphism.

[More explicitly, φ is bijective; also $\varphi(a_1 \vee a_2) = \varphi(a_1) \vee \varphi(a_2)$ and $\varphi(a_1 \wedge a_2) = \varphi(a_1) \wedge \varphi(a_2)$, for all $a_1, a_2 \in A$.]

(a) Show that $\varphi(1_A) = 1_B$ and $\varphi(0_A) = 0_B$.

(b) Show that $\varphi(a') = \varphi(a)'$ for all $a \in A$.

(c) Give an example of a mapping $A \to B$ that satisfies the isomorphism properties except for bijectivity and does *not* satisfy the property in (b).

15. Given the partial ordering of Example 6.115(d), find each of the following. (If the object of interest does not exist, explain why.)

(a) lub$\{2, 6\}$ (b) glb$\{5, 3, 7\}$ (c) lub$\{2, 3, 4\}$

(d) lub$\{3, 6, 9\}$ (e) lub$\{$glb$\{2, 11\}$, lub$\{4, 6\}\}$

16. Let $\leqslant$ be a partial ordering on the set S, let $a \in S$, and let L_a be the set of all lower bounds of the set $\{a\}$.

(a) Determine lub L_a.

(b) What inequality characterizes all the upper bounds of L_a?

17. Let B be a Boolean algebra, let $\leqslant$ be its associated partial ordering (given by 6.118), and suppose $a \leqslant b$. Verify each of the following statements.

(a) $a \vee c \leqslant b \vee c$ for all $c \in B$

(b) $a \wedge c \leqslant b \wedge c$ for all $c \in B$

(c) $b' \leqslant a'$

18. (Refer to Exercise 14.) Let $\varphi: A \to B$ be an isomorphism of Boolean algebras. Show that for all $a_1, a_2 \in A$,

$$a_1 \leqslant a_2 \iff \varphi(a_1) \leqslant \varphi(a_2).$$

(Here we have taken the notational liberty of using the same symbol, $\leqslant$, to denote the partial orderings on both A and B.)

19. Deduce from Theorem 6.122 that if B is any finite Boolean algebra with more than two elements, then the associated partial ordering on B is not a total ordering.

20. Let S be any nonempty set, let $\mathbb{B} = \{0, 1\}$ be the standard two-element Boolean algebra, and let F be the set of all functions from S to $\mathbb{B}$. Define a mapping $\alpha: F \to P(S)$ by the rule

$$f \overset{\alpha}{\mapsto} \{x \in S \mid f(x) = 1\}.$$

(a) Show that α is a bijection from F to $P(S)$.

(b) Show that F is a Boolean algebra under the operations $\vee$, $\wedge$, and $'$ defined by the formulas in Example 6.108(d).

(c) Show that α is an isomorphism from F to $P(S)$, where $P(S)$ is viewed as a Boolean algebra according to Example 6.108(a).

21. (a) Exhibit a Hasse diagram for a lattice with five elements.

(b) Show that no five-element lattice is a Boolean algebra. Do this *without* making use of Corollary 6.124.

22. Let $S = \{1, 2, 3, 4\}$.

(a) Determine the atoms of the Boolean algebra $P(S)$, where as usual we use $\cup$ for $\vee$ and $\cap$ for $\wedge$.

(b) Now view $P(S)$ as a Boolean algebra by using $\cap$ for $\vee$ and $\cup$ for $\wedge$. Under these circumstances determine the atoms of $P(S)$.

23. Exhibit a table showing the functional action of the following Boolean expressions in $\mathrm{Bool}(x_1, x_2, x_3)$.

(a) $(1 \vee x_1) \wedge (x_2 \vee x_3)$

(b) $((x_1 \wedge x_2') \vee (x_1' \wedge (x_2' \vee x_3))) \wedge 1$

24. Suppose we have constructed a table showing the functional action of some Boolean expression $f \in \mathrm{Bool}(x_1, \ldots, x_n)$. Briefly explain how to use this table to construct a new table for f viewed as an element of $\mathrm{Bool}(x_1, \ldots, x_n, x_{n+1})$.

25. Consider the Boolean function $f \in \mathrm{Bool}(x_1, x_2, x_3)$ having the following table:

α_1	α_2	α_3	$f(\alpha_1, \alpha_2, \alpha_3)$
0	0	0	1
0	0	1	0
0	1	0	1
0	1	1	1
1	0	0	0
1	0	1	1
1	1	0	1
1	1	1	0

Write the disjunctive normal form for f.

Hints and Partial Solutions
to Selected Odd-Numbered Exercises

SECTION 1.1

1. (a) false proposition (b) not a proposition (c) false proposition
   (d) not a proposition (e) true proposition

3. *Theorem.* madam
   Proof.   S
           mSm
          maSam
          madam

SECTIONS 1.2 and 1.3

1. (a) $\sim P \wedge \sim Q$: Howard did not fall and Howard did not break his leg.
       $Q \vee \sim P$: Howard broke his leg or Howard did not fall.

3.

P	Q	$P \underline{\vee} Q$
F	F	F
F	T	T
T	F	T
T	T	F

5. The cards with the following markings must be turned over: 1, B.

7. Let P denote the statement "I go to the movies in the afternoon," and let Q
   denote "It is rainy."
   (a) $Q \Rightarrow P$ (b) $P \Rightarrow Q$

9. (b) P: Napoleon is President of the United States.
      Q: Boston is a city.

11. (b)

P	$P \Rightarrow P$	$P \Rightarrow (P \Rightarrow P)$
F	T	T
T	T	T

13. 122

15. The new table has eight times as many rows as the old one.

17. $\sim(P \wedge \sim(\sim(Q \wedge \sim(\sim(R \wedge \sim S)))))$

SECTION 1.4

1. Represent the statements "Dracula seizes power," "democracy is lost," "the use of food additives increases," and "mutations can be expected" by the letters $P, Q, R,$ and $S,$ respectively. Since we are given that P and $P \Rightarrow Q$ are both true, it follows by modus ponens that Q is true. From that and the given truth of $Q \Rightarrow R,$ modus ponens yields the truth of $R.$ Continue the argument in this way.

3. Try a proof by contradiction. If 5 is *not* prime then (by definition of *prime*) 5 can be written in the form $a \cdot b$ for some integers a, b strictly between 1 and 5. Check that that is *not* the case.

5. Compare a side of the diamond to a radius of the circle, and use the fact that all radii of a given circle have the same length.

SECTIONS 1.5 and 1.6

1. Statement forms (a), (b), and (d) are tautologies, but (c) is not.

3. (a) $\sim((P \wedge Q) \wedge \sim R)$ (b) $\sim(\sim P \wedge \sim Q)$

5. $\sim P$: $P \downarrow P$; $P \wedge Q$: $(P \downarrow Q) \downarrow (P \downarrow Q)$

7. Use the fact that $(R \vee \sim R)$ is a tautology to deduce that the given expression is logically equivalent to $(P \wedge Q) \vee (P \wedge \sim Q),$ then contract this further using a distributive law. Finally, conclude that the given expression is logically equivalent to $P.$

9. (a)

SECTION 2.1

1. yes

3. (a) $\{2, 3, 5\}$ (b) $\{2, 3, 5, 7\}$ (c) $\{1, 2, 3\}$ (d) $\{\pm 1, \pm 2, \pm 3, \pm 4, \pm 5\}$

5. (a) true   (b) true   (c) false   (d) false   (e) false   (f) false
   (g) true   (h) false   (i) true   (j) false

7. (a) The set $\{A\}$ has A as a member. But A is *not* a member of A, by the given axiom. Thus A and $\{A\}$ do not have the same members, and so $A \neq \{A\}$.

SECTION 2.3

1. (a) $(\exists x)(x \in \mathbb{N} \wedge x^3 + 15 = 22)$ (c) $\sim(\forall x \in \mathbb{R})(\exists y \in \mathbb{R})(y^2 = x)$
   (e) $(\forall x \in \mathbb{R})(\exists!y \in \mathbb{R})(y^3 = x)$ (g) $(A \subseteq \mathbb{Z}) \;\Rightarrow\; (\exists!x \in \mathbb{R})(x < 0 \wedge x \in B)$

3. (a) With $A = \emptyset$, statements 2(a) and 2(b) are both true.
   (b) There is no set A for which the *hypothesis* of 2(b) is true, so for every set A, *statement* 2(b) is true.

5. (a) $(\forall \epsilon > 0)(\exists \delta > 0)(|x' - x| < \delta \;\Rightarrow\; |f(x') - f(x)| < \epsilon)$

7. P: $x^2 = 5$; Q: $x = 17$

9. (a) true   (c) true   (e) false   (g) true

SECTION 2.4

1. (a) three

3. (a) $\{2, -1\}$ (c) $\{2, 7\}$ (e) $S = \{-2, -1, 0, 1, 2, \{1, 2\}\}$

5. Suppose $\{a\} \subseteq S$. Then every member of $\{a\}$ is a member of S. But the only member of $\{a\}$ is a. This shows that $a \in S$, verifying "$\Rightarrow$".

7. (a) The left-hand set is equal to $\{1, 2\}$, and $1 = 2^0$ and $2 = 2^1$.

9. Let $A = \{1\}$ and $B = \{1, \{1\}\}$.

SECTION 2.5

1. (a) $\{1, 2\}$ (c) $\{3, 4, 5\}$ (e) the set of all the even integers except for 4 and 6; in symbols: $\{0, \pm2, -4, -6, \pm8, \pm10, \pm12, \ldots\}$ (g) the members are 3, 5, and all the even integers: $\{3, 5, 0, \pm2, \pm4, \pm6, \ldots\}$
   (i) $\emptyset$.

3. (a) Observe that the statement $x \in A$ is equivalent to the statement

$$x \in A \vee x \in \emptyset.$$

(c) First note that $A - A = \emptyset$, because no element satisfies the condition "$x \in A \wedge x \notin A$." Then check that $\emptyset - A = \emptyset$.

(e) For every element x, the statements "$x \in A \wedge x \in A$" and "$x \in A$" are equivalent.

5. (a) Every element of $A - B$ is an element of A, but no element of $B - A$ is. Therefore the given intersection contains no elements.

7. Let $z \in Z$. From the hypothesis we then have $z \in X$ and $z \in Y$, as desired.

9. (a) For every element x, the statements

$$x \in A \ \lor \ x \in B \qquad \text{and} \qquad x \in B \ \lor \ x \in A$$

are equivalent. Therefore $A \cup B = B \cup A$. This proves that $A + B = B + A$. Proof of $AB = BA$ is similar, using intersection instead of union.

(c) We have $AB = A \ \leftrightarrow \ A \cap B = A \ \leftrightarrow \ B \supseteq A$. Thus the sets B satisfying the given condition are the sets that contain A.

(e) To prove "$\Leftarrow$", check that the statements "$x \in A$" and "$x \in A \lor x \in \emptyset$" have the same truth value. For "$\Rightarrow$", check that if X contains some element w, and A is a set not containing w, then $A + X \neq A$.

SECTION 2.6

1. (a) $\{1, 2, 3, 4, 0, -1\}$ (b) $\{1\}$ (c) $\mathbb{Z} - \{1\}$ (d) $\mathbb{Z} - \{-1, 0, 1, 2, 3, 4\}$

3. Let $x \in \cup A_i$. Then, by definition of union, $x \in A_i$ for some $i \in I$. But A_i is given to be a subset of A_n, so $x \in A_n$. This proves "$\subseteq$". Conversely, if $x \in A_n$ then since $n \in I$, we have that x is a member of $\cup A_i$, from the definition of union.

5. (a) $\mathbb{R}$ (b) $\{0\}$ (c) $[-5, 5]$

7. Proof of "$\subseteq$":

$$
\begin{aligned}
x \in A - \cup B_i \quad &\Rightarrow \quad x \in A \ \text{and} \ x \notin \cup B_i \\
&\Rightarrow \quad x \in A \ \text{and there is no} \ i \in I \ \text{such that} \ x \in B_i \\
&\Rightarrow \quad x \in A - B_i \ \text{for each} \ i \in I \\
&\Rightarrow \quad x \in \cap(A - B_i)
\end{aligned}
$$

9. (a) union: $\mathbb{R}$; intersection: $\emptyset$ (b) union: the set of all points in the plane whose horizontal coordinate x satisfies the inequality $-1 \leqslant x \leqslant 1$; intersection: $\emptyset$ (d) union: Π; intersection: $\emptyset$

SECTION 2.7

1. (a) $\{\emptyset, \{4\}, \{5\}, \{4, 5\}\}$ (b) $\{\emptyset, \{6\}\}$

3. 128

5. $X \in P(A) \Rightarrow X \subseteq A \Rightarrow X \subseteq A \subset B \Rightarrow X \subset B \Rightarrow X \in P(B)$. This proves "$\subseteq$".

7. $A \cap B = \emptyset$

SECTION 2.8

1. (a) Associate the ordered pair (t, r) with an incoming call to room r on telephone line t.

(b) The subset is a collection of four ordered pairs in $T \times R$, no two of which have the same first coordinate or same second coordinate.

3. This cannot be concluded without further information. For instance, if $a = c = 5$ and $b = d = 6$, then the stated set equality holds, but $c \neq d$.

5. (b) all the points in the plane whose coordinates are both integers
   (d) This set consists of all the points above the x-axis that have integer y-coordinates. Therefore the picture consists of horizontal lines through the points $(0, 1)$, $(0, 2)$, $(0,3)$, and so on.

9. "$\subseteq$": Let $P \in (\cup A_i) \times S$. Then $P = (a, b)$ for some $a \in \cup A_i$ and some $b \in S$. Thus $a \in A_i$ for some $i \in I$, by definition of $\cup A_i$, hence $P \in A_i \times S$, and so $P \in \cup(A_i \times S)$.

SECTION 2.9

1. (a) {the set of negative real numbers, the set of nonnegative real numbers}; {the set of nonzero real numbers, {0}}; {the set of nonnegative rational numbers, the set of negative rational numbers, the set of irrational numbers}

3. (a) false   (b) false   (c) true   (d) false

5. (a) $\Pi_1 = \{\{1\}, \{2\}, \{3\}, \{4\}, \{5, 6\}\}$; $\Pi_2 = \{\{1, 2\}, \{3, 4\}, \{5, 6\}\}$; $\Pi_3 = \{\{1, 2, 3, 4\}, \{5, 6\}\}$; $\Pi_4 = \{\{1, 2, 3, 4, 5, 6\}\}$.

7. $R \cup \{(1, 1), (2, 2), (3, 3), (4, 4)\}$

9. $R \cup \{(1, 1), (4, 2), (4, 3), (4, 4)\}$

11. (a) Example 2.53(c), R_1:

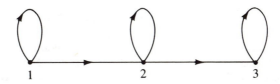

1   2   3

(b) For each pair of vertices p, q for which there is an edge from p to q, there is also an edge from q to p.

(d) For each pair of vertices p, q for which there is path from p to q formed by a succession of directed edges, there is also a single edge leading directly (that is, without passing through any other vertices along the way) from p to q.

15. (b) Reflexive: for each $a \in \mathbb{N}$ we have $a \mid a$, because $a = a \cdot 1$.
Antisymmetric: if $a \mid b$ and $b \mid a$, then $b = ac$ and $a = bd$ for some $c, d \in \mathbb{N}$. Then $a = bd = acd$, therefore $cd = 1$, and so $c = d = 1$. This gives $a = b$.
Transitive: if $a \mid b$ and $b \mid c$ then $b = ad$ and $c = be$ for some $d, e \in \mathbb{N}$. Then $c = a(de)$, and so $a \mid c$.

17. (a) $(1, 1), (2, 2), (3, 3)$ (b) impossible

19. We know that $A \subseteq A \cup B$ and $B \subseteq A \cup B$. This gives

$$R \subseteq A \times B \subseteq (A \cup B) \times (A \cup B).$$

21. The relation $\{(1, 2), (2, 1), (1, 1), (2, 2)\}$ is symmetric and transitive, but not reflexive.

SECTION 2.10

1. When $n = 1$, both sides of the given expression are equal to 1. Having assumed the result to hold for $n = k$, observe that

(*) $\qquad 1^2 + 2^2 + \cdots + (k+1)^2 = (1^2 + 2^2 + \cdots + k^2) + (k+1)^2.$

Now apply the induction hypothesis to expression (*), and use arithmetic to check that the result is equal to

$$\frac{(k+1)[(k+1)+1][2(k+1)+1]}{6}.$$

3. If $n \leq 0$ the result follows from the fact that squares are nonnegative. So it is enough to prove the statement when $n \geq 1$, and for this we use induction. For $n = 1$ the statement is clearly correct. Now suppose we know that $k^2 \geq k$ for some positive integer k. Then we have

$$(k+1)^2 = k^2 + 2k + 1 \geq 3k + 1 \geq k + 1,$$

and this completes the proof.

5. Use the fact that $3k + 3 = 3(k+1)$.

7. For $n = 10$: $1024 > 1000$. Now assume that $2^k > k^3$ for some $k \geq 10$; it must be shown that $2^{k+1} > (k+1)^3$. But $2^{k+1} = 2(2^k) > 2k^3$, so it will suffice to verify the inequality $2k^3 > k^3 + 3k^2 + 3k + 1$, or, equivalently, $k^3 - 3k^2 - 3k - 1 > 0$. This can be done in a variety of ways. For instance, the given inequality is equivalent to the inequality $k(k^2 - 3k - 3) > 1$, so it is enough to show that $k^2 - 3k - 3 > 0$ when $k > 10$.

9. Consider whether the argument is legitimate when $k = 1$.

SECTION 3.1

1. Sets f and g are functions from A to B, but h is not a function; j is not a function from A to B, but j is a function from $\{2, 3\}$ to B.

3. If $x \in X$ and $y_1, y_2 \in Y$, then $(x, y_1), (x, y_2) \in X \times Y$. Consider the consequences if $y_1 \neq y_2$.

5. nine

7. (a) $\chi_\emptyset(x) = 0, \quad \forall x \in A$
   (c) '$\Leftarrow$': Assume that $\chi_A = \chi_B$, and let $x \in A$. To show that $x \in B$, it suffices

to check that $\chi_B(x) = 1$. But $\chi_B(x) = \chi_A(x) = 1$. (Here the first equality follows from the hypothesis $\chi_A = \chi_B$, and the second equality holds because $x \in A$.) This shows that $A \subseteq B$, and a similar argument shows that $A \supseteq B$.

9. The numbers 1 and 6 are the only possible representatives of $\{1\}$ and $\{6\}$, respectively. If 3 were chosen to represent $\{1, 2, 3\}$, then 6 would have to represent $\{3, 6\}$, an impossibility because 6 already represents $\{6\}$. Therefore $\{2\}$ must represent $\{1, 2, 3\}$. Similar reasoning shows that 3 cannot represent $\{3, 4\}$, hence 4 does. Now that the elements 1, 2, 4, 6 have been selected as representatives, the remaining choices are automatic.

11. (a) The answer is 0 if $\alpha \in \mathbb{Z}$, and otherwise the answer is -1.
    (c) Choose β to satisfy the inequality $(1 + 1/[\alpha]) < \beta < 2$.
    (d) First consider the case $m \geq n$, then the case $m < n$.

13. (a) $\{(1, 1), (2, 1)\}$, $\{(1, 1), (2, 2)\}$, $\{(1, 2), (2, 1)\}$, $\{(1, 2), (2, 2)\}$
    (c) Each member of $\{1, 2\}^{(1,2,3)}$ is a set of ordered pairs, and one of those pairs has 3 as its first coordinate. Consider whether the same can be said about the members of $\{1, 2\}^{(1,2)}$.

SECTION 3.2

1. (a) im $f =$ im $g = \{1\}$ (c) the set of nonnegative real numbers   (e) im f is the set of nonnegative real numbers; im g is smaller, consisting of just the set of squares of rational numbers. Thus we can write

$$\text{im } g = \{a^2/b^2 \mid a, b \in \mathbb{Z}, b \neq 0\}.$$

3. (a) injective only   (c) neither injective nor surjective

5. (a) $g^{-1} = \{(5, 1), (-2, 2), (6, 3)\}$ (b) If $f: A \to B$ is not injective, then there exist distinct elements $a_1, a_2 \in A$ and an element $b \in B$ such that (a_1, b), $(a_2, b) \in f$. But then $(b, a_1), (b, a_2) \in f^{-1}$, and hence f^{-1} is not a function.

7. (a) The function f is a bijection if and only if $a \neq 0$.   (b) The necessary and sufficient condition is that $a = \pm 1$.

11. Parts (a) and (b) follow from the fact that the absolute value of any number is nonnegative, and it is zero if and only if the number itself is zero.

13. Compare the current estimate with the preceding one or, if necessary, with the following one. If the current answer agrees with either of those through n decimal places, then it has the desired degree of accuracy.

15. The idea is to obtain a function h such that $h(1) = f(1)$, $h(2) = g(1)$, $h(3) = f(2)$, $h(4) = g(2)$, and so on. Such a function h will take the odd positive integers onto im $f = A$ and the even positive integers onto im $g = B$. Explicitly, define h by

$$h(x) = \begin{cases} f\left(\dfrac{x+1}{2}\right) & \text{if } x \text{ is odd,} \\ g\left(\dfrac{x}{2}\right) & \text{if } x \text{ is even.} \end{cases}$$

Then check that this function works.

17. Go from (1, 1) to (2, 1) to (2, 2) to (1, 2), and continue.

19. Spiral out from the origin.

SECTION 3.3

1. (a) $\mathbb{R}$;   (c) $\mathbb{R} - \{n\pi \mid n \in \mathbb{Z}\}$.

3. (a) If $f(x) = x + 1$ and $g(x) = x/2$, then $(f \circ g)(3) = \frac{5}{2} \neq (g \circ f)(3) = 2$.

7. (b) Let $c \in C$. Because $g \circ f$ is surjective, there exists $a \in A$ such that $(g \circ f)(a) = c$. That is, $g(f(a)) = c$. Thus we have found an element (namely $f(a)$) of B that is taken to c by g.

9. Consider a function f defined by

$$f(x) = \begin{cases} x + 5 & \text{if } x \geq 0, \\ x & \text{if } x < 0. \end{cases}$$

This function "tears a hole" in $\mathbb{Z}$. Now try to patch the hole with a suitable function g.

11. The given function *is* a bijection.

13. (a) Suppose $A = \{1, 2\}$ and $C = \{3, 4\}$. Then, because $\operatorname{im} f^{-1} = A$ and $\operatorname{dom} g^{-1} = C$, it follows that $g^{-1} \circ f^{-1}$ is meaningless, though $(g \circ f)^{-1}$ has meaning.

15. To get the left cancellation law, notice that from the hypothesis we have

$$f^{-1} \circ (f \circ g) = f^{-1}(f \circ h),$$

and proceed from there.

SECTION 4.1

1. (b) We think of two collections as being the same "size" if each object in one can be paired with an object in the other so that when we're done there is nothing left over in either collection. (Demonstrate with a few pears and apples, say.) The statement says that if a first collection and a second collection have the same size in this sense, and also the second collection and a third collection have the same size, then the first and third collections have the same size. (Again demonstrate with actual objects, showing how the pairings for the first and second sets and the pairings for the second and third sets lead naturally to pairings of the elements of the first set with the elements of the third set.)

3. (a) Consider the mapping given by $n \mapsto 5n + 2$.

5. Use differentiation to show that the function is increasing, and hence one-to-one. To show surjectivity it suffices to show that the function is unbounded from above and therefore (why?) also from below.

9. For $n = 1$ there is really nothing to prove, since both sides are equal to $\#A_1$. Corollary 4.16 does the job for $n = 2$. Then rewrite $\cup_{i=1}^{k+1} A_i$ in the form $(\cup_{i=1}^{k} A_i) \cup A_{k+1}$, and use the case $n = 2$ to complete the induction step.

11. According to the definition of "+" in Exercise 10, the job here is to show that if $X \approx (B - A)$ and $Y \approx A$, where $X \cap Y = \emptyset$, then $X \cup Y \approx B$.

13. (a) Apply the pigeonhole principle (4.9) to a function from a given set of eight people to the set of days of the week.

15. (a) Use the product rule (4.22).
    (b) Once the activities have been chosen as in part (a), there remains the choice of *order* for the activities, and the number of possibilities for that is equal to the number of ways of listing the elements of the set {snack, movie, art}. Having determined that number, invoke the product rule.

17. Be careful not to count a given handshake more than once.

19. (a) Each of the n elements of B has m candidates in A that it might be paired with. In all, there are n choices to be made, with m possibilities for each. Invoke the product rule.
    (b) The hint gives the number of subsets of A that can serve as the function's image and, for each of those, part (a) gives the number of functions with domain B that have that image. Use the product rule.

SECTION 4.2

1. (a) "$\Leftarrow$": Assume $A = \emptyset$. The conditional statement

$$\text{If } x \in \emptyset \text{ then there exists } y \in \emptyset \text{ such that } \emptyset(y) = x.$$

is true, because its hypothesis is false. Therefore $\emptyset$ is surjective.

3. This proof is a clone of the proof of the first part of the theorem.

5. (a) The mapping $x \mapsto x$ is an injection from $A - B$ to A.
   (b) Consider the consequences if $\#A > \#B$ and $A \cap B = \emptyset$.

7. A typical element of the square has the form

$$(.a_1 a_2 a_3 \ldots, .b_1 b_2 b_3 \ldots).$$

Now follow the given suggestion.

SECTION 4.3

1. $\mathbb{Z} = \mathbb{N} \cup A \cup B$, where $A = \{n \in \mathbb{Z} \mid n \leqslant 0 \text{ and } n \text{ is even}\}$ and B is the set of odd negative integers

3. (a) By Corollary 4.15, at least one of the blocks must be infinite. Now apply Theorem 4.36(a).

5. Check that the mapping given by $n \mapsto 7n + 3$ is a bijection from $\mathbb{Z}$ to the given set. Then apply Example 4.34.

7. (a) Show that $\mathbb{N} \times \mathbb{N} = \cup_{i \in \mathbb{N}} (\mathbb{N} \times \{i\})$.
   (b) Theorem 4.39(a) will be useful.

9. First check that for each $q \in \mathbb{Q}$, the collection of all intervals having rational endpoints and right endpoint q is a countable set. Then use the result of Exercise 8.

11. Here the word "total" in parts (a) and (b) actually refers to the sum of a geometric series. The idea here is to imagine the rational points on the real line pressing against each other after we have swept away the irrational "debris" from between them. The sums of the series in parts (a) and (b) serve as upper bounds for the length of an interval required to completely surround this huddled mass of rationals.

13. First review the argument in Example 4.43.

SECTION 4.5

1. View x and y as functions from $\mathbb{N}_m$ and $\mathbb{N}_n$ (respectively) into Σ. Then xy is the function from $\mathbb{N}_{m+n}$ into Σ given by

$$(xy)(i) = \begin{cases} x(i) & \text{if } 1 \leqslant i \leqslant m, \\ y(i - m) & \text{if } m < i \leqslant m + n. \end{cases}$$

3. The languages $L(M)$ and $L(N)$ are the same.

5. Let's say that the initial states of M_1 and M_2 are q_0 and q_0', respectively. Erase the arc starting at q_0 that is labelled b, and erase the arc starting at q_0' that is labelled a. Then drag the diagram for M_1 over to the diagram for M_2 in such a way that q_0 and q_0' become superimposed, but so that no other vertices of the two graphs come into contact. The result is the graph of the desired automaton. The point obtained from merging q_0 and q_0' represents the initial vertex of the new automaton, and the new collection of final states is the union of the final state sets of M_1 and M_2. It remains to check that this construction works.

7. (a)

(c)

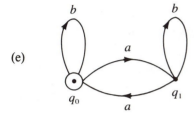

(e)

SECTION 5.2

3. (a) 360
   (b) 144 [First determine how many positive divisors *are* divisible by 10, then apply part (a).]

5. 1320

SECTION 5.3

1. (a) a permutation   (c) not a permutation   (e) a permutation

3. There are 24 members altogether.

5. (a) Given $\lambda \in S_m$, extend λ to an element $\lambda \in S_n$ by the formula

$$\lambda(x) = \begin{cases} \lambda(x) & \text{if } 1 \leqslant x \leqslant m, \\ x & \text{if } m+1 \leqslant x \leqslant n. \end{cases}$$

7. (a) $\begin{pmatrix} 1 & 2 & 3 & 4 & 5 \\ 4 & 2 & 3 & 1 & 5 \end{pmatrix}$ (c) $\begin{pmatrix} 1 & 2 & 3 & 4 & 5 \\ 1 & 3 & 2 & 5 & 4 \end{pmatrix}$

9. Begin by noticing that the sets of elements moved by the given permutations are disjoint.

11. 40,720

13. 840

15. (a) 6   (b) 144   (c) 15!   (d) $7 \cdot 14!$

SECTION 5.4

1. Suppose the triangle's vertices are A, B, C, and let s be a symmetry. Show that if $s(A) = B$ then $s(B) = A$, and therefore $s(C) = C$. Then use this to obtain a contradiction.

3. (a) There are eight members.

   (b) Inscribe a regular triangle in a regular hexagon, and proceed from there.

5. (a) ten symmetries   (c) one symmetry   (e) four symmetries

   (g) one symmetry   (i) the collection of symmetries is countably infinite

   (k) the collection of symmetries is uncountably infinite

7. (a) If $P = Q$ then $R_Q(P) = P$. If $P = (c, d) \neq Q$, then obtain the equation of the line through P and Q and the equation of the circle centered at Q with radius PQ. The intersection of the two geometric figures is obtained by solving these equations simultaneously, and the solution set is $\{P, R_Q(P)\}$.

   (b) Use the formula obtained in (a) to show that for any two points P_1, P_2, the distance between $R_Q(P_1)$ and $R_Q(P_2)$ is equal to the distance between P_1 and P_2.

SECTION 5.5

1. (a) $\begin{pmatrix} 1 & 2 & 3 & 4 & 5 & 6 & 7 \\ 2 & 7 & 1 & 4 & 6 & 5 & 3 \end{pmatrix}$ (c) $\begin{pmatrix} 1 & 2 & 3 & 4 \\ 3 & 1 & 4 & 2 \end{pmatrix}$

3. (a) First notice that once σ^2 has been computed, only one more multiplication is required in order to compute σ^4.

5. They look the same, except that the edge directions are all reversed.

7. (a) The notation tells us that the listed elements are moved cyclically by σ, and that no other elements are moved by σ, but it does not tell us the whole domain.

   (b) $\begin{pmatrix} 1 & 2 & 3 & 4 & 5 & 6 & 7 \\ 5 & 3 & 4 & 2 & 7 & 1 & 6 \end{pmatrix}$

9. (a) $(1 \ \ 5 \ \ 3)(2 \ \ 6 \ \ 7)$ (b) $(1 \ \ 3 \ \ 4 \ \ 5 \ \ 6 \ \ 2)$ (c) $(2 \ \ 4 \ \ 3)$

11. Consider the products $(\sigma\tau)(\tau^{-1}\sigma^{-1})$ and $(\tau^{-1}\sigma^{-1})(\sigma\tau)$.

13. If α satisfies the given equation, it follows that $\alpha = \sigma(2 \ \ 1 \ \ 4)\lambda^{-1}$.

15. Notice that once *some* n is obtained for which S_n contains a non-cycle, then the same is true for all larger values of n.

SECTION 5.6

3. Use the method of Example 5.42.

5. Begin by factoring 2520 into primes, and then use this data to obtain a set of

integers whose sum is 30 and whose least common multiple is 2520. Then apply Theorem 5.45.

7. (b) Every element of S_m has a natural extension to an element of S_n. See Exercise 5 in Section 5.3.

9. 18 (Use the method of Example 5.47.)

SECTION 5.7

1. (a) (1 2)(1 5)(1 4)(1 6)(1 8)(1 7)(1 3)   (b) (1 2)(3 4)

3. (a) two reversals   (b) four reversals

5. Show that the permutation

$$\begin{pmatrix} 1 & 2 & 3 & \cdots & n-1 & n \\ n & n-1 & n-2 & \cdots & 2 & 1 \end{pmatrix}$$

has the desired property.

7. Use Theorem 5.54.

9. Use Corollary 5.36, together with the fact that s moves exactly four elements.

11. (a) -1 (b) 1 (c) -1

13. (a) Observe that a shift of the puzzle pieces to the indicated configuration is associated with the permutation (1 2 6 7 11 12 16), an even permutation.

15. Check each of the cases from $n = 1$ through $n = 5$ to get the idea.

SECTION 5.8

1. (a) 9863   (c) 1001   (e) 39711   (g) 1

5. (a) 336   (b) 231   (c) 2398

7. Use the binomial theorem.

11. Check the cases $n = 1, 2, 3$ individually. For $n \geqslant 4$, use Theorem 5.64.

13. (a) 1120   (b) 1883

15. $\frac{1}{16660} = .00006$

17. (a) $\frac{35}{128}$ (b) $\frac{35}{64}$

19. If a set of n objects is to be assembled in sequence, with only two types of object in the set, consider the consequence of specifying the locations of one of the types.

21. 2,522,520

23. (a) 3,628,800 (b) 30,240 (c) 252

25. -280

SECTION 6.1

1. The left side of the rule becomes $\cdot\,((a, +((b, c))))$.

3. n^3

5. The case $n = 3$: define a 3-ary (more commonly, *ternary*) operation $*_3 : S^3 \to S$ by

$$*_3((s_1, s_2, s_3)) = (s_1 * s_2) * s_3.$$

7. Suppose there *is* an identity element; call it e. Then for each $a \in \mathbb{R}$, the equation $a \div e = a$ holds. But this implies that $e = 1$. (Why?) Check that in fact 1 is *not* an identity element for $\div$.

9. The statement $(x * y) * z = z * (y * z)$ must be checked for each substitution of elements from the set $\{e, a\}$ in place of the symbols x, y, z. There are eight statements to check altogether.

11. First explain how to check the table to see if there is an identity element e for the operation. (If there is none, then no inverses exist.) Assuming that e exists, the element e must appear in every row of the table for inverses to exist. Having checked that, it remains to carry out an appropriate check of the *columns* of the table.

13. (a) yes, yes, no (c) yes, yes, yes (e) yes, no, no

15. If e is an identity for $*$, then

$$ra + se = a * e = e * a = re + sa, \qquad \forall a \in \mathbb{Z}.$$

Consider the consequences when $a \neq e$.

SECTION 6.2

1. (a) a monoid (c) none (e) a group

5. (b) Let e_S and e_T denote the identity elements for S and T, respectively. Check that (e_S, e_T) is an identity for $S \times T$.

7. Most of this is Exercise 7 of Section 5.7 in new terminology.

9. Multiply both sides of the equation in the hypothesis by a^{-1}, and apply the associative law.

11. Refer to Exercise 5 in Section 5.3.

13. Consider the effect of rotating the table by $180°$, using the *main diagonal* (a line drawn from the upper left corner to the lower right corner of the table) as the axis of rotation.

15. (a) This follows from the equation $aa^{-1} = a^{-1}a = e$. (Why?)
    (b) For injectivity, start with the assumption that $a^{-1} = b^{-1}$, multiply both sides of this equation by a, and proceed from there. Surjectivity is a consequence of part (a).

17. (a) Exercise 10(b) becomes this: H is a subgroup of G if and only if $h_1 + h_2 \in H$ for all $h_1, h_2 \in H$ and $-h \in H$ for all $h \in H$.

SECTION 6.3

1. (a) yes   (c) no   (e) yes

3. Let $\psi: G \to H$ be an isomorphism, and let g be a generator for G. Show that $\psi(g)$ is a generator for H.

5. In the argument for $\mathbb{Z}$, remember that an isomorphism must be surjective.

7. (b) Consider the mapping $\mathbb{Z} \to n\mathbb{Z}$ given by $x \mapsto nx$.

9. For associativity, we must check that

$$a \oplus (b \oplus c) = (a \oplus b) \oplus c, \qquad \forall a, b, c \in G.$$

If $a, b, c \in \mathbb{R}$, this equality holds because addition in $\mathbb{R}$ is associative. It remains to consider the other cases. For instance, it follows from the definition of $\oplus$ that

$$(\text{Stanley} \oplus b) \oplus c = b \oplus c = \text{Stanley} \oplus (b \oplus c), \qquad \forall b, c \in G.$$

SECTION 6.4

3. (a) By using the distributive laws we obtain

$$(a + b)(c + d) = (a + b)c + (a + b)d = ac + bc + ad + bd.$$

5. Suppose that for some $a \in \mathbb{Z}$ we have $0 \cdot a = b \neq 0$. Then, from the properties discussed in the text, we deduce that

$$a = 1 \cdot a = (1 + 0) \cdot a = 1 \cdot a + 0 \cdot a = a + b.$$

Now add $-a$ to both sides of this equation to get a contradiction.

7. Use 6.24.

9. Define "$\leq$" in a way modelled after the definition of "$<$" in 6.24.

SECTION 6.5

1. From the hypothesis, we have $b = ax$ and $c = ay$ for some $x, y \in \mathbb{Z}$. This observation will lead to the result.

3. Notice that if $x \in \mathbb{R}$, then

$$[x] = x \iff x \in \mathbb{Z}.$$

7. After the sieving procedure is complete, thirty primes should remain on your list of integers.

9. It suffices to show that n is divisible by 2 and by 3. Achieve these goals one at a time. (The division algorithm (6.30) will be helpful for this purpose, with $k = a$.)

13. The expression on each side of the alleged equation represents a positive integer. It is enough to check that each of these is a divisor of the other.

17. Refer to the proof of Theorem 6.43 for the strategy.

19. Remember Pythagoras!

21. The least common multiple is 9,447,438.

SECTION 6.6

1. (a) true   (c) false   (e) true   (g) false

3. Write n in the form $2k + 1$ for some integer k.

5. (b) Recall the formula $1 + 2 + 3 + \cdots + k = k(k + 1)/2$, and use the test (6.58) for divisibility by 9.

7. Refer to Example 6.52(a).

9. Use Theorem 6.57.

11. (b) As part of the solution, show that if $x, y \in \{0, k, 2k, \ldots, (m - 1)k\}$ and $x \equiv y \pmod{m}$, then $x = y$. Next apply part (a) to each element of the set $\{0, k, 2k, \ldots, (m - 1)k\}$.

15. Use a proof by contradiction. The equation $2^{ab} = (2^a)^b$ will be helpful.

17. (b) 131123031

19. Use Theorem 6.60 as a guide.

SECTION 6.7

1.

$\oplus$	$\emptyset$	$\{1\}$	$\{2\}$	A
$\emptyset$	$\emptyset$	$\{1\}$	$\{2\}$	A
$\{1\}$	$\{1\}$	$\emptyset$	A	$\{2\}$
$\{2\}$	$\{2\}$	A	$\emptyset$	$\{1\}$
A	A	$\{2\}$	$\{1\}$	$\emptyset$

3. (b) Be sure to read the definition of "unit element" carefully.

5. If $f \in F$ is to be invertible, is it possible that $f(x) = 0$ for some $x \in \mathbb{R}$?

7. Consider separate cases according to the number of zeros on the diagonal.

9. Consider the ring of 2×2 diagonal matrices over $\mathbb{R}$.

11. For $x, y \in S$, define $x + y$ to be the image of the sum of their pre-images. (To make this a useful definition, rewrite it as a formula. The bijectivity of f is crucial.) Proceed similarly to define the product xy.

13. Refer to Exercise 1 of Section 6.5.

15. (b) Because $a \mid m$ and $b \mid m$, it follows that $m \in (a) \cap (b)$. Deduce from this that $(m) \subseteq (a) \cap (b)$.

17. (a) Notice that $(5) \supset (10) \supset (20)$. Is there a reasonable way to continue extending this descending chain of ideals?

19. (a) For a start, write two matrices $A, B \in M_2(\mathbb{R})$ such that $AB \neq BA$, and whose entries are 0s and 1s. Then use this as a guide in answering the given question about $M_2(K)$.

    (b) A typical element of $M_2(M_2(\mathbb{R}))$ is a 2×2 matrix, each of whose entries is itself a 2×2 matrix over $\mathbb{R}$.

21. Proceed as for the construction of the tables in 6.63(c) and 6.76.

SECTION 6.8

1. $f + g = 4x^2 + x$.

3. Writing

$$x^{n-1} + x^{n-2} + \cdots + x + 1 = x^{n-1} + (x^{n-2} + \cdots + x + 1)$$

will help the induction step along.

5. First show that the hypothesis $f \mid g$ implies that $\deg f \leqslant \deg g$.

7. Use Lemma 6.90.

9. There are eight polynomials of degree 3.

11. No

13. (a) The elements $1, a, a^2, \ldots, a^q$ cannot all be different, by the pigeonhole principle.

15. (a) Use 6.79 and 6.80 to show that in $F[x]$, the elements $1, x, x^2, x^3, \ldots$ are all different.

    (b) Use the product rule; see page 211. (The fact that F is a field is not needed for this part of the exercise. Any finite set will do.)

17. (a) $x + 2$

SECTION 6.9

1. (a) $5 + 12i$ (c) $-i$ (e) 5 (g) $\dfrac{5}{7} - \dfrac{2}{7}i$

5. $\dfrac{\sqrt{2}}{2} + \dfrac{\sqrt{2}}{2}i$

7. These numbers lie on a spiral; as n grows, the numbers z^n get farther from the origin.

9. The number $\zeta = \cos(\pi/9) + i \sin(\pi/9)$, and suitable powers of it, are the desired roots.

11. (b) Each of the given conditions can be transformed into the equation of a corresponding circle.

13. The average of the real parts of the given numbers is the real part of the midpoint, and similarly for the imaginary part of the midpoint. (The proof of this, which is not required here, uses an argument involving similar triangles.)

15. (b) These numbers occupy the vertices of a regular octagon centered at the origin in the complex plane.

17. (b) Re $z = (z + \bar{z})/2$ (c) 4

19. There *are* examples of the indicated kind.

21. (a) In $\mathbb{R}[x]$:

$$x^6 - 1 = (x^3 - 1)(x^3 + 1)$$
$$= (x - 1)(x^2 + x + 1)(x + 1)(x^2 - x + 1)$$

In $\mathbb{C}[x]$:

$$x^6 - 1 = (x - 1)\left(x + \frac{1 - i\sqrt{3}}{2}\right)\left(x + \frac{1 + i\sqrt{3}}{2}\right) \cdot$$
$$\cdot (x + 1)\left(x - \frac{1 + i\sqrt{3}}{2}\right)\left(x - \frac{1 - i\sqrt{3}}{2}\right)$$

SECTION 6.10

1. (a) The statement

$$a + bc = (a + b)(a + c)$$

is one of the distributive laws.

(b) The distributive law given here as part of the answer to part (a) is not a law in $\mathbb{Z}$. For instance, it fails when $a = b = c = 1$.

3. Refer to the definition of complement, and also to part (a) of Theorem 6.109.

5. Among the ring axioms to be checked is the distributive law $a(b + c) = ab + ac$. Using Boolean algebra notation and the definitions supplied in this exercise, verifying this statement amounts to verifying the following for all $a, b, c \in B$:

(*) $a \wedge ((b \wedge c') \vee (b' \wedge c)) = ((a \wedge b) \wedge (a \wedge c)') \vee ((a \wedge b)' \wedge (a \wedge c))$.

The left side of (*) is equal to $(a \wedge b \wedge c') \vee (a \wedge b' \wedge c)$. To show that the right side of (*) is also equal to this, first bring the complements inside the parentheses using De Morgan's laws; for instance, $(a \wedge c)' = a' \vee c'$. Then

(**) $(a \wedge b) \wedge (a \wedge c)' = (a \wedge b \wedge a') \vee (a \wedge b \wedge c')$.

But $a \wedge b \wedge a' = (a \wedge a') \wedge b = 0 \wedge b = 0$, and so (**) is equal to $a \wedge b \wedge c'$.

7. For one of the laws, observe that

$$a = a \wedge 1 = a \wedge (a \vee a'),$$

then use a distributive law and the fact that $a \wedge a' = 0$.

9. (a) not a partial ordering   (c) not a partial ordering

11. One of the distributive laws to be checked is this statement:

$$(a, [b, c]) = [(a, b), (a, c)].$$

Write each of a, b, c as a product of prime powers; say,

$$a = p_1^{\alpha_1} p_2^{\alpha_2} \cdots p_n^{\alpha_n}, \qquad b = p_1^{\beta_1} p_2^{\beta_2} \cdots p_n^{\beta_n}, \qquad c = p_1^{\gamma_1} p_2^{\gamma_2} \cdots p_n^{\gamma_n}.$$

To check the statement under investigation, it suffices to show that the prime power factors on both sides of the expression are equal. Those prime powers are given by the formulas in 6.45 and 6.46, and the present exercise is thereby reduced to checking that

$$\min\{\alpha, \max\{\beta, \gamma\}\} = \max\{\min\{\alpha, \beta\}, \min\{\alpha, \gamma\}\},$$

for nonnegative integers α, β, γ.

13. (a) A Hasse diagram for a totally ordered set can be drawn as a vertical line.

15. (a) 6   (c) 12   (e) 12

17. (c) $a \leq b \;\Rightarrow\; a \vee b = b$
        $\Rightarrow\; (a \vee b)' = b'$
        $\Rightarrow\; a' \wedge b' = b'$
        $\Rightarrow\; b' = \text{glb}\{a', b'\}$
        $\Rightarrow\; b' \leq a'$

19. The diagram in Example 6.115(b) may help.

21. The Hasse diagram for a finite Boolean algebra must have a unique highest vertex and a unique lowest vertex, corresponding to 1 and 0, respectively. For the present exercise, consider the possible vertex and edge configurations linking the other three vertices to each other and to 0 and 1. The fact that each element has a *unique* complement is crucial.

23. (a) The table has eight lines (in its interior). Here is a start:

α_1	α_2	α_3	$f(\alpha_1, \alpha_2, \alpha_3)$
0	0	0	0
0	0	1	0
0	1	0	1
0	1	1	1

25. The method is demonstrated in Example 6.129.

Index